André König

Temporal Activation Profiles of Gene Sets

André König

Temporal Activation Profiles of Gene Sets

for the Analysis of Gene Expression Time Series

Südwestdeutscher Verlag für Hochschulschriften

Impressum / Imprint
Bibliografische Information der Deutschen Nationalbibliothek: Die Deutsche Nationalbibliothek verzeichnet diese Publikation in der Deutschen Nationalbibliografie; detaillierte bibliografische Daten sind im Internet über http://dnb.d-nb.de abrufbar.
Alle in diesem Buch genannten Marken und Produktnamen unterliegen warenzeichen-, marken- oder patentrechtlichem Schutz bzw. sind Warenzeichen oder eingetragene Warenzeichen der jeweiligen Inhaber. Die Wiedergabe von Marken, Produktnamen, Gebrauchsnamen, Handelsnamen, Warenbezeichnungen u.s.w. in diesem Werk berechtigt auch ohne besondere Kennzeichnung nicht zu der Annahme, dass solche Namen im Sinne der Warenzeichen- und Markenschutzgesetzgebung als frei zu betrachten wären und daher von jedermann benutzt werden dürften.

Bibliographic information published by the Deutsche Nationalbibliothek: The Deutsche Nationalbibliothek lists this publication in the Deutsche Nationalbibliografie; detailed bibliographic data are available in the Internet at http://dnb.d-nb.de.
Any brand names and product names mentioned in this book are subject to trademark, brand or patent protection and are trademarks or registered trademarks of their respective holders. The use of brand names, product names, common names, trade names, product descriptions etc. even without a particular marking in this works is in no way to be construed to mean that such names may be regarded as unrestricted in respect of trademark and brand protection legislation and could thus be used by anyone.

Coverbild / Cover image: www.ingimage.com

Verlag / Publisher:
Südwestdeutscher Verlag für Hochschulschriften
ist ein Imprint der / is a trademark of
OmniScriptum GmbH & Co. KG
Heinrich-Böcking-Str. 6-8, 66121 Saarbrücken, Deutschland / Germany
Email: info@svh-verlag.de

Herstellung: siehe letzte Seite /
Printed at: see last page
ISBN: 978-3-8381-3910-4

Zugl. / Approved by: Dortmund, TU, Dissertation, 2014

Copyright © 2014 OmniScriptum GmbH & Co. KG
Alle Rechte vorbehalten. / All rights reserved. Saarbrücken 2014

Contents

1	Introduction to the Analysis of Gene Expression Time Series Data	3
2	Aspects of Analyzing Gene Expression Time Series Data on a Gene Set Level	5
	2.1 Gene Expression Data Characteristics	6
	2.2 Definitions of Gene Sets	13
	2.3 Experimental Designs, References and Formalization	17
3	Methods for the Analysis of Gene Expression Time Series	21
	3.1 Gene Set Testing Based on Differentially Expressed Genes	23
	3.2 Clustering Approaches	26
	3.3 Other Approaches for the Gene Set Analysis of Gene Expression Time Series	29
4	Estimating Temporal Activation Profiles of Gene Sets	31
	4.1 Adapted Methods from GSA in Two-Condition Experiments	34
	4.2 GSA-Type Algorithms for Determining Gene Set Activation Profiles	40
	4.3 Smoothing of GSA-Type Gene Set Activation Profiles	46
	4.4 Ranking of GSA-Type Gene Set Activation Profiles	52
	4.5 Rotation-Test-Type Algorithm for Determining Gene Set Activation Profiles	53
	4.6 The Modified STEM Algorithm	56
	4.7 The maSigFun Algorithm	59
5	Simulation Study	61
	5.1 Short Characterization of the Four Recycled Example Data Sets	62
	5.2 Construction of a Simulation Study by Recycling of Given Data	65
	5.3 Simulation Study to Compare the Methods	81
	5.4 Simulation Study to Evaluate the Smoothing Algorithms	103
	Simulation Study to Evaluate the Smoothing Algorithms	103

		5.5	Joint Conclusions from Both Simulation Studies	129

6	Application of Gene Set Activation Profile Estimation	135
	6.1 Filtering the Gene Universe	135
	6.2 Summary Over All Profile Algorithms and Data Sets	140
	6.3 Aldosterone Effect on Mouse Heart Gene Expression	146
	6.4 Embryonic Mouse Ovary Development Experiment	150
	6.5 Gene Expression During Wound Healing in Skin and Tongue	155

7	Discussion and Summary	169

List of Figures	193

List of Tables	201

A	FDR q-value in the TS-ABH Procedure	205

B	Figures and Tables	209

List of Symbols	277

CHAPTER 1

Introduction to the Analysis of Gene Expression Time Series Data

In the second half of the 20th century *molecular genetics* was established as a new field in biology. From the *central dogma* of molecular genetics first postulated by Francis Crick in 1958 it lasted another 37 years until the first genome of the bacterium *Haemophilus influenzae* was fully sequenced. This was only possible due to enormous scientific and technological advances like the discovery of the enzyme reverse transcriptase and the proceedings in automation and miniaturization with the PCR procedure as outstanding mile stone. In the decade around the turn of the millennium the advent of high throughput technologies allowed the parallel analysis of tens of thousands of genes or their transcripts with a single experiment on a chip smaller than a fingertip.

This thesis focuses on the analysis of the quantitative gene expression data generated by high throughput time series experiments. The second chapter elucidates the statistical challenges arising from the task to analyze an enormous number of genes with a much smaller sample size for a few time points. Usually analyzing gene sets instead of single genes supports the researcher at least in two ways. The reduction of dimensionality in the data increases statistical power in the findings of significant changes in gene expression and the a priori definition of gene sets in the context of molecular functions or pathways facilitates the interpretation of the findings. There are various methods available for the gene wise analysis of gene expression time course experiments and gene set testing in the time series context. The common approaches are discussed in the third chapter. On the other hand there is a lack of methods for estimating an activation profile of differential expression in the gene set setting, which is the overall objective of this thesis.

Three ways of estimating a group activation profile are introduced in the third chapter. Each algorithm attempts to represent the differential expression of a significant number of the included genes for a parallel consideration of a large number of gene sets. Two standard algorithms from literature – STEM and maSigFun – are presented for comparison purposes. The accuracy of the five approaches is compared in an extensive simulation study in the fifth part of the thesis. A second large simulation study in the same chapter evaluates smoothing procedures for the three activation profile algorithms. The sixth chapter provides the application of the activation profile estimation algorithm on four exemplary data sets and different sources of gene set definition. In the last chapter the findings are summarized and discussed.

CHAPTER 2

Aspects of Analyzing Gene Expression Time Series Data on a Gene Set Level

Gene expression data confronts the statistician with the task of parallel analyzing tens of thousands features representing genes with only a few measurements. In the time course experimental design there are often only a few time points with non-uniform sampling scheme, which is an additional challenge for the researcher. The first section in this chapter elucidates the type of data, which is considered in this thesis; from the biological backgrounds to the symbolic notation used in the following chapters. The second section provides an overview of the ideas and sources of gene set definitions used for gene expression analysis. Currently available methods for analyzing gene expression data resulting from a time course experiment are shown in the third section of this chapter. It has to be noted that there currently no methods are available that explicitly estimate temporal activation profiles of gene sets, which is the objective of the thesis. Nevertheless the established algorithms provide beneficial ideas, which can be modified to achieve an estimation procedure as is demonstrated in chapter 4.

2.1 Gene Expression Data Characteristics

The Biological Background

All cellular components and functions of all living organisms depend directly or indirectly on molecules, called proteins. Proteins consist in their primary structure of one or more chains of aminoacids. The information about the order and type of the 20 different amino acids, which determines each protein is stored in the cells Deoxyribonucleic acid (DNA). In order to build a protein all organisms use a transcription of the double-stranded DNA into a single-stranded Ribonucleic acid (RNA) molecule, which codes the amino acid sequence forming the corresponding protein.

Gene expression data means measurements of the qualities and quantities of the cells RNA. Often these measurements are restricted to the messenger RNA, which is a replicate of the part of the genetic information coding for a single protein in the cells nucleus. *Transcripts* is a synonymously used term for RNA-strands and the whole space of transcripts is called *transcriptome* as can be seen in Figure 2.1 on the facing page, in which the main research areas related to bioinformatics are shown within their biological context.

High Throughput Technologies and Sources of Gene Expression Data

A number of high throughput technologies is available to generate gene expression data corresponding to thousands of transcripts from a small sample of cell plasma. All manufacturers have in common, that they provide chips with an area of 1 - 2 cm^2, which contain a matrix of fixed oligonucleotides complementary to a priori chosen transcripts. A prepared cell medium containing the transcripts of interest in the current experiment is applied on this slide and the (labeled) fragments of the probe RNA hybridize complementary to the corresponding sites on the chip. The measurement of transcript quantity is commonly derived of an optical signal scanning the whole slide with a high resolution.

2.1 Gene Expression Data Characteristics

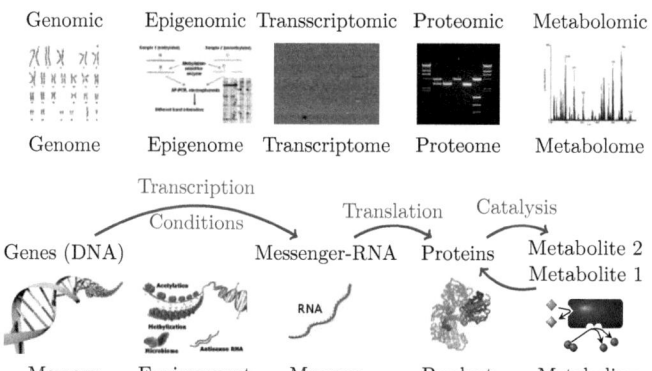

Figure 2.1: Sketch of the main areas of bioinformatics (upper line) in combination with typical data, the term of the entirety of features (the omes), the biological interactions between the cells molecules and a more colloquial terminology lend from informatics.

The Gene Expression Omnibus (GEO, see Edgar, Domrachev, and Lash (2002)) is a data base for high-throughput functional genomic data and a free source of such data. The ArrayExpress data base (Parkinson, Kapushesky, et al. (2009)) is another repository for this type of data, but includes most of the GEO data sets. Though there are other free online data bases for gene expression data, a view on the available technologies on ArrayExpress gives an impression about the distribution of the experiments on the different methods. Table 2.1 on the next page lists the number of experiments for the main high throughput technologies, which provide quantitative information about the presence of transcripts in a cell medium.

Miller and Tang (2009) briefly describe the available array technologies for gene expression experiments. The microarray technologies (chip technologies from Table 2.1) mainly differ in the way how to deposit the complementary oligonucleotides on the chip surface. This can be done by printing copies of these base sequences as a whole on the array (cDNA chip) or generate the sequence base by base with a photolythographic procedure (Affymetrix, NimbleGen) or respectively by contactless

Table 2.1: Gene expression experiments found in the ArrayExpress repository grouped by technology and time series experiments. The numbers were obtained by queries in the *ArrayExpress Browser* on 1st of November 2013. The overall used keyword was "RNA assay". For the Time series value "ef:time" was used. For the microarray technologies the keyword *by array* was combined with the corresponding manufacturer/chip name, i.e "NimbleGen", "Agilent", "Affymetrix", "Illumina", "cDNA". The numbers for the other technologies are obtained by the keyword in the left column.

Manufacturer / Technology	Data Sets	Time Series
cDNA arrays; spotted chips	2330	196
Roche NimbleGen chips	671	53
Agilent one color chips	3991	413
Affymetrix one color chips	17232	1652
Illumina Bead chips	2745	294
RT-PCR + RT-qPCR	915	62
sequencing assay	2924	183

printing (Agilent). Illumina bead arrays use beads with identical oligonucleotides on their surface, which randomly assort to the wells on the array. This necessitates an additional mapping step, which also serves as quality control. Another difference between the array technologies is the number of bases in the oligonucleotides fixed on the array beginning with 20 (Affymetrix) up to several hundreds (spotted arrays). There is a trade-off between sensitivity and specificity related to the length of the sequence. Longer sequences provide higher sensitivity but lower specificity and vice versa (see Miller and Tang 2009, p. 612–614). In contrast to chip technologies the Real-time polymerase chain reaction (RT-PCR) techniques and the up-coming high-throughput sequencing methods (also known as "next generation sequencing") give information about both the transcript quantity and sequence of the probes' oligonucleotides. This allows to quantify even those transcripts, whose sequence was unknown or not yet considered in the expression experiments.

Essentially, all of the technologies shown in Table 2.1 are in principle suited to be analyzed on a gene set level in time. Nevertheless this thesis focuses on the Affymetrix technology due to the fact that microarrays of this manufacturer were used for about two thirds of all gene expression time series experiments listed in the ArrayExpress data base.

Affymetrix GeneChips

Dalma-Weiszhausz, Warrington, et al. (2006) provide a detailed description of the Affymetrix GeneChip technology and the corresponding experimental setup for measuring gene expression from cell RNA. A scheme of the Affymetrix microarray construction and a gene expression experiment is shown in Figure 2.2 on the following page. A transcript on Affymetrix microarrays is represented by a *probe set* of typically 11 oligonucleotides of 25 base pairs length. Each oligonucleotide appears twice on a chip; once with the real target sequence as perfect match (PM) probe and once as mismatch (MM) probe. The MM probe of the *probe pair* differs only by the 13th base from the PM probe and is included due to recognized signal from cross-hybridization, nonspecific binding or technical noise. A photolithographic process applies millions of identical oligonucleotide sequences onto an 11 μm^2-sized feature space on the quartz chip surface corresponding to each probe sequence. The manufacturer Affymetrix sells gene expression arrays for many species with tens of thousands probe sets representing the exon regions (part of a gene sequence transcribed to RNA) of a priori known transcript-sequences.

In order to obtain the gene expression measurements, the RNA extracted from various cells must be amplified. The amplification is achieved by reverse transcription to cDNA, which is thereafter repeatedly transcribed to Biotin-labeled RNA. After a fragmentation step the labeled short RNA-strands are applied onto the Affymetrix array, where the labeled RNA-strands complementary hybridize to their oligonucleotide counterparts. Non-binding and weakly binding RNA-oligonucleotides are washed out by a cleanup step. The Biotin-labeled bases are stained with a fluorescent-dye afterwards and after another cleanup step the hybridized chip is read out by a high definition laser scanner. Only those feature spaces with stained RNA on it send fluorescent light back to the sensor. The corresponding measured light intensity is a measure for the quantity of hybridized RNA and cell RNA, respectively.

RNA from different experimental conditions has to be examined on different chips due to the array design. Furthermore, there are no replicates for the expression probe sets on the array. Hence, replicates for the gene expression measurements

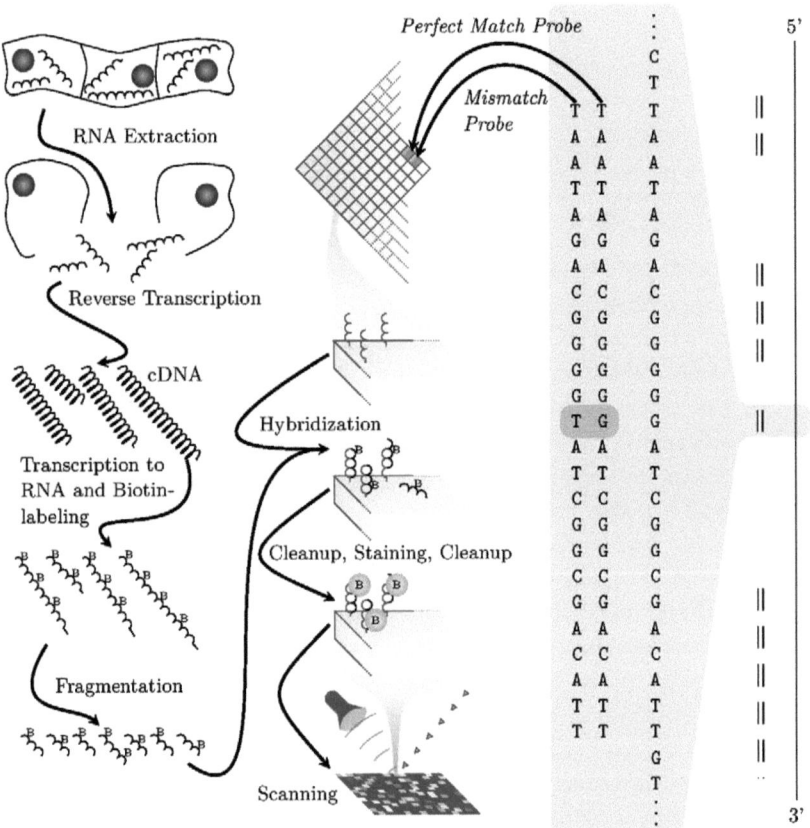

Figure 2.2: General construction of an Affymetrix gene expression microarray and the experimental work flow to obtain quantitative data for the examined transcripts.

are only available, if multiple chips are hybridized under the same condition. The intensity values from the scanner are also called *raw expression values* and commonly stored in .cel files. These intensities are biased due to several reasons and hence have to be preprocessed before applying a differential expression analysis.

Data Preprocessing by Robust Multiarray Analysis

There are several preprocessing algorithms for microarrays proposed in the literature, which can be divided into the three steps: background correction, normalization across arrays and summarization of multiple probes per transcript target. Robust Multiarray Analysis (RMA) is an approach introduced by Irizarry, Bolstad, et al. (2003), which accomplishes these three preprocessing steps.

Optical noise and non-specific binding to the oligonucleotides on the array lead to an unspecific background signal even for absent target transcripts and make a corresponding correction necessary. Z. Wu, Irizarry, et al. (2004) refine the background model of RMA in the sense, that the different binding affinity of the four nucleotide bases is taken into account for the estimation of unspecific binding. The empirical Bayes version of their GC-RMA algorithm is applied as background correction to the raw gene expression data in this thesis.

Each array is processed under individual conditions, which may also include individually prepared RNA samples from different laboratories and read by different Scanners. Those conditions may affect the measured intensities, but do not provide information about the (differential) transcript expression. Therefore, a normalization step is needed to make the expression values comparable across all arrays hybridized in the study. RMA uses quantile normalization to unify the distribution of the probe values across all arrays.

In the Affymetrix array design multiple probes in one probe set quantify the same transcript. The different loci on the sequence and the varying sequence-dependent binding affinity of the probes may lead to a significant variation in the signal strength among the probes within a probe set. A robust summary resistant to outliers is needed in order to combine the probe set intensities to obtain a reliable expression value for the whole transcript. Tukey's median polish procedure is applied by

RMA to estimate the probe set expression value, which is synonymously denoted as transcript or gene expression value in the following.

Time Series Gene Expression Data

Gene expression time series experiments provide insight into the molecular biology processes inside an organism over time. Usually time series experiments focus on the changing transcription after suspending a stress condition (e.g. starvation or a drug) or on the gene activation during a natural process (e.g. embryonic development of a certain organ). The analysis is challenging due to the biological variability between the examined individuals and even between cells of the same individual, which may have different stages in their cell cycles. The analysis is further complicated by the circumstance that most experiments cover only a few time points, which are often not uniformly sampled. In contrast to static experiments the number of biological or technical replicates at the same time point is very restricted as illustrated in Figure 2.3. Since the collection of gene expression data comes along with the destructive testing of several cells no real longitudinal data is available even if the same individuals are tracked across time. All in all the sample size in a typical gene expression time series experiment is small in contrast to the tens of thousands genes analyzed in parallel by microarrays, which is also known as a $p \gg n$ situation. Some of the typical problems facing gene expression time series analysis can be mitigated by considering a priori defined gene sets instead of single genes as is described in the following section.

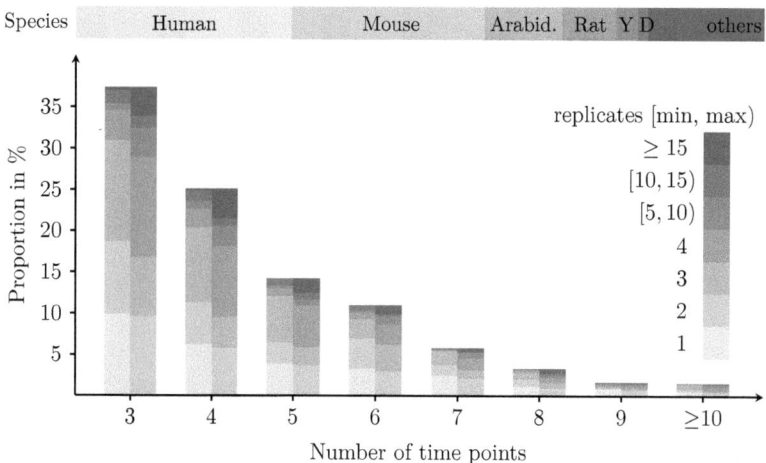

Figure 2.3: The Gene expression experiments on Affymetrix arrays found in the ArrayExpress repository are grouped by the number of time points and available replicates per time point (minimum and maximum for the experimental condition only, i.e. without controls). The numbers were obtained by queries in the *ArrayExpress Browser* at the 1st of November 2013. The overall used query was "RNA assay ef:time Affymetrix". On top the distribution of the species is shown (Y denotes yeast and D drosophilia melanogaster).

2.2 Definitions of Gene Sets

The statistical gene expression analysis is hampered by various problems. As mentioned before, only few replicates per experimental condition are faced with quantitative information for tens of thousands transcripts. The non-experimental variance on the measured light intensities is quite high due to varying technical conditions during the separate hybridization of each chip in the experiment and due to the undoubtedly existing biological differences on the gene expression level, which may result from different cell cycle states or different genetic or epigenetic conditions in the examined individuals or cells. Furthermore, the variance increases proportional to the measured intensity, which makes a detection of differential expression between two or more experimental conditions more difficult for a single transcribed gene. In Affymetrix microarrays often more than one probe set interrogates the quantitative expression of the same gene, which may lead to contradictory signals, because of the different hybridization affinity of these probe sets or because of the abundance of only one of the interrogated transcript variants (e.g. SNPs, splicing variants). All in all the information from a typical microarray gene expression experiment for a single gene alone is not as trustworthy as needed to draw definitive conclusions.

On the other hand, most gene products interact with each other and form networks or can be classified in clusters of similar functions or their contribution to the same biological process. Gene sets can be defined by aggregating the genes corresponding to the gene product classes formed by the functional pathway in the network or their molecular function in biological processes. Other sources of gene set definition might be the position on chromosome, a common regulatory motif (e.g. transcription factor) or the knowledge of preliminary experiments for instance related to certain diseases.

Analyzing gene expression data on a gene set level can provide more reliable biochemical reactions involved in the pathways.

Gene Ontology (GO) Gene Set Definition

The GO project has the goal to provide "a structured, precisely defined, common, controlled vocabulary for describing the roles of genes and gene products in any organism" (Ashburner, Ball, et al. 2000). For this purpose, three ontologies are introduced to organize the biological terms of *biological processes*, *molecular functions* or *cellular components* and describe the relation of the terms to one another with well-defined relationships. Each ontology forms a directed acyclic graph with its terms and relations. Gene products can be assigned to corresponding GO terms regarding to their function or location. The matching genes or respectively the matching transcripts are assigned to the same terms, which allows the definition of gene sets. The gene to GO term annotation is induced by different sources. The evidence codes indicate, whether the annotation comes from an experiment, from a computational analysis, from author statements, from curator statements, or from unsupervised electronic software algorithms. However, it has to be emphasized that these evidence codes cannot be used explicitly as a quality measure of annotation and therefore all evidence types are considered in the following analyses. The inter-term-relations or edges in the GO graph are a subset operator in a mathematical sense. Therefore, all genes annotated to a specific GO term are also assigned to a more general parent term which is linked to the former by a relation. The GO terms and relations are well-defined across species, but the annotation of genes to the terms depends on the understanding of the underlying biology. Due to this, the ontologies and the annotation are a continuously changing work in progress, but the benefit from the current biological interpretation is high.

Kyoto Encyclopedia of Genes and Genomes (KEGG) Pathway Gene Set Definition

The Kyoto Encyclopedia of Genes and Genomes provides daily manually updated functional information about genes and ligand-protein-reactions (Kanehisa and Goto 2000). Kanehisa, Goto, et al. (2012) give an overview about the 15 main databases managed by KEGG. The pathway maps include "graphical diagrams representing knowledge on molecular interaction and reaction networks for metabolism, genetic information processing, environmental information processing, cellular processes, organismal systems, human diseases and drug development"(Kanehisa, Goto, et al. 2012).. The cross-species gene annotation by KEGG allows to define gene sets according to the pathway maps and makes the underlying biological information about the molecular reactions usable in the analysis of gene expression experiments.

BioCarta Pathways as Gene Set Definition

BioCarta LLC is a profit oriented US company, but provides the freely available *Proteomic Pathway Project*. The "open-source" project invites the research community to share their knowledge about the "molecular or cell-to-cell interactions" (www.biocarta.com). Newly submitted pathway annotations are checked by a peer-review-system. Currently pathway information is available for human and mice. For each pathway the reference, the review history, and the contributing protein/gene list is given, which allows for creating gene sets based on the submitted pathway annotation.

Reactome Pathway Driven Gene Sets

The Reactome data base was originally restricted to human reactions and pathways (Vastrik, D'Eustachio, et al. 2007). However, the need to compare the knowledge on the human level with such from model organisms like mice, which are more feasible for invasive experiments, leads to more than 20 data bases displaying the reactions and interactions of proteins and ligands across different species. The possibility to cross-reference the pathway proteins with the corresponding genes yield gene set definitions according to the peer-reviewed expert knowledge of biological processes and reactions inside and outside of cells (Matthews, Gopinath, et al. 2009).

BioCyc Data Bases as Gene Set Source

The BioCyc data base collection includes more than 1500 Pathway/Genome data bases for the most species whose genome is completely sequenced. The collection is divided into three quality classes (Tier 1–3) regarding the curation intensity for the originally computationally created networks. The PathLogic program uses existing genome annotations and combines the found genes with the metabolic pathway information already understood in other organisms or respectively existent in the species-transcending MetaCyc data base (Karp, Ouzounis, et al. 2005). A likelihood-score is used to predict the presence probability of the pathway in the species. This automatic creation of biochemical networks has a special charm, since due to the common descendants the metabolic pathways of all species are highly correlated. On the other hand annotation mistakes and changes during the evolution can lead to a misunderstanding of the underlying biology not only for a single but also for a set of different organisms. Hence, the contribution of the research community in the evaluation of the networks is essential. The genes involved in each pathway can be used to construct a gene set.

2.3 Experimental Designs, References and Formalization

The classical statistical time series analysis examines a long list of a single value (e.g. temperature) observed within a time window, which could (theoretically) be expanded into both directions, the past and the future. The *Markovian assumption*, that every time points depends only on its direct predecessor(s), often facilitates the modeling of states, trends and periodicity. The wide field of time series analysis allows to take additional model coefficients or structural breaks into consideration, but is not directly applicable to cope with the very different nature of gene expression time series, which is briefly described in the last paragraph of section 2.1.

Figure 2.4 on the next page shows typical data collected for a single gene in a time course experiment. Different experimental designs are illustrated. The simplest case is one expression value per time point. This design of experiment does not allow to consider the variability of the gene expression data in contrast to a replicated time course experiment. A distinction should be made between technical replicates (from the same individual) and biological replicates (different individuals). Only the latter allows to consider the biological variation in the analysis, while the amount of measurement errors can be estimated by the first one. Some approaches discriminate between longitudinal (gene expression from the same individual observed over time) and cross-sectional (different individuals observed at different time points) time course experiments. The idea behind longitudinal designs is to mitigate the biological variance. However, the measurement of gene expression destroys the cells from which the RNA is taken. Hence, it is uncertain, whether the longitudinal design brings real benefits for the analysis and does not falsify the results because of the intervention in the living organism or tissue. Sometimes the (replicated) time series are divided by a single factor as the health or the treatment status. If there is only one factor, the proposed methods in chapter 4 can be applied. If a multi-factorial design is used or an even more complex design (e.g. continuous cofactors), then simple gene set activation profiles would not be the aim of the study and the proposed methods are not applicable.

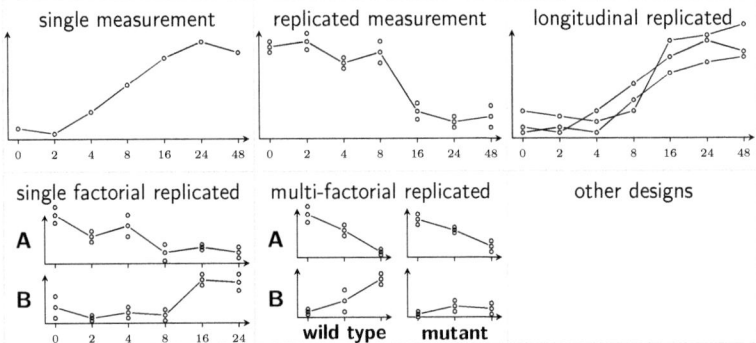

Figure 2.4: Most common time series types resulting from the design of experiment shown for a single gene of a gene expression microarray time series experiment. The time scale is plotted on the horizontal axis and the gene expression on the vertical axis. The measurements are shown as circles and the interpolated gene expression trajectories are drawn by straight lines. The number of time points is typically small as well as the number of replicates. Additional factors are usually discrete or categorized. The number of genes in the experiment is normally much greater than 10,000.

The approaches proposed in this thesis allow to find differential expression profiles over time for a priori defined gene sets with respect to a reference. This reference can be a mean or median expression profile for each gene on the chip over all time points or alternatively depending on the experimental design a reference presented in Figure 2.5 can be used. Another, here not considered approach would be to use the direct previous time point as reference to focus on changes within the studied time interval.

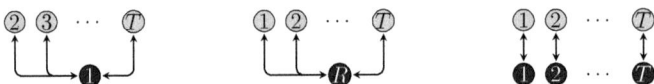

Figure 2.5: Three possible reference types: A reference time point within the time series (left), a single reference measurement (center) or a reference time series with an identical sampling scheme (right).

2.3 Experimental Designs, References and Formalization

In microarray time series experiments there is often an additional effect besides time included, for instance the tissue, the health status or one treated and one untreated group. For the control or reference experiment there are usually less replicates available than for the time points under experimental condition. Rarely, there is data available in form of a whole control time series experiment.

Formalization and Notation

This section presents the mathematical notation used in the method chapters. Since this thesis does not focus on preprocessing the analysis is based on the preprocessed and hence logarithmized gene expression values. A single gene expression value of the mth measurement for gene g at time point t is denoted by x_{tjm}, where

$$t \in \{1, \ldots, T\}, \quad g \in \check{G} = \{g_1, \ldots, g_G\} \quad \text{and} \quad m \in \{1, \ldots, M\}.$$

$\check{G}$ is the whole gene universe and it is assumed, that every gene g is annotated to at least one gene set $s \in \check{S} = \{s_1, \ldots, s_S\}$. Hence, it holds $\check{G} = \bigcup_{i=1}^{S} s_i$ and $s_i \subseteq \check{G} \ \forall s_i \in \check{S}$. The capital letters T, G, M and S describe the total numbers of time points, genes, replicates and gene sets in the analysis.

A symbol list of all used notations in the thesis is given on page 277.

CHAPTER 3

Methods for the Analysis of Gene Expression Time Series

The statistical analysis of typical gene expression time course experiments is in general an explorative analysis, though the most methods provide a significance measure like a p-value or promise to hold an a priori determined significance border, for instance a family wise error rate (FWER) or a false discovery rate (FDR). Hence, it is proven standard to evaluate findings from a microarray study in an external experiment or an independent study. The main reason for this is the small sample size, which hampers to distinguish between the effects of time, experimental condition and the omnipresent biological (and technical) variance. Another reason is the adjustment for multiple experiments (e.g. tens of thousands genes), which may result in no or only few significant findings and poor power of the method. Therefore, the hard significance border is often mitigated to a ranked list of findings, which includes some random items.

There is an uncounted variety of published methods to analyze gene expression time series experiments. Bar-Joseph (2004) gives an early review on analyzing this type of data. X. Wang, M. Wu, et al. (2008) list analysis strategies with an emphasis on clustering and incorporation of multi-source information. *Gene-module level analysis*, hence the identification of gene groups and the group interactive dynamics is the topic of a review by X. Wang, Dalkic, et al. (2008). A more recent review by Coffey and Hinde (2011) focuses on the functional data analysis (FDA) of gene expression microarray time courses. It is not the intention of this work to explain all these procedures in detail, but to give a survey about the main fields in this research area, which will shed a light on the great importance of this data type.

Sima, Hua, and Jung (2009) provide a survey about the inference of gene regulatory networks (GRN) based on gene expression time course experiments. A large number of samples is needed to estimate GRNs while the total number of genes is typically restricted (e.g. $G = 9$ genes in Bansal, Gatta, and Di Bernardo (2006)). The inference of GRN does not make use of the strength of micoarrays to provide information about tens of thousands genes, but yields relevant insights into the microbiological gene-to-gene interactions. The outcome of a GRN analysis can be used as a gene set definition in the framework of a comprehensive gene set analysis.

A second way to analyze gene expression time series is to identify genes, which are differentially expressed across the examined time period. This single gene analysis is described in more detail in section 3.1. The influence of the noise on the single gene results is quite large and hence a gene set analysis is often applied as second stage in the analysis, but the temporal information is in general lost on this level and hard to reconstruct.

The idea that a similar gene expression trajectory over time indicates a common function or contribution to the same biological process is the foundation of the various existing clustering approaches for the analysis of gene expression time series. Section 3.2 gives a brief overview of the methods proposed in the literature. Most clustering approaches provide an exhaustive partition of the gene universe and hence allow formulation of hypotheses for the function of genes, where this information is not yet available. Conversely, the hard cut between the clusters seems not to be a good assumption for modeling the complex interaction between the true biologic processes, because one gene product may contribute to different biological processes in different cell states.

There are methods which does not fit into the three former categories. Those, which are related to gene set analysis are summarized in section 3.3. Some others are designed to identify genes jointly regulated by so called transcription factors (TFs) in time series experiments (Das, Nahlé, and M. Q. Zhang 2006; L. Wang, G. Chen, and Li 2007). Holter, Maritan, et al. (2001) attempt to explore the causal relationship among genes by deducing a time translational matrix on the characteristic modes specified by singular value decomposition (SVD). Gene expression time series data is also used to identify regulatory modules and condition-specific regulators (Segal, Shapira, et al. 2003). The discovery of genes with periodic gene expression may also be an analytic goal, see J. Chen and Chang (2008) for a review. Hirose, Yoshida, et al. (2008) attempt to simultaneously identify transcriptional modules (gene sets) and analyze the resulting module-based networks within a state space model, whose applicability is limited to time series with more than ten time points and a comparatively small number of genes (several thousands). Still other approaches face the task of classifying time series profiles in categories according to prototypes of trajectories over time. Such a model gene expression pattern can be acquired from previous experiments with respect to molecular reaction to toxins (Hafemeister, Costa, et al. 2011) or in relation to the progress of a disease (Costa, Schönhuth, Hafemeister, et al. 2009). Finally, there are methods to analyze gene expression time series in order to identify the different timing of gene expression under certain conditions (Yoneya and Mamitsuka 2007) or to distinct between a consensus gene expression trajectory and an individual response to a drug in clinical studies (Kaminski and Bar-Joseph 2007).

In the following the focus lies on published methods, which allow to analyze time series experiments on the level of a priori defined gene sets.

3.1 Gene Set Testing Based on Differentially Expressed Genes

In principle, all methods providing a list of significant genes can be used for an enrichment analysis on the gene set level. By using just the list of genes significantly changing their expression over time has the drawback of loosing all information about the form of differential expression across the examined time points. Another disadvantage is the large influence of noise on the resulting gene list and hence its bad performance in terms of reproducibility.

Nevertheless, methods identifying significant differences in the gene expression values across time are largely used. They can be roughly grouped by the way they assign significance to the gene trajectories.

Three statistics to identify significantly differentially expressed genes between two equally sampled gene expression time courses are proposed by Di Camillo, Toffolo, et al. (2007). The method uses the maximal difference or the area between the linear or spline interpolated gene expression measurements across the given time points. The significance assessment follows from various null distributions estimated by using replicate measurements.

Some approaches provide a gene ranking according to differential expression across time. Chuan Tai and Speed (2009) use a multivariate empirical Bayes approach to sort genes according to their differential expression within one or between two or more gene expression time series (i.e. two or more experimental conditions besides time). Kalaitzis and Lawrence (2011) rank differentially expressed genes through a likelihood ratio quotient or a Bayes factor after modeling gene trajectories by Gaussian process regression. The approach of Cheng, X. Ma, et al. (2006) investigates the changes in an estimated gene neighborhood by the mean absolute rank difference (MARD) and hence relies completely on gene expression ranks per time point.

Other methods directly model the gene expression under various conditions and experimental designs directly on the discrete sampled time series. Xu, Olson, and Zhao (2002) use a regression based statistical modeling approach in combination with permutation tests to find significantly differentially expressed genes. *Limma* is a very popular procedure fitting linear models to the gene expression values and using moderated tests in the analysis of variance (ANOVA) framework to assign significance to its findings (G. K. Smyth 2004; G. Smyth 2005). Park, Yi, et al. (2003) apply ANOVA models in combination with F– or permutation tests to identify significant time-group-interactions or effects of experimental groups. ElBakry, Ahmad, and Swamy (2012) propose a modified repeated measure ANOVA, which removes the variance caused by individual differences and assigns significance per permutation of columns (i.e. within time points and replicates). The idea of

M. J. Nueda, Conesa, et al. (2007) is to utilize a principal componant analysis (PCA) for a dimension reduction of the estimated parameters from an ANOVA model in multiple series time course experiments. Genes contributing to a significant effect in this so called ANOVA-SCA model are identified by permutation tests on the leverage or squared prediction error.

Hidden Markov models (HMM) are a further class of statistical tools applied in the gene-wise analysis of gene expression time-course experiments. Yuan and Kendziorski (2006) estimate non-homogeneous HMMs to distinguish between the two states *equally expressed* and *differentially expressed* at each time point. Hidden spatial-temporal Markov random fields are used by Wei and Li (2008) to identify genes, which are differentially expressed at each time point in the context of known biological pathways.

A wide range of methods originates from the field of functional data analysis (FDA). They have in common, that the measured gene expression trajectory is modeled as continuous function in time. Bar-Joseph, Gerber, Simon, et al. (2003) apply gene-wise hypotheses testing on the integral of the quadratic difference between the B-spline curves of two aligned gene expression time series experiments. Storey, Xiao, et al. (2005) introduce the popular *Extraction of Differential Gene Expression* (EDGE) software and method for the identification of differentially expressed genes. The procedure fits a natural cubic spline representation of the gene expression trajectory under the alternative hypothesis and a constant mean curve under the null hypothesis. Permutation testing based on the residual sums of squares of both models assigns significance to the detected differentially expressed genes. A functional hierarchical model on the p-dimensional B-spline basis gene expression curves is utilized by Hong and Li (2006). *Bayesian Analyisis of Time Series* (BATS) is a software tool proposed for one-sample time series (Angelini, De Canditiis, et al. 2007; Angelini, Cutillo, et al. 2008). The functional Bayesian approach expands the gene temporal profiles over an orthonormal basis and assigns significance for differential gene expression in the form of Bayes factors. P. Ma, Zhong, and J. Liu (2009) utilize a functional ANOVA mixed-effects model to identify either non-parallel differentially expressed genes or parallel differentially expressed genes. X. Liu and Yang (2009) apply a functional principal component analysis to test for changes in the temporal gene expression under different conditions.

The advantage of gene wise approaches is the consideration of all transcripts measured on the chip. Hence, hypotheses about unknown gene-functions can be made. The power of gene-wise approaches is low, because smaller effects caused by an experimental factor will not show up on a top position in the list of significant genes and are therefore not considered for a detailed view. Although a subsequent gene set testing step based on the gene-wise tests mentioned above would be straightforward and would give additional insights into the underlying biology, it is not very often applied in the literature. The clustering approaches in the next section are more often combined with an analysis of predefined gene sets.

3.2 Clustering Approaches

Clustering approaches intend to identify groups of genes with a coherent gene expression over time. The underlying idea of these procedures is, that the similar gene expression trajectory results from a common cause (e.g. a transcription factor) or a common function. The clustering methods can be discriminated in three main fields the similarity based approaches, the model based procedures and template based methods, which attempt to recognize genes with a gene expression time profile similar to predefined patterns.

Similarity based gene clusters can for instance be found by simple visual inspection (Cho, Campbell, et al. 1998). Some clustering algorithms need a predetermined total number of clusters as the k-means procedure (Tavazoie, Hughes, et al. 1999) or in the self organizing map (SOM) framework (Tamayo, Slonim, et al. 1999). The standard agglomerative hierarchical clustering has been applied by Eisen, Spellman, et al. (1998) or Magni, Ferrazzi, et al. (2008), the latter provide the software package `TimeClust`, which implements several clustering techniques (e.g. Bayesian clustering). Brown, Grundy, et al. (2000) propose to use a supervised learning algorithm, which is based on support vector machines (SVMs), to group genes with

3.2 Clustering Approaches

unknown function to clusters with a priori known function. The CLICK algorithm (Sharan, Shamir, et al. 2000) combines graph-theoretical and statistical techniques to find homogeneous gene expression clusters. J. Kim and J. H. Kim (2007) define the similarity on symbolic vectors representing the first and second order differences between adjacent time points. *Gene Shaving* is introduced by Hastie, Tibshirani, et al. (2000) as a cluster algorithm, which applies sequential PCA techniques to identify those genes, which are both largely varying across time and coherent to each other. Tchagang, Bui, et al. (2009) propose clustering in a *rank order preserving matrix* framework or by identifying *minimum mean squared residue clusters*.

The second type of cluster procedures groups the genes either based on the model fit of their gene expression trajectory in time, by the application of a specific clustering model, or a combination of both. Ben-Dor, Shamir, and Yakhini (1999) propose a corrupted clique graph model for the non-hierarchical clustering of genes. Fitting a mixture of multivariate Gaussian distributions to the gene expression values is the approach of Yeung, Fraley, et al. (2001) available in the MCLUST software package. Šášik, Iranfar, et al. (2002) attempt to cluster genes involved in a specific biological process based on a biological kinetic model. The expectation maximization (EM) algorithm is used by Bar-Joseph, Gerber, Gifford, et al. (2002) to cluster genes on the basis of their cubic spline representation in a predefined number of sets. Ramoni, Sebastiani, and Kohane (2002) introduce CAGED (Cluster analysis of gene expression dynamics), a pseudo-Bayesian agglomerative clustering approach applied on auto-regressive gene expression models. An extension of this procedure based on polynomial models for describing the gene expression trajectory in the framework of a Bayesian hierarchical mixture model was published by L. Wang, Ramoni, and Sebastiani (2006). Luan and Li (2003) utilize the EM algorithm to fit a mixed effects model on the B-spline representations of the gene expression profiles. The EM algorithm is also used by Chudova, Hart, et al. (2003) for modeling a mixture of simplified differential equations in order to cluster genes according to their gene expression time course. Schliep, Schonhuth, and Steinhoff (2003) applied model based clustering based on linear HMMs, which is implemented in the Graphical Query Language (GQL) (Costa, Schönhuth, and Schliep 2005). An approach to infer gene clusters from finite mixtures of HMMs while using prior informations in a semisupervised learning framework is proposed by Schliep, Costa, et al. (2005). The approach called maSigPro (microarray significant profiles) identifies gene clusters of differentially expressed genes by a two-step regression approach, where the algorithm is based on the similarity of the gene-wise regression model coefficients (Conesa, M. J. Nueda, et al. 2006). Heard, Holmes, and Stephens (2006) introduces a Bayesian hierarchical clustering of nonlinear regression spline representation of the gene expression trajectories. A mixture of mixed-effects models, which applies a rejection controlled EM algorithm to estimate the class assignment and the corresponding mean expression curves is used by P. Ma, Zhong, and J. Liu (2009). Scharl, Grün, and Leisch (2010) find clusters using the EM algorithm to fit mixtures of linear models or linear mixed models.

Template based methods attempt to identify predefined patterns in the temporal gene expression trajectories. In contrast to most other clustering approaches significance can be assigned to the various template clusters by permutation or resampling procedures. Peddada, Lobenhofer, et al. (2003) propose an order-restricted inference methodology defining candidate temporal profiles in terms of inequalities among the mean expression levels at the time points. T. Liu, N. Lin, et al. (2009) introduce the ORICC algorithm, which groups the gene trajectories according to an order-restricted information criterion to pre-specified candidate inequality profiles. A hierarchy of trend temporal abstraction profiles (consisting of steady, increasing, decreasing) serve as prototypes for the clustering of the gene expression profiles in the approach of Sacchi, Bellazzi, et al. (2005). The EPIG method uses a multi-step filtering procedure to generate representative candidate patterns from the data (Chou, Zhou, et al. 2007). StepMiner software (Sahoo, Dill, et al. 2007) identifies genes with one or more binary transitions across the gene expression time series by modeling segment-wise constant adaptive regression. Ramakrishnan, Tadepalli, et al. (2010) introduce the GOALIE procedure, which uses linear time logics to identify segments in the time series and separate gene clusters with coherent gene expression behavior within the segments. Ernst and Bar-Joseph (2006) implemented their method in the Short Time-series Expression Miner (STEM) software. The STEM procedure matches the gene expression profiles on data-independent chosen model profiles and applies a time point permutation test to assign significance to the corresponding gene clusters. A Fisher test is used to identify GO gene sets enriched with genes from significant clusters. Springer, Ickstadt, and Stoeckler (2011) propose a data-driven selection of model profiles, which gains a better fit to the data structure, but with the drawback of loosing the significance assessment for the identified clusters.

An own implementation of the STEM method based on the median gene expression profiles is used to compare the performance of the proposed methods (see chapter 4) in a simulation setting (see chapter 5) and for the example data sets (see chapter 6).

3.3 Other Approaches for the Gene Set Analysis of Gene Expression Time Series

There are published methods, which do not fit in the former two main categories, but attempt to analyze gene expression time series on a gene set level. Hvidsten, Lægreid, and Komorowski (2003) propose a systematic supervised learning approach to generate hypotheses about the function of genes not yet annotated to any considered predefined GO gene set. The procedure is based on learning a classification rule model within the rough set framework, which is evaluated by cross validation. Other approaches assume that the gene expression of all genes in a predefined gene set follows the same trajectory, which is hence estimated by regarding the genes as multiple measurements of the same pattern in time. The *GlobalANCOVA* procedure by Hummel, Meister, and Mansmann (2008) fits a linear model to the gene expression values for every gene set and identifies those groups, in which a design factor (e.g. treatment-time interaction) is significant in contrast to a reduced model. L. Wang, X. Chen, et al. (2009) construct a unified mixed effects model for the mean trajectory of every gene set and claim that their algorithm detects gene sets, in which at least 20 – 50 % of the genes follow the same linear trend. A nonparametric Wald-type test statistic in combination with a permutation based test is used by K. Zhang, H. Wang, et al. (2011) to detect treatment effects or treatment-time interactions in predefined gene sets. The principal components analysis through conditional expectation (PACE) is proposed by Yao, Muller, and J. Wang (2005) to estimate the mean trajectory function for sparse longitudinal data and apply it on yeast cell cycle gene sets. The idea of fitting regression models to the gene set gene expression values in the time series or to the corresponding principal components is introduced by M. Nueda, Sebastián, et al. (2009), who denote their method as `maSigFun` or respectively as `PCA-maSigFun`. The first approach assumes, that all group genes follow the same underlying trajectory, but the second allows for more than one model profile per group, but due to the regression on the principal components, the group profile is hard to reconstruct. The `maSigFun` is used as competitor in the simulation study (see chapter 5 and the data analysis (see chapter 6). A brief introduction to the method is given in section 4.7. The implementation maintained in the original article is used.

None of the approaches for the analysis of gene expression time series known to the author intends to estimate an activity profile for an a priori defined gene set. The assumption that all or at least a majority of genes in a gene set defined by its general function follows a common gene trajectory is very strict and counterexamples may easily be found. Imagine complex functions like embryonic development or immune response, whose gene expression in healthy organisms needs to be well balanced and fine-tuned. Hence, it is implausible that all genes related to those general processes show the same (functional) expression trajectory at all possible time points. This thesis focuses on the idea to identify a gene set activity with respect to a reference at the examined time points and the therefor proposed methods are the subject of the next chapter.

CHAPTER 4

Estimating Temporal Activation Profiles of Gene Sets

In contrast to the numerous approaches proposed in the literature to analyze gene expression time series the objective of this thesis is to estimate a temporal activation profile with respect to a reference for an a priori defined gene set. Most existing methods which intent to make use of the knowledge provided by reasonably defined gene sets focus on the clustering of genes with a common differential expression profile or use a significance measure on the gene wise differential expression before a common gene set enrichment test is applied (see chapter 3). Most common gene set definitions are mentioned in section 2.2. It seems to be a promising approach to estimate an activation profile over the examined time points for every gene set of interest in order to obtain an exploratory view into the molecular functions summarized by the gene set of the current definition. For instance, it allows the researcher in a way of *helicopter view* to obtain a sight on the bigger picture of biological functions or protein pathways without getting lost in the list of hundreds of significant genes.

Since genes interact in a way that gene products of an early stage gene may suppress the gene expression of their own genetic information, but promote the gene expression of other genes (e.g. those in a following stage of the pathway), an average gene expression profile seems not to be a good representation for the activation of a gene set. The proposed approaches in this chapter therefore are derived from ideas from enrichment testing for the analysis of two condition gene expression microarray experiments, which are described in section 4.1.

The proposed algorithms follow a three stage procedure, which is illustrated in Figure 4.1 on the next page. The input of the algorithms includes the preprocessed gene expression values for the time points and reference, the gene set definitions, and parameters for testing and smoothing. The first step provides the information of differential expression with respect to the reference for all genes and every time point either qualitative (up, down, not different), ordinal (ranked gene list) or quantitative (shrunken t-test values). The second step identifies gene sets which are enriched with up or down expressed genes. The three algorithms use different statistical tests to detect enrichment, either as over representation of differential expressed genes in the sets (Fisher test) or as asymmetric distribution of the gene set genes across the ranked list (segment test and GSEA like test). These profile algorithms are described in sections 4.2 and 4.5. Smoothing of non-reliable alternating profiles, which are hard to biologically interpret is the final step of the algorithms. The smoothing algorithms, which are evaluated by a simulation study in chapter 5, are explained in detail in section 4.3. A proposal to rank the resulting temporal activation profiles is given in section 4.4.

All methods described in chapters 4 and 5 are implemented in R-code (using R-version 2.14.3, see R Development Core Team (2012)) programmed by the author and is available on request.

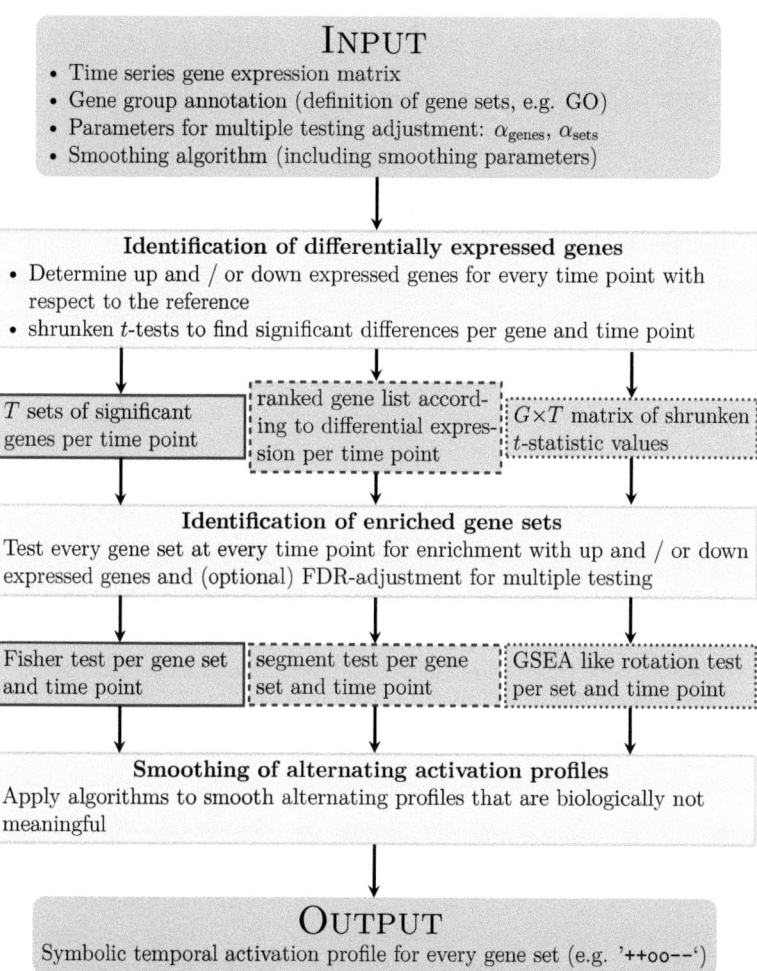

Figure 4.1: Schematic illustration of three possible algorithms for the estimation of gene set activation profiles. The differences of the three methods are shown in the intermediate steps enclosed by solid (threshold variant of the GSA-type algorithm), dashed (non-threshold GSA-type algorithm), or dotted (GSEA by rotation testing) line.

4.1 Methods Adapted from Gene Set Testing For Two Condition Gene Expression Experiments

Microarray studies conducted to compare only two conditions or respectively a single bivariate experimental factor (e.g. healthy and diseased or stem cell and differentiated cell) often provide much more replicates per condition than time series experiments. Hence, in the gene set analysis (GSA) of such experiments permutation and resampling approaches (either for the condition label or the genes) are quite common to estimate the null distribution of an enrichment test. The order of measurements is not arbitrary in time series experiments and the assumption of the exchangeability of genes seems not to be justified, since the gene expression is obviously correlated. Due to these reasons only those methods can be regarded, which can abandon permutation and resampling.

Detection of Differentially Expressed Genes

Cui and Churchill (2003) give a review of most common statistical tests for differential expression in gene expression experiments. In case of no replicates the relative gene expression of case and control samples – the so called *fold change* – is applied. If the expression values are assumed to be logarithmized to base two, the *fold change* for a gene g is defined as:

$$\text{FC}_g = 2^{x_g^B - x_g^A},$$

where A and B denote the two different experimental conditions. If the assumption of equal variances across the gene expression intensities is made (which is usually not true), an average fold change can be applied.

The t-test is a standard method to detect differential expression if replicates are available. Despite its optimality for normal distributed data it is also applicable for other distributions. The test statistic is calculated as:

4.1 Adapted Methods from GSA in Two-Condition Experiments

$$D_g^{t-\text{test}} = \frac{\bar{x}_g^B - \bar{x}_g^A}{\text{sd}_g},$$

where sd_g denotes the sample estimate of the standard deviation, provided by

$$\text{sd}_g = \sqrt{v_g^A/M^A + v_g^B/M^B} \qquad v_g^{A/B} = \frac{1}{M^{A/B} - 1} \sum_{m=1}^{M^{A/B}} \left(x_{gm}^{A/B} - \bar{x}_g^{A/B}\right)^2$$

and the sampling means for gene g are given by

$$\bar{x}_g^{A/B} = \frac{1}{M^{A/B}} \sum_{m=1}^{M^{A/B}} x_{gm}^{A/B}.$$

These test statistic values take into account the variance of the measurements across the samples. Due to the usually small sample size the estimated variances per gene and condition are not stable and may lead to extreme large test statistic values because sd_g is small by chance for some genes. To account for this problem some approaches were published to modify the standard t-test. Tusher, Tibshirani, and Chu (2001) propose to add a constant sd_0 in the denominator of the test statistic

$$D_g^{\text{SAM}} = \frac{\bar{x}_g^B - \bar{x}_g^A}{\text{sd}_g + \text{sd}_0}$$

and call the new statistic Significance Analysis of Microarrays (SAM). With some computational costs it is possible to find the optimal value of sd_0. Opgen-Rhein and Strimmer (2007) apply a distribution-free shrinkage estimate to detect differential gene expression. The gene wise sample estimate of the variance v_g is replaced by a shrinkage estimator

$$v_g^{\text{shrink}} = \hat{\lambda} v^{\text{median}} + (1 - \hat{\lambda}) v_g,$$

with

$$\hat{\lambda} = \min\left(1, \frac{\sum_{g=1}^{G} \widehat{\text{Var}}(v_g)}{\sum_{g=1}^{G} (v_g - v^{\text{median}})^2}\right).$$

The term v^{median} denotes the median of the v_g across all genes and $\widehat{\text{Var}}(v_g)$ is calculated as follows

$$\widehat{\text{Var}}(v_g) = \frac{M}{(M-1)^3} \sum_{m=1}^{M} (w_{gm} - \bar{w}_g) \quad w_{gm} = (x_{gm} - \bar{x}_g) \quad \bar{w}_g = \frac{1}{M} \sum_{m=1}^{M} w_{gm}.$$

Opgen-Rhein and Strimmer (2007) propose two versions for determining $\hat{\lambda}$, namely a separate calculation for both conditions ($M = M^A$ or $M = M^B$) and a common calculation $M = M^A + M^B$. In this work the variant with the separate estimation across all conditions is preferred and referred to as

$$D_g^{\text{shrink}} = \frac{\bar{x}_g^B - \bar{x}_g^A}{\sqrt{\frac{v_{g,A}^{\text{shrink}}}{M^A} + \frac{v_{g,B}^{\text{shrink}}}{M^B}}}.$$

The activation profile algorithms (see section 4.2) use this last modified t-test statistic to detect differentially expressed genes for the comparison of each time point in the experimental setting with a common reference.

Detection of Gene Set Enrichment with Differentially Expressed Genes

Testing functional gene groups in two conditions gene expression experiments reduces the dimensionality of the underlying statistical problem and typically increases power compared with a single gene analysis. There are various approaches of gene set testing in the literature (e.g. parametric analysis of gene enrichment (PAGE) by S.-Y. Kim and Volsky (2005)). Ackermann and Strimmer (2009) describe the standard framework of gene set enrichment analysis with a significance assessment by sampling approaches. They list the typical gene set statistics used in a framework with resampling significance assessment in the field of gene set testing and compare their performance in combination with different gene level statistics and their transformations. According to Al-Shahrour, Carbonell, et al. (2008) Gene Set Enrichment Analysis (GSEA) is one of the most popular enrichment tests. GSEA is based on a Kolmogorov-Smirnov-type statistic for each considered gene set across a ranked gene list (Subramanian, Tamayo, et al. 2005). The significance is evaluated by repeating the algorithm on permutations on the phenotype assignments. Since significance assessment by permutation and resampling methods seems not to be a good choice in the context of gene expression time series experiments due to the dependencies among time points and genes explained at the beginning of chapter 4, the focus of this thesis lies on gene set methods avoiding such techniques. There are a few approaches (e.g. Y.-T. Huang and X. Lin 2013) attemting to take the covariance structure into account for the gene set analysis, but this methods are not directly applicable to the time series setting due to the lack of observations in relation to the large number of model parameters.

A contingency table is defined (see Table 4.1) by separating the gene universe $\breve{G}$ in differentially expressed (DE) genes and those which are not on the one hand and in genes annotated to a gene set s and those which are not on the other hand. Since the marginal totals are fix due to the a priori definition of the gene sets (threshold definition for DE), the exact null distribution of $N_{DE \in s}$ – the unrealized number of DE genes in gene set s – is hypergeometric (Rivals, Personnaz, et al. 2007). An enrichment is considered if the probability of the random value $N_{DE \in s}$ to be equal

Table 4.1: Contingency table for Fisher's exact test in the gene set enrichment analysis.

	$\in$ DE	$\notin$ DE	$\sum$
$\in$ gene set s	$n_{DE \in s}$	n_{12}	$\lvert s \rvert$
$\notin$ gene set s	n_{21}	n_{22}	$G - \lvert s \rvert$
$\sum$	n_{DE}	$n_{\setminus DE}$	G

or higher than the observed $n_{\text{DE}\in s}$ is very small. This probability is calculated as

$$p^E = P(N_{\text{DE}\in s} \geq n_{\text{DE}\in s}) = 1 - \sum_{i=0}^{n_{\text{DE}\in s}-1} \frac{\binom{|s|}{i}\binom{G-|s|}{n_{\text{DE}}-i}}{\binom{G}{n_{\text{DE}}}}.$$

In some applications the detection of a depletion with DE genes is desirable. The probability

$$p^D = P(N_{\text{DE}\in s} \leq n_{\text{DE}\in s}) = \sum_{i=0}^{n_{\text{DE}\in s}} \frac{\binom{|s|}{i}\binom{G-|s|}{n_{\text{DE}}-i}}{\binom{G}{n_{\text{DE}}}}$$

is small in case of a depletion. The values p^E and p^D are also known as the p-values of a one-sided Fishers's exact test. There exist different approaches for determining a p-value for the simultaneous test for enrichment and depletion (Rivals, Personnaz, et al. 2007; Hosack, Dennis, et al. 2003). The activation profile algorithms proposed in this thesis use only the enrichment probabilities for up and down regulated genes.

Assessment of significance in a multiple experiment setting

The parallel hypothesis testing for many genes and many groups has to account for an overall error rate for the assessment of significance to the single tests. Accepting a usual significance level in the single tests yields many type-I-errors (i.e. reject true null hypotheses). One strategy to account for multiple testing is the family wise error rate (FWER), which limits the probability for the rejection of at least one true null hypothesis. The strong control of the FWER for instance by the Bonferroni-correction (Bonferroni 1936) is in many situations too restrictive and will lead to many missing findings, in particular if the number of tests is large.

An approach of identifying as much true significant features as possible, while limiting the proportion of false positives (rejection of true null hypotheses) among all rejected hypotheses was introduced by Benjamini and Hochberg (1995). The limit of the expected proportion of falsely rejected null hypotheses is called false discovery rate (FDR). The FDR became the standard significance measure for the simultaneous testing of multiple hypotheses in gene expression studies and various procedures have been proposed to control this quantity. The parallel testing of

4.1 Adapted Methods from GSA in Two-Condition Experiments

gene expression differences makes FDR procedures desirable, which account for the potential dependence between the genes and the corresponding test results. K. I. Kim and Wiel (2008) conducted a simulation study analyzing the behavior of various FDR controlling algorithms applied in the gene expression context under dependence. The two-stage adaptive linear step-up procedure (TS-ABH) proposed by Benjamini, Krieger, and Yekutieli (2006) shows the at least conservative estimation of the true FDR under dependence without being anti-conservative.

The TS-ABH procedure uses the linear step-up procedure of Benjamini and Hochberg (1995) to estimate the number of true null hypotheses n_{H_0} and is briefly described in the following with the terms of Table 4.2. The term FP denotes the erroneously rejected null hypotheses and $n_{\hat{H}_1}$ is the total number of rejected null hypotheses. The false discovery rate (FDR) is defined as expected proportion of falsely rejected null hypotheses among all rejections

$$\text{FDR} := E(Q), \qquad Q = \begin{cases} \text{FP}/n_{\hat{H}_1}, & \text{if } n_{\hat{H}_1} > 0 \\ 0, & \text{if } n_{\hat{H}_1} = 0 \end{cases}.$$

The step-up procedure makes use of the ordered observed p-values of the n hypothesis tests

$$p_{(1)} \leq p_{(2)} \leq \ldots \leq p_{(n)}$$

and rejects the first k hypotheses in this order for

$$k = \max\{i \colon p_{(i)} \leq qi/n\}$$

to control the FDR at level q if such a k exists (otherwise accept all null hypotheses).

Table 4.2: Possible results of multiple testing for significance.

	accept	reject	$\sum$
true null hypotheses	TN	FP	n_{H_0}
true alternatives	FN	TP	n_{H_1}
$\sum$	$n_{\hat{H}_0}$	$n_{\hat{H}_1}$	n

The TS-ABH procedure uses the linear step-up procedure at the level $q/q+1$ in the first stage and adopts the number of rejected hypotheses k^{I} to estimate an upper bound of true null hypotheses

$$\hat{n}_{H_0} := n - k^{\text{I}}$$

The realization of this estimate is utilized in the second run of the linear step-up procedure, but replacing the FDR threshold q by

$$q^{\text{II}} := \frac{q}{q+1} \cdot \frac{n}{\hat{n}_{H_0}}.$$

The k^{II} of the second stage still controls the FDR at level q, but since $k^{\text{II}} \geq k^{\text{I}}$ the two-step procedure provides a higher power (see Definition 6 in Benjamini, Krieger, and Yekutieli 2006).

In accordance with Storey and Tibshirani (2003) a single-hypothesis significance measure is denoted as q-value $q_{(i)}$ and calculated by choosing the smallest possible FDR threshold q yielding a rejection of the hypothesis corresponding to p-value $p_{(i)}$ (using the TS-ABH procedure; details are shown in Appendix A). This adjusted p-value is the smallest FDR controlled by the adaptive two-stage procedure if the ith test and all tests with a smaller p-value than $p_{(i)}$ would be rejected.

The following sections in chapter 4 provide three algorithms to assign a gene set s with a simple activation profile AP, indicating the enrichment with up or down regulated genes in the gene set with respect to a reference.

4.2 GSA-Type Algorithms for Determining Gene Set Activation Profiles

In this section three algorithms are proposed for the estimation of gene set activation profiles, which are closely related to the standard functional gene set testing described in section 4.1. The first algorithm is based on the gene set enrichment idea and uses

4.2 GSA-Type Algorithms for Determining Gene Set Activation Profiles

two FDR thresholds to combine the enrichment test results at every time point to an activation profile per gene set. The second method is inspired by the threshold free segmentation test proposed by Al-Shahrour, Díaz-Uriarte, and Dopazo (2005). The third introduced method restricts the segmentation test approach of the second procedure to the significant part of the differential expressed genes (with respect to FDR threshold α_{genes}) and hence combines the two previous approaches.

Threshold Variant of the GSA-Type Activation Profile Estimation

The first step of the algorithm identifies seperately at every time point such genes, which are significantly differentially up expressed or down expressed with respect to a reference. In principle, every procedure yielding information about significant up and down differential expression is suitable for this task. The information about the *direction* of differential expression with respect to the reference in the time series experiment is important due to the intention to estimate a gene set activation profile indicating an enrichment with up or down expressed genes.

If replicates are available the shrinkage t-test of Opgen-Rhein and Strimmer (2007) is applied on the differences to the reference

$$D_g^{\text{shrink}(t)} = \frac{\bar{x}_g^{(t)} - \bar{x}_g^{(\text{Ref})}}{\text{sd}_g^{\text{shrink}}}$$

(for the possible reference types see section 2.3). This yields $G \cdot T$ test statistics. A gene g is considered as up expressed at time point t if the test statistics results in a positive value and denoted as down expressed for a negative value of the test statistic. The significance of these differential expression tests is assigned by transforming the test statistics to the q-values of the adaptive step-up procedure of Benjamini, Krieger, and Yekutieli (2006) over all tests done on the gene level. The single test p-values for each test are needed to apply the q-value procedure. These are determined in the same way as in the one-sided standard (two sample) t-tests with respect to the reference type. Let α_{genes} be the upper limit for the FDR across all shrinkage t-tests for the gene expression difference of a gene against the reference. The two sets

$$\mathrm{DE}_+^{(t)} = \left\{ g \colon D_g^{\mathrm{shrink}(t)} > 0 \wedge q_g^{(t)} \leq \alpha_{\mathrm{genes}} \right\}$$
$$\mathrm{DE}_-^{(t)} = \left\{ g \colon D_g^{\mathrm{shrink}(t)} < 0 \wedge q_g^{(t)} \leq \alpha_{\mathrm{genes}} \right\}$$

include all significantly differentially expressed genes per time point t (with $q_g^{(t)}$ as the q-value of gene g at time point t). If there are no replicates available, it is possible to define $\mathrm{DE}_+^{(t)}$ and $\mathrm{DE}_-^{(t)}$ by a fold change (FC)-threshold α_{genes}.

The second step identifies predefined gene sets, whose genes are overrepresented among the significantly up or down expressed genes. Two enrichment tests per gene set and time point are performed; one for the enrichment with up regulated genes and one for the enrichment with down expressed genes. The one-sided Fisher's exact test is used to verify enrichment as shown in section 4.1. If $n_{\mathrm{DE}_+^{(t)} \in s}$ and $n_{\mathrm{DE}_-^{(t)} \in s}$ denote the observed numbers of significantly up and down expressed genes

$$n_{\mathrm{DE}_+^{(t)} \in s} = \left| s \cap \mathrm{DE}_+^{(t)} \right| \qquad n_{\mathrm{DE}_-^{(t)} \in s} = \left| s \cap \mathrm{DE}_-^{(t)} \right|,$$

the Fisher test enrichment p-values for gene set s at time point t are given by

$$p_{s+}^{E(t)} = 1 - \sum_{i=0}^{n^\star - 1} \frac{\binom{|s|}{i} \binom{G - |s|}{n^{\star\star} - i}}{\binom{G}{n^{\star\star}}} \qquad n^\star = n_{\mathrm{DE}_+^{(t)} \in s} \quad n^{\star\star} = n_{\mathrm{DE}_+^{(t)}}$$

$$p_{s-}^{E(t)} = 1 - \sum_{i=0}^{n^\star - 1} \frac{\binom{|s|}{i} \binom{G - |s|}{n^{\star\star} - i}}{\binom{G}{n^{\star\star}}} \qquad n^\star = n_{\mathrm{DE}_-^{(t)} \in s} \quad n^{\star\star} = n_{\mathrm{DE}_-^{(t)}}$$

(compare the contingency table terms in Table 4.1).

The corresponding q-values ($q_{s+}^{E(t)}$, $q_{s-}^{E(t)}$) from the adaptive step-up procedure (see section 4.1 or Benjamini, Krieger, and Yekutieli 2006) allow to control the FDR over all $2 \cdot S \cdot T$ tests with respect to an in advance determined threshold α_{sets}. The combination of the two enrichment tests per gene set s and time point t is

4.2 GSA-Type Algorithms for Determining Gene Set Activation Profiles

transformed into a symbol characterizing the gene set activation

$$AP_s(t) := \begin{cases} + & \text{if } q_{s+}^{E(t)} \leq \alpha_{\text{sets}} \text{ and } q_{s-}^{E(t)} > \alpha_{\text{sets}} \\ - & \text{if } q_{s+}^{E(t)} > \alpha_{\text{sets}} \text{ and } q_{s-}^{E(t)} \leq \alpha_{\text{sets}} \\ \pm & \text{if } q_{s+}^{E(t)} \leq \alpha_{\text{sets}} \text{ and } q_{s-}^{E(t)} \leq \alpha_{\text{sets}} \\ \circ & \text{if } q_{s+}^{E(t)} > \alpha_{\text{sets}} \text{ and } q_{s-}^{E(t)} > \alpha_{\text{sets}}. \end{cases}$$

Concatenating the symbols across all time points in the experiment yields the gene set activation profile

$$AP_s := [AP_s(1), \ldots, AP_s(T)].$$

A profile like '++oo--' for a gene set s can easily be interpreted as overrepresentation of up expressed genes for the first two time points in that gene set and enrichment with down expressed genes for the last two time points, while the middle two time point are not enriched with any of the two kinds of differentially expressed genes. The direction (i.e. up or down expression) has always to be regarded as change with respect to the reference and not to the previous time point in the examined time series. In the following, this gene set activation profile algorithm is abbreviated with 2T-GSA.

Threshold-free Variant of the GSA-Type Activation Profile Estimation

The choice of the two thresholds α_{genes} and α_{sets} has a substantial impact on the resulting gene set activation profiles. It would be desirable if the researcher does not have to choose a specific significance threshold. Further the activation profile should be reasonable. Al-Shahrour, Carbonell, et al. (2008) propose for their Babelomics web tool a threshold free enrichment test referred to as *segmentation test*. The procedure proposed in the following is similar to the one in the article by Minguez, Al-Shahrour, and Dopazo (2006), but the application in the article focuses on identifying time intervals, where the genes in gene sets according to GO biological processes are overexpressed. The method described in the following estimates a symbolic activation profile for enrichment with down regulation and up regulation in the considered gene sets.

The procedure is based on an ordered list of features (e.g. genes or proteins) according to their significance. This list is divided into two sets: on the one hand the d most significant features and on the other hand the $J - d$ other features, for a varying d. Fisher's exact test is applied on each partitioning of the gene universe $\breve{G}$ to identify a priori defined feature sets enriched with significant features. The control of the FDR over all enrichment tests ensures a limited proportion of false positive (FP) findings. The authors propose to test each feature set for 20 to 50 different values for d and assign the q-value of the most significant partition as significance measure for the enrichment of the tested feature set. Al-Shahrour, Díaz-Uriarte, and Dopazo (2005) present a more detailed description of the original procedure.

The approach of the segmentation test is adopted to obtain an activation profile procedure without the need to choose a specific threshold on the gene level. The starting point is an ordered gene list according to the observed values of the shrinkage t-test statistics

$$L_{\text{DE}}^{(t)} := \left\{ g_{(1)}, \ldots g_{(G)} \colon D_{g_{(1)}}^{\text{shrink}(t)} \leq \ldots \leq D_{g_{(G)}}^{\text{shrink}(t)} \right\}.$$

Hence, the most down expressed genes with respect to the reference at time point t are at the beginning and the most up regulated are at the end of the ordered list. Since the number of enrichment tests is increased by the factor T (number of time points), only 20 partitions are applied for the segmentation test. These are defined by the $1, 4, \ldots, 28$ percentiles of the ordered gene list for the enrichment with down expressed genes and analogously as $72, 75, \ldots, 99$ percentiles for the up regulation. The FDR is controlled across all $40 \cdot T \cdot S$ enrichment tests by the adaptive step-up procedure of Benjamini, Krieger, and Yekutieli (2006) at the level α_{sets}. In order to enable smoothing on the activation profile every gene set s is assigned with two q-values from the step-up procedure, which are defined as

$$\breve{q}_{s+}^{E(t)} := \min q_{s,d}^{E(t)}, \qquad d \in \{72, 75, \ldots, 99\},$$
$$\breve{q}_{s-}^{E(t)} := \min q_{s,d}^{E(t)}, \qquad d \in \{1,, 4, \ldots, 28\},$$

4.2 GSA-Type Algorithms for Determining Gene Set Activation Profiles

where $q_{s,d}^{E(t)}$ denotes the q-value of the enrichment test using d as percentile for partitioning the gene universe. This q-value definition assigns the q-value of the most significant of the partitions considered. The activation profile is derived analogously to the threshold variant of the approach

$$\mathbb{P}_s^{\mathrm{tl}}(t) := \begin{cases} + & \text{if } \check{q}_{s+}^{E(t)} \leq \alpha_{\mathrm{sets}} \text{ and } \check{q}_{s-}^{E(t)} > \alpha_{\mathrm{sets}} \\ - & \text{if } \check{q}_{s+}^{E(t)} > \alpha_{\mathrm{sets}} \text{ and } \check{q}_{s-}^{E(t)} \leq \alpha_{\mathrm{sets}} \\ \pm & \text{if } \check{q}_{s+}^{E(t)} \leq \alpha_{\mathrm{sets}} \text{ and } \check{q}_{s-}^{E(t)} \leq \alpha_{\mathrm{sets}} \\ \circ & \text{if } \check{q}_{s+}^{E(t)} > \alpha_{\mathrm{sets}} \text{ and } \check{q}_{s-}^{E(t)} > \alpha_{\mathrm{sets}} \end{cases}$$

and concatenated across all time points in the experiment the gene set activation profile for the non-threshold variant is defined as

$$\mathbb{P}_s^{\mathrm{tl}} := \left[\mathbb{P}_s^{\mathrm{tl}}(1), \ldots, \mathbb{P}_s^{\mathrm{tl}}(T) \right].$$

This approach does not apply a single threshold for significance on the gene level, but uses several thresholds for the partitioning of the gene universe. If the number of tested partitions is not too large the power of detecting gene sets with an asymmetric distribution across the ordered gene list may be higher than in the variant with a single fixed significance threshold, despite the FDR control for a larger number of tests. This approach is referred to as 1S-GSA algorithm in the following.

Threshold-Segmentation GSA-Type Activation Profile Estimation Considering the Significance of Differential Expressed Genes

The simple threshold-free approach of the previous section using quantiles of the test statistics for differential gene expression does not account for the significance of the differences. Hence, the number of genes regarded for the gene set enrichment tests is too large in case of only few genes changing their expression. In order to benefit from both approaches, controlling the significance of the differential expression with a threshold α_{genes} and the increased power from the segmentation test for the identification of asymmetries in the gene set proportion of differential expressed genes, the two methods are combined to a third method, which is called Two-threshold segmentation GSA-type activation profile algorithm (2S-GSA) method. Only those genes, which are significantly differential expressed according to a TS-ABH FDR threshold of α_{genes} are separated into ten segments at each extreme in order of their test statistics $D_g^{\text{shrink}(t)}$ at every time point. An enrichment test is applied for each of these ten segments in up and ten segments in down regulation considering all genes in the actual and in the more extreme segments as differential expressed. The corresponding activation profile AP_s^{ts} for a gene set s is constructed analogously to the activation profiles AP_s^{tl} and AP.

4.3 Smoothing of GSA-Type Gene Set Activation Profiles

Due to the separate testing of the experimental time points activation profiles GSA-type algorithms sometimes have an alternating structure (e.g. '+o++o-o'). Such a profile is hard to interpret and unlikely the biological truth for a process, especially if the time points are close to each other. For instance in an embryonic development microarray experiment it is not very likely, that a development process is turned on and off repeatedly, but it is more likely, that the process has a starting point and an end and is thereafter inactive in an adult individual. The reasons for alternating and noncontinuous activation profiles are diverse. In a gene set including just a few genes, the unavoidable variance in the differential expression of a single gene may cause the misdetection of the enrichment with differentially expressed genes at a time point if the gene is differentially expressed, but to a smaller amount than for the other time points. The opposite case may also occur when one or more genes are randomly differentially up expressed to a large extent and causes an overrepresentation with up expressed genes in this gene set at the random time point. The small number of replicates per gene increases this effect, but summarizing the information for all genes of a gene set mitigates the influence of the single gene variation in particular for gene sets of reasonable size. The enrichment approach provides a protection against the influence of extreme outliers. Nevertheless, it is desirable to estimate a reasonable non-fluctuating activation profile for gene sets of any size, in particular

4.3 Smoothing of GSA-Type Gene Set Activation Profiles

since the small sets are often the most specialized and hence more interesting for the interpretation than the larger ones. A smoothing of the gene set activation profiles of GSA-type yields more reasonable and interpretable profiles.

This section describes the methods considered for smoothing activation profiles with an alternating unreasonable structure. The methods are originally designed for the threshold variant of the GSA-type algorithm, but the summarizing methods are also suitable for the non-threshold variant. The subprofiles considered for smoothing are shown in Table 4.3. For each position considered for smoothing one of the algorithms explained in the following is applied. Smoothing is only considered if a non significant positions are adjacent to enriched positions with the same direction of enrichment. Hence, in the profile '+o+++−' the second position is considered for smoothing, but the last one is not. Smoothing is not considered if the lack of significance (or the surplus) covers more than one position (e.g. in '++oo++') or for smoothly changing profiles (e.g. in 'ooo−−−'). In case of long time series those subprofiles may be reasonably considered for smoothing and the methods proposed in the following can easily be adapted.

The strength of smoothing for the different algorithms is controlled by two smoothing parameters λ_{fill} and λ_{wipe}, which have opposite influence of the smoothing extent in dependence on the particular algorithm.

Table 4.3: Subprofiles in a gene set activation profile considered for smoothing. Two directions of smoothing are distinguished. The fist smoothing direction in the second and third column fills a enrichment gap in the profile (either for up expressed genes in the second column or down expressed genes in the third column). The other smoothing direction wipes out a significant enrichment for a single position in the activation profile. The exact position considered for smoothing is emphasized by fading.

| direction | fill | | wipe | |
position	original	smoothed	original	smoothed
begin	o++..	+++..	+oo..	ooo..
	o−−..	−−−..	−oo..	ooo..
middle	..+o+..	..+++..	..o+o..	..ooo..
	..−o−..	..−−−..	..o−o..	..ooo..
end	..++o	..+++	..oo+	..ooo
	..−−o	..−−−	..oo−	..ooo

The smoothing methods can be divided into five categories, from which only the first is also applied to the segmentation test variants (1S-GSA and 2S-GSA) of the GSA-type gene set activation profile algorithm. All smoothing methods have in common that they use the two adjacent profile positions to decide whether a considered position is smoothed or not. The categories are

1. Summarization of the Fisher test q-values

2. Determination of gene continuity

3. Shift of genes in the Fisher test contingency table

4. Evaluating the distance from significance

5. Utilization of the sequential information in the time series.

Smoothing by Summarizing the Fisher Test q-Values
The first idea is to calculate a weighted mean of the (transformed) q-values of the enrichment tests for the three time points around the position considered for smoothing. The smoothing position is denoted as t. For simplification, we assume in the following that t is neither the first nor the last time point. Let further be $A\!\!P_s(t) = {\tt +}$ or $A\!\!P_s(t-1) = A\!\!P_s(t+1) = {\tt +}$, F the cumulative distribution function of a standard normal distribution and F^{-1} its inverse. The *weighted inverse normal score mean* (INSM) denoted by $^{\text{INS}}q_{s+}^{(t)}$ is obtained as follows:

$$\text{INS}_{s+}^{(t)} := F^{-1}(q_{s+}^{(t)})$$

$$^{\text{INS}}q_{s+}^{(t)} = F\left(\frac{0.5\,\text{INS}_{s+}^{(t-1)} + w\text{INS}_{s+}^{(t)} + 0.5\,\text{INS}_{s+}^{(t+1)}}{1+w}\right).$$

Hence, the weight is used for the inverse normal score of the q-value at the smoothing position in contrast to the adjacent positions, whose weights sum up to 1. The cumulative distribution function F projects the weighted mean back in the range of q-values. The weight w is chosen according to the parameter $\lambda_{\text{wipe}}^{\text{INS}}$ if $A\!\!P_s(t) = {\tt +}$ (and $A\!\!P_s(t-1) = A\!\!P_s(t+1) = {\tt o}$) or as $\lambda_{\text{fill}}^{\text{INS}}$ if $A\!\!P_s(t) = {\tt o}$ (and $A\!\!P_s(t-1) = A\!\!P_s(t+1) = {\tt +}$).

4.3 Smoothing of GSA-Type Gene Set Activation Profiles

This procedure is similar but different to Stouffer's Z-score method with weights due to the smoothing intention instead of an agreement for independent test decisions. A natural alternative for the distribution function F would correspond to the χ^2 distribution with one degree of freedom (due to testing in a 2×2 contingency table), which is denoted by $^{\text{IXS}}q_{s+}^{(t)}$. Applying a weighted geometric or arithmetic mean directly to the q-values leads to similar results ($^{\text{GM}}q_{s+}^{(t)}$ and $^{\text{AM}}q_{s+}^{(t)}$).

Switching the position from $P_s(t) = \text{o}$ to $P_s(t) = +$ occurs only if $^{\text{INS}}q_{s+}^{(t)} \leq \alpha_{\text{sets}}$, and for the other direction smoothing occurs only if $^{\text{INS}}q_{s+}^{(t)} > \alpha_{\text{sets}}$.

Smoothing by Determination of Gene Continuity

The second idea for smoothing does not make use of a smoothing parameter. A correction term is used to count all genes that are differentially up expressed in the current gene set at time points $t-1$ and $t+1$ but not for t, if $P_s(t) = \text{o}$. In case of $P_s(t) = +$ in the example setting above the correction term is given by the negative number of genes in the gene set up expressed only for time point t but not for $t-1$ or $t+1$:

$$\text{ct}^{\text{conti}} := \begin{cases} \left| s \cap \left(\left[\text{DE}_+^{(t+1)} \cap \text{DE}_+^{(t-1)} \right] \setminus \text{DE}_+^{(t)} \right) \right|, & \text{if } P_s(t) = \text{o} \\ -\left| s \cap \left(\text{DE}_+^{(t)} \setminus \left[\text{DE}_+^{(t+1)} \cup \text{DE}_+^{(t-1)} \right] \right) \right|, & \text{if } P_s(t) = + \end{cases}.$$

The enrichment test is recalculated for the new $n_{\text{DE}_+^{(t)} \in s}^{\text{conti}} = n_{\text{DE}_+^{(t)} \in s} + \text{ct}^{\text{conti}}$ and compensating the other contingency table entries under the condition of fixed marginal sums.. The resulting p-value is transformed into a FDR-adjusted p-value $^{\text{conti}}q_{s+}^{(t)}$ as if all other enrichment p-values would have the same value as without smoothing. The smoothing decision is made analogously to the inverse normal score mean (INSM) method

$$^{\text{conti}}P_s(t) = \begin{cases} \text{o}, & \text{if } ^{\text{conti}}q_{s+}^{(t)} > \alpha_{\text{sets}} \wedge P_s(t) = + \\ +, & \text{if } ^{\text{conti}}q_{s+}^{(t)} \leq \alpha_{\text{sets}} \wedge P_s(t) = \text{o} \\ P_s(t), & \text{otherwise}. \end{cases}$$

Thus, smoothing occurs only if the correction term leads to a significant q-value in the case of filling the gap of significance in the profile or if the recalculation leads to a non-significant q-value if the *wipe out* smoothing direction is used.

Smoothing by Shift of Genes in the Fisher Test Contingency Table

A shift of genes within the contingency table is the idea of the third method. A correction term ct^{shift} is used, similar to the gene continuity approach. Here, in the case of potential smoothing, ct^{shift} genes from the gene set are moved from "differential expression" to "non-differential expression" or vice versa. The smoothing parameter α_{wipe} and α_{fill} determine the proportion of genes shifted. In the exemplary situations from above for a '+o+' subprofile or respectively a single '+' position between two 'o' positions the correction term is calculated as follows

$$ct^{shift} := \begin{cases} -\max\left\{\left\lceil \alpha_{wipe} \cdot n_{DE_+^t \in s} \right\rceil, n_{DE_+^{(t)}} \right\}, & \text{if } \mathbb{P}_s(t) = + \\ \max\left\{\left\lceil \alpha_{fill} \cdot (|s| - n_{DE_+^t \in s}) \right\rceil, G - n_{DE_+^{(t)}} \right\}, & \text{if } \mathbb{P}_s(t) = o \end{cases}$$

with $\lceil \cdot \rceil$ denoting upper Gaussian brackets and time point t as the middle time point, which is considered for smoothing. It is bounded by the total numbers of genes in the gene set and the total numbers of differentially up regulated genes at t and the total number of the complementary not up expressed genes. Analogously to the gene continuity smoothing the next step is the recalculation of the enrichment test with

$$n^{shift}_{DE_+^{(t)} \in s} = n_{DE_+^{(t)} \in s} + ct^{shift}$$

and compensating the other contingency table entries under the condition of fixed marginal sums. The resulting p-value is transformed into a q-value $^{shift}q_{s+}^{(t)}$ as if all other enrichment p-values would have the same value as without smoothing. The smoothed position $^{shift}\mathbb{P}_s(t)$ is determined analogously to the gene continuity and q-value summarizing approaches.

4.3 Smoothing of GSA-Type Gene Set Activation Profiles

Smoothing by Evaluating the Distance from Significance

The fourth method calculates the "relative distance from significance" for the number of differentially expressed genes in a gene set s at a time point t, where smoothing is considered. A small distance should be an incentive for smoothing. In the example situation of above, in which the smoothing method has to decide between enrichment with up regulated genes ($\textbf{P}_s(t) = +$) or no enrichment ($\textbf{P}_s(t) = \text{o}$) the term

$$\check{n}_{\text{DE}_+^{(t)} \in s} := \min\left\{n_{\text{DE}_+^{(t)} \in s} : q_{s+}^{E(t)} \leq \alpha_{\text{sets}}\right\}$$

denotes the boundary for significant enrichment for gene set s at time point t. The expression

$$d_{s+}^{(t)} := \frac{n_{\text{DE}_+^{(t)} \in s} - \check{n}_{\text{DE}_+^{(t)} \in s}}{\min\{|s| - n_{\text{DE}_+^{(t)} \in s},\ n_{\text{DE}+}^{(t)} - n_{\text{DE}_+^{(t)} \in s}\}}$$

gives a distance from significant enrichment with up regulated genes, with respect to the residual number of genes in group s and with respect to the residual number of up regulated genes. This distance determines the smoothing decision depending on the two threshold values α_{wipe} and α_{fill}:

$$^{\text{dist}}\textbf{P}_s(t) := \begin{cases} \text{o}, & \text{if } d_{s+}^{(t)} \in [0, \alpha_{\text{wipe}}) \wedge \textbf{P}_s(t) = + \\ +, & \text{if } d_{s+}^{(t)} \in [-\alpha_{\text{fill}}, 0) \wedge \textbf{P}_s(t) = \text{o} \\ \textbf{P}_s^{(t)}, & \text{otherwise.} \end{cases}$$

Smoothing by Using the Sequential Information in the Time Series

The last smoothing algorithm utilizes the sequential structure of the time series to perform smoothing. For a gap time point t, associated for instance with a sub profile 'o+o' (at time points $t-1$, t, $t+1$) we test for differential gene expression between time points $t-1$ and t as well as for differential gene expression between time points t and $t+1$. More explicit, the corresponding gene set s is analyzed with respect to an enrichment with differentially expressed genes, comparing time point t with its two neighbors, respectively. This analysis can be performed in an undirected or directed manner. In the directed case, in the example, we test twice for an enrichment with up regulated genes at time point t, once compared with time point $t-1$ and once with time point $t+1$. For testing, the same combination of shrinkage t-test and Fisher's exact test as in the standard profile algorithm before smoothing is used, while controlling the FDR at both the gene level and the gene set level. If both comparisons yield no significant enrichment with respect to the FDR thresholds α_{wipe} (if $\textit{\textbf{P}}_s(t) = $ +) or α_{fill} (for $\textit{\textbf{P}}_s(t) = $ o), then smoothing is applied. In the undirected manner the direction of differential expression is not taken into account and therefore two sided shrinkage t-tests are applied before the enrichment tests. In order to control the FDR all gene sets considered for smoothing at any of the examined time points have to be tested parallel for the enrichment with sequential differential expression with respect to the corresponding neighboring time points.

4.4 Ranking of GSA-Type Gene Set Activation Profiles

The outcome of the GSA-type gene set activation profile algorithms is an activation profile per gene set. The combination of the gene set defining knowledge and the estimated profile can be used to identify interesting findings and to generate new hypotheses about the molecular genetic processes in the examined time series. However, if the number of examined gene sets S is large, a ranking of the non-constant-'o' profiles would facilitate the analysis of the results. In order to identify gene sets with a conspicuous enrichment with up expressed genes and discriminate those from noticeable extreme findings with down regulated genes with respect to the reference a score is proposed, which summarizes the shrunken t-statistics for the significantly differentially expressed genes in the direction indicated by the current profile position. The summarization is done by calculating the mean over the medians of the statistics at the enriched time points and the score is denoted by

D_s^{med} for gene set s. It is defined by

$$D_s^{\text{med}} := \frac{1}{|\{t: \textit{\textbf{P}}_s(t) \notin \{\text{'o'},\text{'}\pm\text{'}\}\}|} \sum_{t:\, \textit{\textbf{P}}_s(t)\notin\{\text{'o'},\text{'}\pm\text{'}\}} \text{median}\left\{D_g^{\text{shrink}} : g \in \text{DE}_{\textit{\textbf{P}}_s(t)}^{(t)} \bigcap s\right\}.$$

Both extreme tails of the D_s^{med} score ranking are interesting. Low values for D_s^{med} indicate a clear under expression with respect to the reference at the stated positions in the profile and a high value gives a hint towards clear over expression. It has to be noted, that an activation profile, which has both enrichment types for instance '+++−−' would not achieve an extreme D_s^{med} score. A second score accounts for this by using the absolute test statistic values, which causes a loss of the direction information, but allows a reliable ranking for *mixed* profiles

$$D_s^{|\text{med}|} := \frac{1}{|\{t: \textit{\textbf{P}}_s(t) \neq \text{'o'}\}|} \sum_{t:\, \textit{\textbf{P}}_s(t)\notin\{\text{'o'},\text{'}\pm\text{'}\}} \text{median}\left\{\left|D_g^{\text{shrink}}\right| : g \in \text{DE}_{\textit{\textbf{P}}_s(t)}^{(t)} \bigcap s\right\}.$$

In praxis, both scores tend to rank small gene sets higher than larger ones. $\text{DE}_{\textit{\textbf{P}}_s(t)}^{(t)}$ is used as the gene set corresponding to the most significant partition in the direction indicated by the profile position $\textit{\textbf{P}}_s(t)$ in the segmentation test variants of the GSA-type activation profile algorithm.

4.5 Rotation-Test-Type Algorithm for Determining Gene Set Activation Profiles

GSEA is one of the most popular gene set enrichment test approaches, but its significance assessment is based on permutations, which cannot be applied even on the time point wise comparison to reference testing, if the number of replicates is small as is standard for time course microarray experiments. Dørum, Snipen, et al. (2009) applied the rotation testing approach on the GSEA analysis of two color microarray experiments for direct and indirect comparisons on the same chip (i.e. each array is hybridized with a common reference for both conditions or the conditions are compared directly on the same wafer). The rotation test approach was rediscovered and expanded in the linear model framework by Langsrud (2005). Here it is used to apply the GSEA approach on the time point wise comparison of gene expression with respect to reference in order to identify significant gene sets, in which the up regulated or down expressed genes are overrepresented, while the significance is obtained from rotation of the data instead of permutation.

The input for the GSEA algorithm is a ranked gene list of association measures (correlation, signal-to-noise ratio etc.) with the phenotype (Subramanian, Tamayo, et al. 2005). Here, the shrinkage t-statistics per time point t similar to the proposal of Dørum, Snipen, et al. (2009) are considered directly. In order to aggregate the differential expression information for a gene set s two statistics similar to the *maxmean statistic* proposed by Efron and Tibshirani (2007) are used.

$$\mathrm{ES}_{s-}^{(t)} := \frac{1}{|s|} \sum_{g \in s} \min \left\{ D_g^{\mathrm{shrink}(t)}, 0 \right\}$$
$$\mathrm{ES}_{s+}^{(t)} := \frac{1}{|s|} \sum_{g \in s} \max \left\{ D_g^{\mathrm{shrink}(t)}, 0 \right\}.$$

The significance assessment is done in contrast to the original permutation approach by a rotation test approach, in which the $J \times T$ matrix of shrunken t-statistics

$$\mathbf{D}^{\mathrm{shrink}} = \begin{bmatrix} D_1^{\mathrm{shrink}(1)} & \cdots & D_1^{\mathrm{shrink}(T)} \\ & \vdots & \\ D_J^{\mathrm{shrink}(1)} & \cdots & D_J^{\mathrm{shrink}(T)} \end{bmatrix}$$

across all time points is rotated by a random $J \times J$ rotation matrix $\mathbf{Q}$ and the enrichment scores ($\mathrm{ES}_{s-}^{(t),k}$, $\mathrm{ES}_{s-}^{(t),k}$) are recalculated for each set s and time point t.

4.5 Rotation-Test-Type Algorithm for Determining Gene Set Activation Profiles

This is repeated for $K = 1000$ random rotations. The normalized enrichment scores

$$\text{NES}_{s-}^{(t)} := \frac{\text{ES}_{s-}^{(t)}}{1/K \sum_{k=1}^{K} \text{ES}_{s-}^{(t),k}}$$

$$\text{NES}_{s+}^{(t)} := \frac{\text{ES}_{s+}^{(t)}}{1/K \sum_{k=1}^{K} \text{ES}_{s-}^{(t),k}}$$

are determined for all gene sets and time points and analogously for the rotated enrichment scores (yielding $\text{NES}_{s-}^{(t),k}$, $\text{NES}_{s+}^{(t),k}$) in order to remedy gene set size effects.

The calculation of a gene set enrichment q-value at time point t according to Subramanian, Tamayo, et al. (2005) uses the distribution of $\text{NES}_{s+}^{(t),k}$ (and $\text{NES}_{s+}^{(t),k}$ for the up expression direction) over all $K \cdot T \cdot S$ values as null distribution. The q-values are computed by

$$\check{q}_{s-}^{E(t)} := \min\left\{ \frac{\left|\text{NES}_{s-}^{(t),k} \leq \text{NES}_{-}^{\star}\right|/K}{\left|\text{NES}_{s-}^{(t)} \leq \text{NES}_{-}^{\star}\right|/S}, 1 \right\}$$

$$\check{q}_{s+}^{E(t)} := \min\left\{ \frac{\left|\text{NES}_{s+}^{(t),k} \geq \text{NES}_{+}^{\star}\right|/K}{\left|\text{NES}_{s+}^{(t)} \geq \text{NES}_{+}^{\star}\right|/S}, 1 \right\}$$

across all values for $s \in \{1, \ldots, S\}$, $t \in \{1, \ldots, T\}$, $k \in \{1, \ldots, K\}$, while $\text{NES}_{-}^{\star} = \text{NES}_{s-}^{(t),k}$ ($\text{NES}_{+}^{\star} = \text{NES}_{s+}^{(t),k}$) denotes the fixed normalized enrichment score value for the gene set s at time point t. The rotation test type activation profile $\mathbb{P}_s^{\text{rot}}$ is estimated analogously to the GSA-type profiles (see section 4.2):

$$\mathbb{P}_s^{\text{rot}}(t) := \begin{cases} + & \text{if } \check{q}_{s+}^{E(t)} \leq \alpha_{\text{sets}} \text{ and } \check{q}_{s-}^{E(t)} > \alpha_{\text{sets}} \\ - & \text{if } \check{q}_{s+}^{E(t)} > \alpha_{\text{sets}} \text{ and } \check{q}_{s-}^{E(t)} \leq \alpha_{\text{sets}} \\ \pm & \text{if } \check{q}_{s+}^{E(t)} \leq \alpha_{\text{sets}} \text{ and } \check{q}_{s-}^{E(t)} \leq \alpha_{\text{sets}} \\ \circ & \text{if } \check{q}_{s+}^{E(t)} > \alpha_{\text{sets}} \text{ and } \check{q}_{s-}^{E(t)} > \alpha_{\text{sets}}. \end{cases}$$

A potentially necessary smoothing step is applied according to the smoothing of the non-threshold GSA-type algorithm in section 4.3. The ranking is obtained by considering those genes as significant, which have a more extreme shrinkage t-statistic value than the corresponding significant enrichment score $\mathrm{ES}_{s-}^{(t)}$ or $\mathrm{ES}_{s+}^{(t)}$. This promising approach is not given any further consideration in this thesis due to its expected huge computational effort in the conducted simulation studies.

4.6 Modified STEM Algorithm for the Identification of Active Gene Sets

The Short Time-series Expression Miner (STEM) is originally a web-tool for clustering gene expression time trajectories according to their similarity with model profiles (Ernst and Bar-Joseph 2006). The model profiles are determined independent from the data, which allows for significance assessment by permutation testing. Independently defined gene sets can be tested for enrichment with genes from one or more (significant) model clusters by the standard Fisher test (see section 4.1). The standard STEM procedure clusters only the dynamics of gene expression in time (i.e. changes of gene expression over time), but not the level of gene expression. The STEM procedure was originally designed for simple time series experiments without replicates and therefore the variance of replicated measurements is not considered.

The original STEM algorithm by Ernst, Nau, and Bar-Joseph (2005) is modified in this thesis in order to not only capture the dynamic of gene expression in time, but also the (relative) level of gene expression at the time points in the time series experiment. The main differences are the used similarity metric and the slightly different generated model profiles.

The input for the STEM algorithm is a matrix with one value per gene and time point. Here, the median of the replicates of gene expression differences with respect to the available reference for each gene and time point is used. The median gene expression differences may vary strongly between the genes under consideration. Therefor, the median gene expression difference values are rescaled to the interval $[-1, 1]$ per time point in order to preserve the information about both the relative level of gene expression and the dynamic in the course of time. This rescaled median gene expression difference matrix is the input in the clustering step.

4.6 The Modified STEM Algorithm

The model profiles are generated in a similar way as in the original STEM algorithm. The step width sw and the number of time points T determines the number of potential model profiles. The original STEM procedure started at 0 (first time point as reference for all other time points) and allowed at each time point a whole numbered change of up to sw. In the modified version, which is used here, the difference to an external reference is considered (compare section 2.3), hence the first time point may have already a high difference to the reference. Therefor, the starting point of the potential model profiles is an arbitrary value of the set

$$\left\{\frac{2(i-(T+1)/2)}{T-1} : i = 1, \ldots, T\right\}$$

and the step width allowed to change the value in the model profile between two adjacent time points is $2/T-2$. The total number of generated model profiles is given by

$$T \cdot 3^{T-1}.$$

Only such generated profiles are considered as model profiles, which are non-constant and do not exceed the admissible interval of $[-1, 1]$. This number is still high, if T grows (e.g. 702 for $T = 6$). Ernst, Nau, and Bar-Joseph (2005) propose to use only a small selection of model profiles as cluster prototypes. Here, the number of cluster prototypes is fixed to 40. The selection is chosen with a greedy algorithm, which maximizes the minimum of the pairwise measured distance d_{STEM} between two profiles of the selection. The applied distance function between two vectors y_1 and y_2

$$d_{\text{STEM}}(y_1, y_2) = \begin{cases} 0.5 \cdot \sum_{t=1}^{T} \frac{|y_1^{(t)} - y_2^{(t)}|}{T}, & \text{if } y = \text{const} \lor y_2 = \text{const} \\ 0.25 \cdot (1 - \text{cor}(y_1, y_2)) + 0.5 \cdot \sum_{t=1}^{T} \frac{|y_1^{(t)} - y_2^{(t)}|}{T}, & \text{else} \end{cases}$$

sets more attention on the profile level in relation to the profile shape in contrast to the just correlation based proposal by Ernst, Nau, and Bar-Joseph (2005).

In the following clustering step each gene is attached to the model profile, which minimizes the distance d_{STEM} between the corresponding model profile and gene vector from the gene expression difference matrix. A significance assessment for the resulting clusters is achieved by a permutation test, which repeats the cluster attachment for all permutations of the time points and calculates a p-value for each cluster cl_i

$$p_{\text{STEM}}^{\text{cl}_i} = P(X \leq |\text{cl}_i|) \qquad X \sim \text{Bin}\left(G, \frac{\sum_{p=1}^{T!}|cl_i^p|}{J \cdot T!}\right),$$

where $|cl_i^p|$ denotes the number of genes annotated to the ith model profile in the pth permutation. An adjustment for multiple testing is done with the TS-ABH FDR procedure and an FDR bound of 0.01 is controlled.

The last step is an enrichment test analogously to the test in section 4.1. Each gene set is tested for enrichment of genes attached significant (FDR q-value below 0.01) gene clusters. The resulting p-values are adjusted for multiple testing with the TS-ABH FDR method. Only those gene sets are considered as significantly enriched with genes from one or more significant model clusters, if the corresponding FDR q-value is below 0.01.

The major disadvantage of the STEM procedure is the circumstance, that constant profiles (indicating no difference or constant difference to the reference) cannot be captured in a constant model profile, since the time point permutation would not affect such profiles. Ernst, Nau, and Bar-Joseph 2005 face this issue with an extensive gene filtering procedure, which only selects dynamically changing genes. This type of filtering is not used in this work.

4.7 The maSigFun Algorithm for the Estimation of Regression Profiles for Gene Set Activity

M. Nueda, Sebastián, et al. (2009) describe three algorithms to analyze gene expression time series data on gene sets. The most simple approach is termed maSigFun. It is the only one, which directly allows to report a gene set activation profile. The method is based on a two-stage regression approach considering the gene set genes as different observations of an underlying gene set expression profile. In this thesis the dependent variable includes the differences in gene expression to the reference and the time is the only independent variable in the regression model. In general, other covariates can be included in the maSigFun procedure. The expression values of genes within a set may vary strongly. Hence the data matrix is gene-wise standardized to mean 0 and variance 1. Subsequently to the standardization the procedure follows roughly the maSigPro algorithm of Conesa, M. J. Nueda, et al. (2006).

In the first stage a full linear regression model of polynomial degree 3 is fitted. The gene sets are selected for the second stage if the F-test for the significance of the link between gene expression differences to reference and the time factor can be controlled by a TS-ABH FDR limit of 0.01.

The second stage uses variable selection (i.e. backward elimination) to improve the model of the first stage in the selected gene sets. In this thesis, the variable selection refers to the determination of the polynomial degree of the model. If the unadjusted measure of determination R^2 is larger than a moderate threshold, the corresponding gene sets are considered as significant. Here the threshold is fixed at 0.5.

The maSigFun procedure identifies gene sets with a high level of co-expression among the included genes. The fitted regression model represents the theoretical temporal activation profile for the gene set based on the regression coefficients. This can be compared with the median gene expression values in the gene set over time. The main disadvantage of this method is the assumption that all genes included in a particular gene set follow the same expression profile, which is not necessarily true for all gene sets.

CHAPTER 5

Simulation Study

In time-course microarray experiments the truth is neither known for the activation on the gene level nor for the activation on the more general level of pathways or biological processes. Spike-in and dilution data sets were used for the validation of microarray preprocessing methods. In the field of gene expression time series no such study is known to the author and this approach would hardly be able to properly model the complex reality of molecular genetic changes in living organisms. Hence, simulations are an often used procedure to validate and compare analysis methods for microarray time series. Due to the dependencies in the gene regulatory networks and the interaction of genes, the commonly used approach to model a simple pattern with normally distributed error seems not to capture the true molecular biology in real cells.

The idea of Törönen, Pehkonen, and Holm (2009) to utilize the original gene expression value from a real gene expression experiment in the case of gene set testing in the two classes design is introduced in section 5.2. This idea is modified for the situation of gene expression time series in two ways. The results for the two conducted simulation studies are presented in sections 5.3 and 5.4. The next section introduces the four gene expression time series experiments, whose data was recycled in the two simulation studies.

5.1 Short Characterization of the Four Recycled Example Data Sets

Four gene expression time series experiments are used in this thesis; on the one hand for applying the proposed methods and on the other hand as data basis for the two large simulation studies. The data sets are all publicly available for free from the above mentioned GEO data base. Mice are the subjects of all experiments, although there is a large variety of species in gene expression time series (see Figure 2.3). The species *mus musculus* is the higher mammal, which is widely used in biology as model organism for analyzing the molecular genetics of mammals in general and human diseases in particular. The commonly used inbred strains have a very homogeneous genome. Therefore, they promise more reproducible gene expression values than wild strains or humans. Particularly, the fixed species and technology allows to draw conclusions from the data characteristics and different methods instead of confounding effects from different species or different Affymetrix arrays.

The original data of four different gene expression time series is reused for the two simulation studies in this chapter. The data sets differ in main characteristics like the number of time points, the type of reference, the availability of replicates and slightly with respect to the gene set universe due to filtering as is displayed in Table 6.1 on page 136. All data sets result from mouse experiments with hybridized Affymetrix mouse 430 2.0 arrays. The common preprocessing, the gene filtering, the gene annotation, a detailed description including number of differentially expressed genes and significant sets are given in chapter 6. Each data set is briefly described in the following.

Aldosterone Heart Data Set (AH)

The GEO series GSE3440 provides data from a time series study conducted to analyze the effect of Aldosterone on the gene expression of mouse heart cells. Two groups of mice were examined in the experiment. The treatment group received an injection with a single dose of 10 µg/kg Aldosterone, whereas the control group

5.1 Short Characterization of the Four Recycled Example Data Sets

received the vehicle only. The mixed tissue of the hearts of five individuals per group was hybridized on a single microarray for the time points 0.5, 1, 2, 3, 4, 5, and 12 hours after injection. Hence, there are two measurements per time point and gene available, one for the treatment group and one for the control group. The lack of replicates requires to define significant differential expression by fold change. The differences in gene expression between the two conditions are relatively small across all time points. Therefore, exceeding a fold change of 1.5 is used as definition for significant differential expression on basis of the gcRMA gene expression values. This definition yields a moderate number of 173 to 903 up or down expressed genes per time point as shown in Figure 6.5. The simple design of the experiment reduces the expected variety of different gene set activation profiles. The missing replicates hamper the validation of the differential expression, which cannot be compensated by the complete reference time series. Nevertheless, this is a typical study design and the simulations may reveal the abilities of the competing algorithms to explore significant activation profiles in such a data situation.

Ovary Development Data Set (OD)

The embryonic development of mouse ovaries is the focus of the second experiment. The GEO series GSE5334 includes 19 raw data .cel-files of hybridized Affymetrix mouse 430 2.0 arrays. The RNA from the ovaries of three individual female mice was examined at six time points: gestational days GDs 11, 12, 14, 16, 18 and postnatal day PN 2. In contrast to the three replicates per time point only a single chip is available as reference, whereas a mixed tissue sample of whole body RNA from four male and four female mice 12 hours after birth was hybridized. Despite the particular type of tissue in the focus of the experiment, a large number of genes show significant differential expression as is demonstrated in Figure 6.6. This relatively strong expression activity is observed due to the early stage in the ontogenesis and leads to a high number of significant gene sets. Hence, it can be expected that not all identified sets are closely related to the research question, but those sets are very likely involved in the general embryonic development. The interest in the performance of the profile algorithm in such a situation of overwhelming signals is one reason for selecting this data set for the simulation studies.

Skin Healing Data Set (SH)

The skin healing data set is one part of the gene expression experiment stored in the GEO series GSE23006. The study was originally conducted to identify the genetic differences between wound healing on skin and mucosa (on tongue). Here, the gene expression information is split in two time series for each type of wounded tissue. The first data set considers only the chips related to mice with wounded skin. Three biological replicates are available before wounding and at time points 6 h, 12 h, 24 h, 3 d, 5 d, 7 d and 10 d after wounding. In contrast to the question of the original study, here the focus lies on the changes during the healing of the wounded skin in relation to the unwounded skin. The number of TS-ABH FDR adjusted differentially expressed genes is small, between 2 and 30 for down regulation and 8 to 132 for up expression (see Figure 6.7). However, the design of experiment provides comparatively many measurements with three replicates for the reference and each time point. The profile algorithms have to prove in the two simulation studies whether the resulting profiles for the gene sets are robust and reliable in a study with a very limited number of differentially expressed genes.

Tongue Healing Data Set (TH)

The second gene expression time series data set extracted from the GEO series GSE23006 is focused on the molecular genetic changes during the healing of the mucosa on mice tongues. The experimental design is identical to the SH data set yielding a total of 24 hybridized Affymetrix mouse 430 2.0 arrays – three replicates for the reference and each of 7 time points. Although originating from the same study, the number of differential expressed genes is much higher for the tongue healing group than for the skin healing cohort. The exact values are reported in Figure 6.8 (97 to 1072 genes). The total numbers are significantly smaller than in the ovary development data set, but regarding the fact that replicates are available for each measurement this is almost an ideal data situation among the available studies in data bases. Recycling those data and creating simulated gene sets with corresponding inputed activation profiles should lead to signals that can be found by all considered algorithms.

5.2 Construction of a Gene Set Simulation Study by Recycling of Given Gene Expression Time Series Data

There is no commonly accepted approach for simulating gene expression time series data in the context of gene sets. The proposed simulation strategies adopt the idea from Törönen, Pehkonen, and Holm (2009). It is proposed to use a real gene expression data matrix for the generation of Gene Ontology benchmark datasets with various types of positive signal. The simulation approach is called POSGODA. It is based on the gene expression data from an experiment with two classes and the GO gene set definitions. Briefly described, the algorithm in the first step randomly chooses gene sets, which are not highly correlated (the intersection of the two sets is empty or includes only few genes) and suitable to assume an enrichment p-value in the parameter interval for positive enrichment signal. In the second step the determined gene set signal p-value is approximately obtained by declaring significance for the corresponding number of genes in the gene set.

The approach for the two conditions gene set simulations is adapted and expanded to the time course design in two different ways. The intention is to stay as close as possible to the real gene expression time series situation while enabling the control of the gene set temporal activity. This activation profile has either the form of a symbolic enrichment with up or down expressed genes for the time points, it follows a regression profile found by the maSigFun approach (see section 3.3, page 30) or comes from a model profile by the STEM approach (see section 3.2, page 28).

The construction of two simulation studies is described in the following sections. The first evaluates the performance of the five profile algorithms in a setting of an artificial gene set universe with fixed gene set size. A controlled number of independent gene sets includes an a priori defined proportion of genes, which contribute to a significant activation profile in the analysis of the four chosen data sets. The *spiked-in* genes and the *non-informative* genes are recycled as complete vector across all time points to create the data matrix in each simulation step. Hence, the correlation between time points is preserved in the first simulation study.

The second simulation is conducted for the evaluation of the smoothing algorithm in the GSA-type algorithms. This is done in the framework of a gene set analysis within the same gene set universe as used in chapter 6. The complete data matrices from the four original experiments are recycled in each simulation step, in a way that allows to input enrichment for chosen gene sets per permuting the gene expression values per time point. Since the permutation occurs per time point, the original dependence structure between the time points is lost. The gene sets defined by GO are not independent, since all genes annotated to a GO term in a lower level are also annotated to their ancestor terms in the directed acyclic graph representation. This results in false positive results of the form, that a set not preset to a certain activation pattern is active due to the high number of genes, which are shared with its active neighbor gene set. Therefore, the second simulation type with dependent gene sets is used for the comparison of the different smoothing algorithms in contrast to the accuracy comparison of activation profile procedures on independent artificial gene sets in the first simulation.

Simulation to Compare the Profile Algorithms

The first simulation study is conducted to compare the five competing gene set analysis methods: the Two-threshold GSA-type activation profile algorithm (2T-GSA), the One-threshold segmentation GSA-type activation profile algorithm (1S-GSA), the Two-threshold segmentation GSA-type activation profile algorithm (2S-GSA), the modified STEM algorithm and the maSigFun procedure with each other. The GSA-type methods are applied without smoothing. The simulation results are shown in section 5.3.

The simulation is used to reveal the differences of the profile algorithms with respect to their sensitivity to gene set size, their proportion of *active* genes in the gene sets and the proportion of active gene sets in the whole gene set universe. Independent gene sets are used to compare the performance of the five competitors in order to avoid effects resulting from gene set overlap. In accordance with the idea from Törönen, Pehkonen, and Holm (2009), which inspired the simulation study for the evaluation of the smoothing methods, the data from original gene

5.2 Construction of a Simulation Study by Recycling of Given Data

expression studies is recycled. In contrast to the second simulation study, the time points are not considered independently, but the gene expression trajectories in time are considered as units. Those gene expression trajectories which fit well to the significant gene set activation profile of their own gene set are spiked-in into artificial gene sets together with genes whose time trajectory does not show any clear differential expression with respect to the reference. The algorithm for one complete simulation turn is illustrated in Figure 5.1 and is described in more detail in the following.

Input for the Simulation Algorithm for the Comparison of Profile Algorithms

The simulation study conducted for comparing the five competing profile algorithms (maSigFun, STEM, 2S-GSA, 1S-GSA, 2T-GSA) is based on the real gene expression time series experiments briefly described in section 5.1 (more details in chapter 6). As shown in nodes one to three in Figure 5.1 a profile analysis applying each of the five competing algorithms is carried out separately for the underlying gene expression matrix of the selected data set (one of AH, OD, SH or TH).

The resulting significant gene sets of GSA-type are filtered to omit too small or too large sets as prototypes for *active* gene sets in the simulation study. The gene sets are limited to include at least 10 and at most 500 genes. Moreover, the activation profiles of significant sets are filtered to include a minimum number of non-zero positions in the activation profile in order to include only those significant sets as prototypes, which are probably not identified by chance. In contrast to the second simulation study the prototypes are not selected for their continuous activation profile, though this type of profile results more often from the profile analysis of experimental data. The significant gene sets resulting from STEM are only filtered according to the gene set size limits mentioned above, whereas the few significant maSigFun gene sets are not filtered at all. The gene set numbers per data set resulting from filtering (block four in Figure 5.1) are listed in Table 5.1. Unfortunately, there are no significant gene sets arising from the 2T-GSA and maSigFun algorithms after applying the filtering procedure for the AH data. However, the data set investigating the effect of aldosterone on the mouse heart gene expression experiment is included in the simulation study.

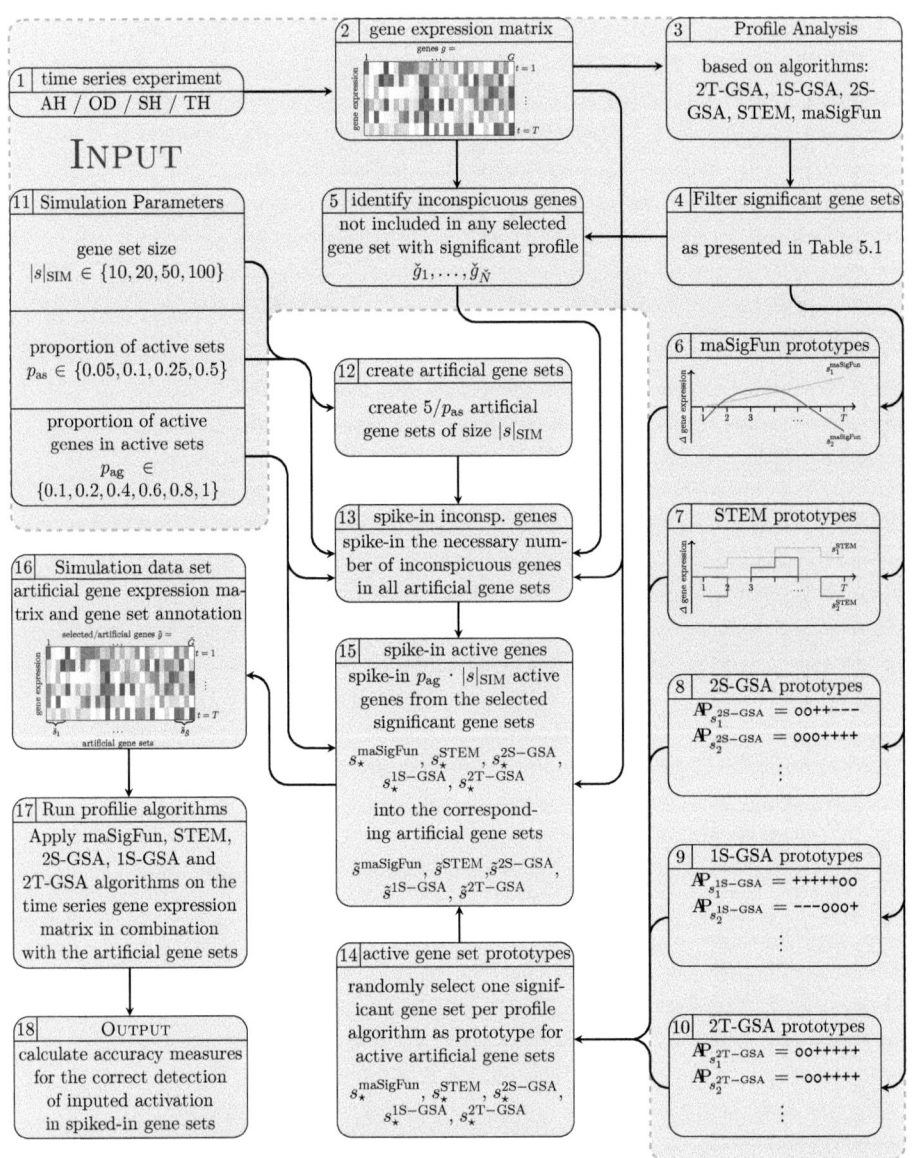

Figure 5.1: Simulation algorithm flow chart of the simulation for the comparison of the competing profile algorithms.

5.2 Construction of a Simulation Study by Recycling of Given Data

Profile analyses and filtering yields a number of S^{maSigFun} maSigFun prototypes, S^{STEM} STEM prototypes, $S^{\text{2S-GSA}}$ 2S-GSA prototypes, $S^{\text{1S-GSA}}$ 1S-GSA prototypes and $S^{\text{2T-GSA}}$ 2T-GSA prototypes. These are used as input in the simulation algorithm (nodes six to ten in Figure 5.1) to select one prototype per algorithm for the artificial *active* gene sets in the simulation data set.

Since only a certain proportion of gene sets in the simulation is supposed to obtain an *active* differential expression profile with respect to the reference, non-differential genes are needed. It would be too optimistic, if this set of genes would consist only of those genes with a small agglomerated difference to the expression values of the reference measurements. In the simulation all those genes are regarded as *uninformative* or *inconspicuous*, which are not annotated to any of the prototype gene sets. It has to be noted that some of these genes might be differential at some time points, in particular if they are included in gene sets skipped by the filtering process. This definition adds noise and allows for the occurrence of false positive findings by the profile algorithms in the simulation study. The *inconspicuous* genes are denoted by $\check{g}_1, \ldots, \check{g}_{\check{N}}$ as illustrated in block five of Figure 5.1.

Table 5.1: Gene set numbers resulting from the filtering for the simulation study to compare the five profile algorithms. All STEM and maSigFun gene sets are kept. Only gene sets with a minimum of non-zero position are used as prototype in the simulation study for the GSA-type algorithms and additionally the obtained groups must fulfill the gene set size limits to be selected.

	AH	OD	SH	TH
min set size	10	10	10	10
max set size	500	500	500	500
min non-zero pos (GSA algorithms)	3	4	4	4
selected sets				
2T-GSA	0	538	3	155
1S-GSA	3	705	205	413
2S-GSA	1	681	12	247
STEM	1	85	87	9
maSigFun	0	3	4	3
genes not in selected sets	12,439	3288	5861	5982

The simulation parameters shown in field eleven of Figure 5.1 control the common gene set size $|s|_{\text{SIM}}$ of all gene sets, the proportion of *active* gene sets p_{as} and the proportion of *active* (i.e. differentially expressed) genes p_{ag} in *active* gene sets in the artificial gene set universe of the simulation. The parameter values are given in Figure 5.1.

Generation of the Simulation Data Set

The simulation algorithm for the comparison of the five competing profile algorithms generates an artificial gene expression matrix and the corresponding annotation to artificial equal sized gene sets. The number of gene sets is determined by the number of prototypes (one for each algorithm) and the proportion of active sets. Assuming that five prototypes are available (for AH data the prototypes for maSigFun and 2T-GSA algorithms are missing, see Table 5.1) the number of gene sets in the simulation data set is given by

$$S_{\text{SIM}} = 5/p_{\text{as}}$$

and the gene set size for each set is fixed at $|s|_{\text{SIM}}$ genes as shown in block 12 of Figure 5.1.

The next step (node 13 in Figure 5.1) determines the inconspicuous genes for the simulation study. The total number (assuming five prototypes) of

$$G_{\text{SIM}}^o = (S_{\text{SIM}} - 5 + 5 \cdot [1 - p_{\text{ag}}]) \cdot |s|_{\text{SIM}}$$

inconspicuous genes is selected randomly from the $\check{N}$ inconspicuous genes identified in step 5 of Figure 5.1, if $G_{\text{SIM}}^o \leq \check{N}$ is fulfilled. The probability for selecting gene g is given by

$$\frac{r_g^o}{\sum_{g=1}^{G_{\text{SIM}}^o} r_g^o},$$

where r_g^o denotes the rank of sum of absolute gene expression differences to reference for gene g. If $G_{\text{SIM}}^o > \check{N}$, a number of $\check{N} - G_{\text{SIM}}^o$ genes is created by randomly

5.2 Construction of a Simulation Study by Recycling of Given Data

selecting gene expression values per time point from the G^o_{SIM} genes and used as artificial uninformative gene expression trajectories in addition to the available G^o_{SIM} inconspicuous genes.

In order to obtain a proportion of p_{as} gene sets with spiked in differentially expressed genes a single prototype gene set from each of the five profile algorithms is selected in each simulation turn. Beginning with selecting $s^{\text{maSigFun}}_\star$ from the S^{maSigFun} prototypes resulting from the maSigFun algorithm, a single set is randomly chosen and deleted from the prototype candidates for the subsequent algorithms. The selection order follows the top-down order of Figure 5.1 (maSigFun, STEM, 2S-GSA, 1S-GSA, 2T-GSA). The probability for selecting a set as prototype depends on its size and the simulation parameter $|s|_{\text{SIM}}$. It is given for the maSigFun algorithm (analogously for the other algorithms) by

$$\frac{r^{\text{maSigFun}}_{s^{\text{maSigFun}}}}{\sum_{s^{\text{maSigFun}}=1}^{S^{\text{maSigFun}}} r^{\text{maSigFun}}_{s^{\text{maSigFun}}}}, \quad r^{\text{maSigFun}}_{s^{\text{maSigFun}}} := 1 + S^{\text{maSigFun}} - \text{rank}\left(\left|\frac{|s|^{\text{maSigFun}}}{|s|_{\text{SIM}}} - 1\right|\right)$$

resulting in higher selection probability, when the prototype gene set size is similar to the gene set size defined by the simulation parameter $|s|_{\text{SIM}}$. Hence, the selection process symbolized in block 14 of Figure 5.1 results in five prototype sets $s^{\text{maSigFun}}_\star$, $s^{\text{STEM}}_\star$, $s^{\text{2S-GSA}}_\star$, $s^{\text{1S-GSA}}_\star$, $s^{\text{2T-GSA}}_\star$, which have shown significant profiles in the analysis of the original gene expression time series experiment.

In step 15 of the flow chart in Figure 5.1 each *active* gene set obtains a total number of $p_{\text{ag}} \cdot |s|_{\text{SIM}}$ genes from the corresponding prototype selected in the previous step in the simulation algorithm. The selection of spiked in gene expression values depends on the underlying profile algorithm of each prototype set and the number of available genes in the prototype set:

maSigFun

if $|s|_{\text{SIM}} \cdot p_{\text{ag}} \leq |s|^{\text{maSigFun}}_\star$:

Include a random sample of $|s|_{\text{SIM}} \cdot p_{\text{ag}}$ genes from the prototype set $s^{\text{maSigFun}}_\star$ into the artificial *active* gene set $\tilde{s}^{\text{maSigFun}}$. The sample probability for a gene g from $s_\star^{\mathrm{maSigFun}}$ is chosen as

$$\frac{r_g^{\mathrm{maSigFun}}}{\sum_{g=1}^{|s|_\star^{\mathrm{maSigFun}}} r_g^{\mathrm{maSigFun}}}, \quad r_g^{\mathrm{maSigFun}} := |s|_\star^{\mathrm{maSigFun}} + 1 - \mathrm{rank}(\mathrm{RSS}_g)$$

where RSS_g denotes the residual sum of squares for the single gene g and the model given from the maSigFun algorithm for the prototype gene set $s_\star^{\mathrm{maSigFun}}$. This allows for favoring genes with a good fit to the determined gene set regression model.

if $|s|_{\mathrm{SIM}} \cdot p_{\mathrm{ag}} > |s|_\star^{\mathrm{maSigFun}}$:
Sample λ_i^{SIM} from the interval $(0, 1)$ and the two genes $g_i^{\mathrm{SIM}-1} \neq g_i^{\mathrm{SIM}-2}$ from $s_\star^{\mathrm{maSigFun}}$ for $i = 1, \ldots, |s|_{\mathrm{SIM}} \cdot p_{\mathrm{ag}} - |s|_\star^{\mathrm{maSigFun}}$. Include the convex combinations

$$\lambda_i^{\mathrm{SIM}} x_{g_i^{\mathrm{SIM}-1}} + (1 - \lambda_i^{\mathrm{SIM}}) x_{g_i^{\mathrm{SIM}-2}}$$

as artificial gene expression vectors in addition to the $|s|_\star^{\mathrm{maSigFun}}$ available informative genes in the artificial gene set $\tilde{s}^{\mathrm{maSigFun}}$. This generation of artificial genes provides intermediate gene expression trajectories similar to those in the set. They will fit the maSigFun regression model if the two sampled genes from the set $s_\star^{\mathrm{maSigFun}}$ fit the model, too.

STEM

if $|s|_{\mathrm{SIM}} \cdot p_{\mathrm{ag}} \leq |s|_\star^{\mathrm{STEM}}$:
Include a random sample of $|s|_{\mathrm{SIM}} \cdot p_{\mathrm{ag}}$ genes from the prototype set $s_\star^{\mathrm{STEM}}$ into the artificial *active* gene set $\tilde{s}^{\mathrm{STEM}}$. The sample probability for a gene g from $s_\star^{\mathrm{STEM}}$ is chosen as

$$\frac{r_g^{\mathrm{STEM}}}{\sum_{g=1}^{|s|_\star^{\mathrm{STEM}}} r_g^{\mathrm{STEM}}}, \quad r_g^{\mathrm{STEM}} := |s|_\star^{\mathrm{STEM}} + 1 - \mathrm{rank}\left(d_{\mathrm{STEM}}(d_g, \hat{m}_\star^{\mathrm{STEM}})\right),$$

where d_g denotes the vector of differential expression across time points ($\log_2 \mathrm{FC}$ or D_g^{shrink}). This ensures that genes with a smaller distance to the prototype model profile $\hat{m}_*^{\text{STEM}}$ annotated to s_*^{STEM} is selected more likely than a gene with a larger distance.

if $|s|_{\text{SIM}} \cdot p_{\text{ag}} > |s|_*^{\text{STEM}}$:

Spike in the gene expression values from all genes included in s_*^{STEM} and include additional $|s|_{\text{SIM}} \cdot p_{\text{ag}} - |s|_*^{\text{STEM}}$ expression vectors from genes annotated to the same model profile as s_*^{STEM}.

2S-GSA if $|s|_{\text{SIM}} \cdot p_{\text{ag}} \leq |s|_*^{\text{STEM}}$:

Include a random sample of $|s|_{\text{SIM}} \cdot p_{\text{ag}}$ genes from the prototype set $s_*^{\text{2S-GSA}}$ into the artificial *active* gene set $\tilde{s}^{\text{2S-GSA}}$. The sample probability for a gene g from $s_*^{\text{2S-GSA}}$ is chosen as

$$\frac{r_g^{\text{2S-GSA}}}{\sum_{g=1}^{|s|_*^{\text{2S-GSA}}} r_g^{\text{2S-GSA}}}$$

with

$$r_g^{\text{2S-GSA}} := \text{rank}\left(d_g^T \widetilde{AP}_{s_*^{\text{2S-GSA}}}\right) \quad \text{and} \quad \widetilde{AP}_s(t) = \begin{cases} 1 & , \text{if } AP_s(t) = + \\ -1 & , \text{if } AP_s(t) = - \\ 0 & , \text{if } AP_s(t) = \circ \end{cases}.$$

This choice prefers genes, which follow the activation profile of $s_*^{\text{2S-GSA}}$ to a more extreme extent, though ignoring non-differential positions.

if $|s|_{\text{SIM}} \cdot p_{\text{ag}} > |s|_*^{\text{2S-GSA}}$:

Spike in the gene expression values from all genes included in $s_*^{\text{2S-GSA}}$ and include additional $|s|_{\text{SIM}} \cdot p_{\text{ag}} - |s|_*^{\text{2S-GSA}}$ expression vectors from genes annotated to gene sets with the same activation profile as $s_*^{\text{2S-GSA}}$. If the number of these additional genes is not sufficient, add all genes from sets with the same activation profile and add the missing genes by successively reducing the necessary number of positions agreeing with the activation profile of $s_*^{\text{2S-GSA}}$. It has to be noted that the gene choice depends on the activation profiles of the gene sets and not on the differential expression of the genes included in those sets.

1S-GSA analogously to 2S-GSA

2T-GSA analogously to 2S-GSA

The steps 12 to 15 in Figure 5.1 result in a new gene expression matrix and annotation information for $5/p_{as}$ artificial gene sets of size $|s|_{SIM}$ as shown in field 16. It has to be emphasized that the rows in the new matrix have whenever possible a corresponding row with identical expression values in the data matrix of the original experiment.

Run Profile Algorithms on Simulation Data Set

The artificial gene expression matrix, which mainly includes gene expression vectors of the underlying gene expression time series experiment, is the basis for the application of all five profile algorithm (see block 17 in Figure 5.1). The new artificial data matrices includes less genes (rows) than the full gene expression matrix from the original experiment. The significance limits for each profile algorithm are set to the same values as in the original data analysis. Simulation data sets including a small number of genes may identify a higher proportion of significant genes than expected from the p_{ag} parameter, since the same FDR adjustment is applied on a lower number of tests. This adds another kind of uncertainty to the simulation when validating the performance in identifying the inputed gene sets in each simulation turn.

The five competing algorithms are applied to the gene expression matrix generated in each simulation turn. For each resulting significant gene set it is checked whether it was generated from a prototype of the same algorithm. Moreover, for all algorithms except maSigFun the information is kept whether the resulting profile match exactly the input profile of the corresponding prototype set. Additionally, the simulation algorithm notes, which other *active* sets were recognized and the number of false positive sets, i.e. those artificial sets without inputed genes from originally significant gene sets. These values are used to calculate accuracy measures as stated in step 18 of the flow chart in Figure 5.1.

Four overall accuracy measures are used for the validation of the five profile algorithms in the simulation study. *True positives* (TP) is the term for the total of

5.2 Construction of a Simulation Study by Recycling of Given Data

identified active gene sets, i.e. those genes which were generated from a significant prototype set. *True negatives* (TN) is the designation for the number of those sets, which are not identified as significant by the algorithm and were not inputed with differentially expressed genes from a prototype. The total of gene sets which has spiked in genes, but was not identified by the algorithm is denoted by *false negatives* (FN). The number of incorrect identifications without any spiked in differentially expressed genes is called *false positives* (FP). The true positive rate (TPR) is the ratio of the correctly identified number of the five gene sets with active profile and the total number of sets with spiked in profiles (TPR = TP/(TP + FN)) over all 1000 simulation turns with the same simulation parameters. The second measure is the false discovery rate (FDR). The FDR is the ratio of gene sets identified as significant but without spiked in information and all significant sets found by the algorithm (FDR = FP/(FP + TP)). The accuracy (ACC) is the ratio of all correctly identified sets (TP + TN) and all sets in the simulation study. The negative predictive value (NPV) denotes the proportion of gene sets correctly identified as non significant on all sets declared as non-significant by the certain algorithm. In addition to this summarizing measures the TPR is calculated for each kind of spiked in profile, i.e. the ratio is determined separately for each algorithm in identifying the active profiles derived from each of the five algorithms.

The results of the simulation study to compare the five profile algorithms are shown in section 5.3 beginning on page 81.

Simulation to Evaluate the Smoothing Algorithms

The smoothing methods proposed for smoothing the gene set activation profiles of the GSA-approaches, which were introduced in section 4.3 (see pages 46ff.), are evaluated by a simulation study described in the following. In the simulation setting, the detection of differentially up and down expressed genes with respect to the reference is controlled by a FDR of $\alpha_{\text{genes}} = 0.05$ and for simplification a strict unadjusted significance level threshold α_{sets} is used for the Fisher tests (enrichment with differentially expressed genes). This second threshold is selected as the p-value equivalent to the FDR limit of 1 %, which is applied in the application chapter 6. The idea is similar to the one from Törönen, Pehkonen, and Holm (2009). A flow chart of the simulation algorithm with its twelve key stages is illustrated in Figure 5.2. Gene sets are randomly chosen (while considering the gene set overlap of identical genes) and a fixed smooth symbolic activation profile (e.g. '+++ooo') is assigned to these gene sets. In a second step the gene set genes obtain test statistic values (from the tests of differential gene expression with respect to the common reference), which result in a significant enrichment at the time points according to the preset profile. In this controlled scenario it can be evaluated, which smoothing algorithm best recovers the preselected gene sets and their activation profiles.

More precisely, a data set dependent number of gene sets (see Table 5.10) limited in size between 10 and 200 are selected under the constraint that all pair-wise gene set correlations lie within the interval of $[-0.1, 0]$ (stage 4 in Figure 5.2). The set-to-set correlation is defined as the correlation of the two corresponding binary vectors of length G. These vectors turn out to be 1 if the gene is part of the gene set and 0 otherwise. This prevents to choose gene sets with a large part of overlapping genes, which would hinder the successful assignment of different preset profiles. Different preset activation profiles are used at the same simulation run in accordance to the results from analyzing real gene expression time course experiments. Each selected gene set is associated with a randomly chosen prototype activation profile from a pool, which is shown in Table 5.10. These profiles are denoted in the following as *preset profiles* and true patterns in the simulation study. Each of the preset profiles is used multiple times in the same simulation run in order to yield a more realistic scenario with 10 to 90 gene sets with *known* activation profiles. To add noise in the preset profiles, every non-significant position (i.e. the 'o' potions in '+++ooo') in the preset profiles is changed to a randomly chosen significant position with probability 2.5 % (node 3 in Figure 5.2). The "mutated" profiles are used as prototypes for the assignment of gene statistics from the original study to the selected gene sets (step 5 in Figure 5.2).

Subsequently, for each profile gene set pair and each time point, in which the profile shows an enrichment with up or down expressed genes, all genes in the gene set are annotated to one of the three types of significance gene sets per time

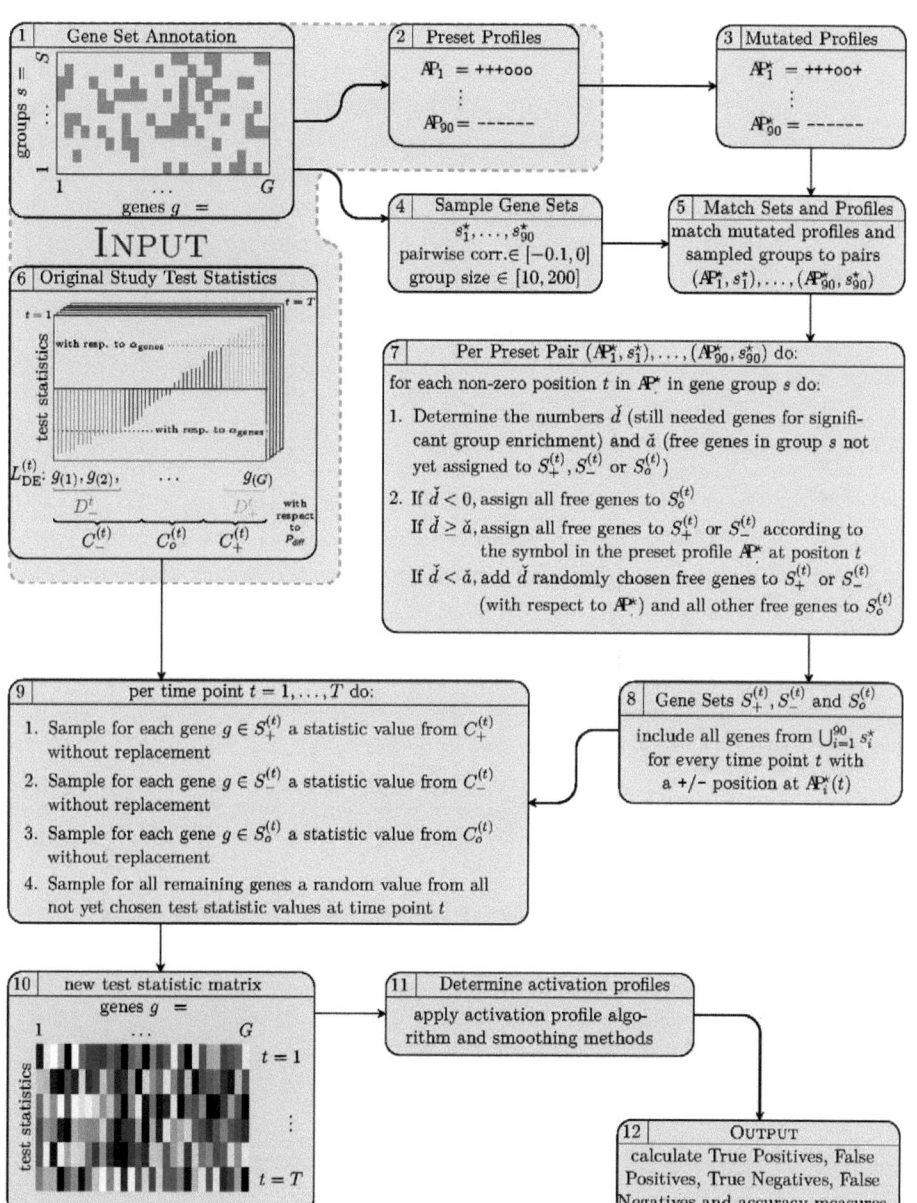

Figure 5.2: Simulation algorithm flow chart of the simulation for validation of profile smoothing algorithms. Here, 90 gene sets are assumed to be set to a preset activation profile as intended for the OD data set.

point. These sets include genes, which are designated to obtain a test-statistic value signaling up expression ($S_+^{(t)}$), down expression ($S_-^{(t)}$), or no differential expression ($S_o^{(t)}$) at time point t. The number of up and down regulated genes is defined in accordance with the data analysis in chapter 6 by the control of a (TS-ABH) FDR value of $\alpha_{\text{genes}} = 0.05$ or in case of the aldosterone heart data set exceeding a fold change of 1.5. The lowest number yielding enrichment with respect to the unadjusted significance threshold α_{sets} (unadjusted Fisher test p-value corresponding to the q-value threshold in application) in gene set s at time point t is denoted with

$$\check{n}^{(t,s)}_{\text{DE}^{(t)}_{+/-} \in s} := \min\left\{ n^{(t,s)}_{\text{DE}^{(t)}_{+/-} \in s} : P\left(N^{(t,s)}_{\text{DE}^{(t)}_{+/-} \in s} \geq n^{(t,s)}_{\text{DE}^{(t)}_{+/-} \in s} \right) \leq \alpha_{\text{sets}} \right\}$$

depending on the direction of the differential expression (i.e. '+' or '−') given by the significant position in the profile. This number is the target of up or down expressed genes in the gene set s at significant position t in order to obtain a not too strong enrichment. The procedure is complicated by the fact that gene sets are treated iteratively and a set can share genes with a set treated earlier. Consider a gene group s and assume that the preset profile at time point t is '+' (up regulation). Let $\check{n}^{(t,s)}_{\text{DE}^{(t)}_+ \in s}$ stand for the lowest number of up regulated genes in the group in order to obtain significant enrichment with respect to given α_{genes} and α_{sets}. The two numbers

$$\check{d} = \check{n}^{(t,s)}_{\text{DE}^{(t)}_+ \in s} - \left| s \cap S_+^{(t)} \right| \quad \text{and} \quad \check{a} = \left| \left(s \setminus (S_+^{(t)} \cup S_-^{(t)} \cup S_o^{(t)}) \right) \right|,$$

denote the number of genes missing to achieve significant enrichment ($\check{d}$) and the number of genes available to obtain a different significance status ($\check{a}$), i.e. not yet included in the union $S_+^{(t)} \cup S_-^{(t)} \cup S_o^{(t)}$. Three cases concerning the relation between the two terms determine the division of the free (not yet assigned) genes in s to the three significance sets:

5.2 Construction of a Simulation Study by Recycling of Given Data 79

$\check{d} < 0$: There are already more genes assigned to be up regulated than needed for significant enrichment. The remaining group genes not yet included in $S_+^{(t)}$, $S_-^{(t)}$ or $S_o^{(t)}$ are added to the list $S_o^{(t)}$ (not differentially expressed).

$\check{d} \geq \check{a}$: In the case of less available group genes than needed for significant enrichment, all free genes are annotated to $S_+^{(t)}$ in order to be as close as possible to enrichment (smoothing could correct this missed enrichment).

$\check{d} < \check{a}$: The number of d genes is sampled from the set of available group genes if there are more free genes available than needed for significant group enrichment. If some available genes were already among the up regulated genes at the previous time point, their sampling probability is 75 %, otherwise it is equal for all possible genes (uniform sampling). This sampling is the major difference to the deterministic algorithm by Törönen, Pehkonen, and Holm (2009) and the following steps extend their procedure.

The sampling (if $\check{d} < \check{a}$) prefers in case of a preset continuous activation profile (e.g. '+++...') those genes in relation 3:1, which have a significant expression value at neighboring previous time point. This procedure ensures that the same group genes obtain more likely a significantly differential expression value if neighboring time points are preset to enrichment with differential expression.

The gene sets $S_+^{(t)}$, $S_-^{(t)}$ and $S_o^{(t)}$ include all genes from the sampled gene sets for those time points with a '+' or '-' in the corresponding "mutated" preset activation profile at the current stage of the simulation algorithm. In the following step (step 9 in Figure 5.2) the original test statistic values from the time course experiment are rearranged for every time point such that the estimated gene set activation profiles match their "mutated" preset profiles with high probability. For every time point in the original time series data an index list $L_{\text{DE}}^{(t)} = \{g_{(1)}, \ldots, g_{(G)}\}$ is created, ordered increasingly according to the test-statistics of the tests for differential gene expression. Three candidate sets are derived from this list, which are later used to assemble the preselected gene groups with test statistic values such that enrichment with up or down regulated genes according to their preset profiles is achieved with high probability. These candidate sets (see step 6 in Figure 5.2) are defined as

$$C_-^{(t)} := \{g_{(1)}, \ldots, g_{(\lfloor n_{\text{DE}+}^{(t)}/P_{\text{diff}}\rfloor)}\}$$
$$C_+^{(t)} := \{g_{(J-\lfloor n_{\text{DE}-}^{(t)}/P_{\text{diff}}\rfloor)}, \ldots, g_{(G)}\}$$
$$C_o^{(t)} := L_{\text{DE}}^{(t)} \setminus \left(C_+^{(t)} \cup C_-^{(t)}\right).$$

For a randomly sampled item from the sets $C_+^{(t)}$ or $C_-^{(t)}$, by construction the probability for significant up or down regulation of this item is greater than or equal to P_{diff}, since $n_{\text{DE}+}^{(t)}$ is the number of significantly up expressed genes and $\lfloor n_{\text{DE}+}^{(t)}/P_{\text{diff}}\rfloor$ is the smallest integer succeeding this number by the factor $1/P_{\text{diff}}$. In the conducted simulation studies P_{diff} is fix with $P_{\text{diff}} = 0.95$. Each gene in the gene sets $S_+^{(t)}$, $S_-^{(t)}$ and $S_o^{(t)}$ obtains a randomly chosen test statistic value from the corresponding candidate sets $C_+^{(t)}$, $C_-^{(t)}$ and $C_o^{(t)}$ by sampling without replacement. All not yet used statistic values (significant and non-significant) are pooled per time point and randomly assigned to the genes from the set $\breve{G} \setminus \left(S_+^{(t)} \cup S_-^{(t)} \cup S_o^{(t)}\right)$ at each time point. This procedure yields a new $T \times G$ matrix (step 10 in Figure 5.2), which includes all original test statistic values from the real study in a reordered way to spike-in positive signals of gene set activation. Depending on the original study results there are gene sets, which have a non-zero activation profile by chance, since only the preset gene sets are controlled (and only for non-zero positions).

The last two steps (11 and 12 in Figure 5.2) include the algorithms to estimate and smooth the gene set activation profiles and the calculation of accuracy measures. The output of a simulation consists of average numbers of true positives, true negatives, false positives and false negatives regarding the originally preset profiles. Here, true positives (TP) and false negatives (FN) are the number of profiles belonging to the preselected gene groups that are correctly identified or not identified. True negatives (TN) and false positives (FP) are the number of profiles belonging to the remaining gene groups that fit or do not fit to a continuous non-significant profile ('oooooo'). These values are used to derive sensitivity (TPR) as TP/(TP+FN), false positive rate (FPR) as FP/(FP+TN) and accuracy (ACC) as (TN + TP)/(TP+FP+TN+FN) to assess the quality of the overall algorithm and in particular in combination with the smoothing algorithm. The false discovery rate (FDR) defined as FP/(TP+FP) would be high due to the construction of the simulation, in which all signals are recycled and only a small proportion is used to assign a preset gene set activation profile. Hence, this value would be misleading for the analysis of the simulation results. An additional accuracy criterion is the positive predictive value restricted to the preset profile types and a maximum similarity of 90% with preset gene sets (PPV_{p90}). Therein, false positive profiles are only those, which fit to one of the used preset profiles and do share a maximum of 90% of their genes with any preset gene set. Hence, false positive gene sets with a preset profile, which have more than 90% of their genes in common with a preset gene set do not count as FP. However, this value ignores FPs that turn out to have another activation profile than the prototypes in the input of the simulation study, but it reports the proportion of correctly identified profiles, if the set of preset profiles would be the objective of the activation profile analysis. The results of the simulation study are presented in section 5.4.

5.3 Simulation Study to Compare the Methods

This section shows the results of the first simulation study, which compares the five proposed profile algorithms in a realistic simulation scenario using recycled original data from published experiments. The three algorithms based on gene set analysis methods (2T-GSA, 1S-GSA, 2S-GSA) are applied without smoothing. The STEM and maSigFun procedures are used as described in section 4.6 and section 4.7. The simulation algorithm described in section 5.2 and an overview of all inputs and outputs is given in Figure 5.1. 1000 simulation runs per parameter setting are applied for each of the four gene expression time series data sets presented in section 5.1 and analyzed in chapter 6. Each data set has its own characteristics and therefore the results of the simulation study are reported separately for each data set in the following.

Aldosterone Heart Experiment

Two experimental groups are compared at seven time points without any replicates in the gene expression time series experiment analyzing the effect of Aldosterone on the mouse heart. The exceeding of a fold change of 1.5 is used to determine differential expression due to the lack of replicates. The application of the five profile algorithms and the filter rules for suitable prototype gene sets lead to the very low number of four candidates (compare Table 5.1, while the 2S-GSA set is among the 1S-GSA sets). Due to the missing prototype sets, there is no spiked-in information from a significant maSigFun set or a 2T-GSA set. Nevertheless, these algorithms are applied to identify the active sets spiked-in with gene expression information from the 1S-GSA, 2S-GSA and STEM prototypes as described in section 5.2.

Table 5.2 reports the overall true positive rate (TPR) and false discovery rate (FDR) considering all identified gene sets with spiked-in information as true positive (TP) and all gene sets found with a significant activation profile without spiked-in differentially expressed genes as false positive (FP). It was not possible to spike in genes from a significant maSigFun set or a significant 2T-GSA for the simulation based on the Aldosterone heart data set. While the maSigFun procedure is not able to identify any significant gene set (TPR=0 and FDR=0), the 2T-GSA method leads to significant detections. The STEM algorithm has in general a low TPR far below the levels of the GSA-type procedures. This observation is mitigated for larger sets ($|s|_{\text{SIM}}$) and fewer sets in the gene set universe ($p_{\text{as}} = 0.5$). Table B.1 in the Appendix reveals that the STEM algorithm only succeeds in detecting significant gene sets with spiked-in genes from a former significant STEM gene set, but the TPR for detecting STEM sets is even higher applying GSA-type algorithms.

In general, at least for the three GSA-type procedures the TPR is increasing with increasing p_{ag}, e.g. the sensitivity is increased if more differentially expressed genes are spiked-in in the chosen gene set. The TPR is higher for larger set sizes ($|s|_{\text{SIM}}$) and it seems to be advantageous if the proportion of active gene sets p_{as} is not set to 0.5, whereas STEM shows best sensitivity for the smallest gene set universe ($p_{\text{as}} = 0.5$). The FDR of the GSA-algorithms decreases for a rising number of genes in the set, for an augmented p_{as} and for larger values of p_{ag}.

Table 5.2: TPR and FDR for identifying spiked-in activation profiles in the simulation to compare the five profile algorithms on basis of the AH data.

	Aldosterone heart data		TPR per p_{ag}						FDR per p_{ag}				
$|s|_{\mathrm{SIM}}$	p_{as}	algorithm	0.1	0.2	0.4	0.6	0.8	1	0.1	0.2	0.4	0.6	0.8	1
10	0.05	2T-GSA	4.60	7.07	10.60	16.73	22.80	26.57	93.81	90.50	85.94	79.08	72.84	68.41
		1S-GSA	19.10	22.50	27.97	36.13	45.73	57.03	94.37	93.44	91.86	89.66	87.04	84.20
		2S-GSA	7.87	10.83	16.40	23.50	31.30	37.90	93.59	91.38	86.92	82.21	77.18	73.10
		STEM	0.00	0.00	0.00	0.00	0.07	0.20	100.00	100.00	100.00	100.00	75.00	40.00
		maSigFun	0.00	0.00	0.00	0.00	0.00	0.00						
	0.1	2T-GSA	5.30	7.03	10.63	15.30	20.60	25.87	85.90	81.41	72.26	65.75	56.08	49.90
		1S-GSA	25.77	31.57	39.33	48.77	59.07	69.93	89.19	86.90	83.85	80.25	76.49	73.18
		2S-GSA	7.77	10.87	15.63	22.33	30.10	35.83	86.90	81.96	74.36	68.02	59.34	55.36
		STEM	0.00	0.03	0.00	0.00	0.07	0.20	100.00	50.00	100.00	100.00	50.00	40.00
		maSigFun	0.00	0.00	0.00	0.00	0.00	0.00						
	0.25	2T-GSA	3.00	3.90	6.97	8.60	13.50	16.13	67.51	59.23	41.13	33.33	21.66	18.10
		1S-GSA	26.13	29.93	36.80	44.87	55.33	62.77	71.63	68.84	61.79	55.93	48.73	42.92
		2S-GSA	4.67	6.07	10.53	14.23	20.83	24.43	66.59	61.11	42.86	34.81	22.74	21.10
		STEM	0.00	0.00	0.03	0.07	0.07	0.17	100.00	100.00	50.00	0.00	33.33	44.44
		maSigFun	0.00	0.00	0.00	0.00	0.00	0.00						
	0.5	2T-GSA	0.23	0.30	0.70	1.53	2.83	4.60	41.67	25.00	8.70	4.17	4.49	1.43
		1S-GSA	15.63	17.67	23.03	26.43	31.93	39.90	47.24	41.82	31.24	25.75	19.29	17.22
		2S-GSA	0.27	0.47	0.93	1.93	3.93	6.27	38.46	17.65	9.68	3.33	4.07	1.57
		STEM	0.03	0.00	0.03	0.10	0.13	0.33	0.00	100.00	0.00	25.00	42.86	16.67
		maSigFun	0.00	0.00	0.00	0.00	0.00	0.00						
20	0.05	2T-GSA	6.67	9.50	17.60	27.20	37.30	44.53	92.50	89.55	81.58	73.38	66.09	61.02
		1S-GSA	21.13	26.87	39.70	54.00	68.23	80.00	94.34	92.90	89.48	86.15	82.87	80.06
		2S-GSA	7.33	10.10	18.37	28.50	39.23	46.47	92.60	90.00	82.65	74.58	67.68	63.24
		STEM	0.00	0.00	0.03	0.00	0.43	0.43	100.00	100.00	66.67	100.00	13.33	13.33
		maSigFun	0.00	0.00	0.00	0.00	0.00	0.00						
	0.1	2T-GSA	7.47	9.77	17.17	26.20	34.27	41.30	84.44	80.81	68.86	57.63	47.34	41.58
		1S-GSA	27.23	32.60	43.30	59.40	73.50	84.63	88.36	85.80	81.94	75.76	71.38	67.39
		2S-GSA	8.23	10.87	18.43	28.03	37.50	45.83	84.62	80.81	69.86	59.02	48.82	43.28
		STEM	0.00	0.00	0.00	0.07	0.23	0.50		100.00	100.00	66.67	30.00	21.05
		maSigFun	0.00	0.00	0.00	0.00	0.00	0.00						
	0.25	2T-GSA	5.30	6.47	11.13	18.50	24.50	32.60	64.98	58.90	42.71	26.78	18.60	13.53
		1S-GSA	21.57	24.50	35.63	49.43	61.33	75.80	72.03	68.39	57.19	45.70	39.33	31.96
		2S-GSA	5.83	7.93	13.23	22.57	29.63	37.57	66.09	58.10	42.46	26.49	19.18	15.45
		STEM	0.00	0.10	0.03	0.27	0.20	0.73	100.00	50.00	75.00	38.46	40.00	29.03
		maSigFun	0.00	0.00	0.00	0.00	0.00	0.00						
	0.5	2T-GSA	0.77	1.70	3.10	7.13	10.07	13.20	46.51	19.05	13.08	7.36	2.58	0.75
		1S-GSA	24.97	28.73	35.90	45.27	55.37	68.07	44.64	38.87	30.47	22.62	17.17	12.59
		2S-GSA	1.33	2.57	4.90	10.30	14.80	20.07	47.37	20.62	15.03	8.31	2.63	0.99
		STEM	0.03	0.03	0.17	0.33	0.80	1.27	66.67	0.00	37.50	28.57	22.58	13.64
		maSigFun	0.00	0.00	0.00	0.00	0.00	0.00						
50	0.05	2T-GSA	10.17	15.07	30.97	47.77	60.30	68.73	90.94	87.32	75.48	65.91	59.27	54.46
		1S-GSA	27.37	38.43	61.33	81.73	94.37	99.50	93.96	91.67	87.31	83.16	80.69	79.16
		2S-GSA	12.57	17.80	37.37	55.23	69.70	84.27	91.21	88.17	77.30	68.87	62.72	57.22
		STEM	0.03	0.07	0.23	0.77	1.93	2.23	50.00	66.67	36.36	17.86	7.94	5.63
		maSigFun	0.00	0.00	0.00	0.00	0.00	0.00						
	0.1	2T-GSA	8.73	16.03	30.57	47.93	57.90	68.07	83.78	73.76	56.10	42.50	36.07	29.95
		1S-GSA	26.90	36.63	59.03	80.27	93.37	99.03	87.59	83.36	74.69	67.42	63.04	60.09
		2S-GSA	10.87	18.80	36.10	55.53	67.27	82.70	84.09	75.77	58.25	46.81	41.08	33.79
		STEM	0.00	0.00	0.10	0.30	1.73	1.63	100.00	62.50	35.71	27.78	13.33	19.67
		maSigFun	0.00	0.00	0.00	0.00	0.00	0.00						
	0.25	2T-GSA	8.90	13.70	26.47	35.63	51.17	60.30	61.30	48.82	26.48	13.16	10.08	7.33
		1S-GSA	34.03	42.83	63.13	76.93	92.57	97.77	69.69	62.39	48.66	34.28	32.96	28.46
		2S-GSA	16.60	23.63	42.70	44.97	70.90	83.83	63.08	54.14	32.11	15.63	17.91	12.67
		STEM	0.07	0.13	0.43	0.97	1.67	1.80	75.00	55.56	35.00	17.14	13.79	10.00
		maSigFun	0.00	0.00	0.00	0.00	0.00	0.00						
	0.5	2T-GSA	3.83	6.70	13.90	22.40	27.93	39.27	38.50	22.09	7.95	5.49	1.53	0.93
		1S-GSA	30.40	35.37	51.93	67.57	75.00	91.70	41.73	34.79	23.14	17.03	6.33	7.99
		2S-GSA	8.10	14.63	24.37	38.50	47.60	65.23	37.69	19.60	10.85	6.40	1.59	1.26
		STEM	0.17	0.20	0.30	1.23	3.00	2.60	37.50	40.00	25.00	13.95	13.46	17.89
		maSigFun	0.00	0.00	0.00	0.00	0.00	0.00						
100	0.05	2T-GSA	12.03	21.87	46.47	67.23	98.33	99.83	91.21	84.55	70.10	61.51	50.45	47.77
		1S-GSA	35.40	49.13	77.00	93.90	99.87	100.00	93.34	90.91	86.15	83.03	81.42	81.02
		2S-GSA	19.23	32.47	60.60	90.47	99.27	99.97	91.82	86.60	76.71	67.76	63.97	62.93
		STEM	0.07	0.07	0.60	0.53	3.40	12.00	80.00	75.00	14.29	27.27	8.93	2.44
		maSigFun	0.00	0.00	0.00	0.00	0.00	0.00						
	0.1	2T-GSA	12.57	24.00	48.83	71.07	98.30	99.80	81.06	66.97	47.32	34.88	25.51	21.83
		1S-GSA	33.57	49.60	77.60	95.23	99.83	100.00	86.63	81.33	71.84	66.19	62.71	60.47
		2S-GSA	18.60	32.20	60.83	91.87	99.20	99.90	82.29	71.16	54.35	41.16	36.64	32.86
		STEM	0.00	0.13	0.53	0.47	3.17	12.83	100.00	66.67	30.43	36.36	8.65	2.28
		maSigFun	0.00	0.00	0.00	0.00	0.00	0.00						
	0.25	2T-GSA	10.70	20.30	40.30	61.03	93.63	98.83	59.82	40.70	16.45	7.38	4.65	2.88
		1S-GSA	39.80	54.60	81.03	94.37	99.90	100.00	67.50	58.39	44.01	31.02	29.25	24.15
		2S-GSA	15.23	28.67	53.67	84.43	97.50	99.53	60.91	41.42	20.02	9.92	7.55	5.00
		STEM	0.03	0.17	0.87	0.57	5.47	15.87	85.71	54.55	18.75	29.17	13.23	7.75
		maSigFun	0.00	0.00	0.00	0.00	0.00	0.00						
	0.5	2T-GSA	8.30	14.17	27.87	43.33	61.37	88.30	30.64	16.34	6.28	1.74	0.11	0.41
		1S-GSA	32.10	41.03	65.70	81.80	92.20	99.53	39.74	30.41	16.80	9.38	2.61	4.63
		2S-GSA	16.80	26.43	46.83	74.63	82.30	95.90	33.07	18.75	7.63	2.82	0.32	0.72
		STEM	0.07	0.23	1.17	0.73	7.30	20.77	71.43	30.00	16.67	33.33	27.00	19.51
		maSigFun	0.00	0.00	0.00	0.00	0.00	0.00						

The 1S-GSA method identifies all gene sets with spiked-in information in simulation runs with a high p_{ag} and a sufficient gene set size ($|s|_{SIM} \geq 50$), but at the cost of very high FDRs. The 2S-GSA algorithm performs similar or with slightly worse TPR for the ($|s|_{SIM} = 100$) simulations, but with a much lower FDR. The FDR and TPR of the STEM procedure show in opposite to its competitors a rising tendency if p_{as} increases.

The accuracy (ACC) and negative predictive value (NPV) are reported in Table B.2 in the appendix. The NPV is only for the GSA-type algorithms clearly above the trivial limit of $1 - p_{as}$, whereas for STEM the NPV stays near to this border and is constantly on this limit for maSigFun due to its *non detection property*. The overall ACC for the GSA-methods lies below the trivial border of $1 - p_{as}$ if the gene set size is smaller than 100 and $p_{as} = 0.05$. In the case of a high p_{as} (e.g. 0.5) the accuracy of 1S-GSA is higher than the values of 2T-GSA and 2S-GSA, but for smaller values this relation turns around due to the higher rate of false discoveries for the 1S-GSA algorithm.

Despite to the weak signals found in the Aldosterone heart data set, the simulation reveals clear differences between the performance of the five examined profile algorithms. While 1S-GSA shows in general the highest sensitivity its false discovery rate is clearly worse than the two other GSA algorithms. The STEM and maSigFun procedures show no or only very weak sensitivity for the detection of gene sets spiked-in with a moderate or even high proportion of genes from significant sets found by GSA-type algorithms. At least for larger gene sets the algorithms 1S-GSA and 2S-GSA are able to identify gene sets spiked-in with genes from significant sets reported by STEM.

Ovary Development Study

The gene expression during embryonic development of the female reproductive glands (ovaries) is examined at six time points with three replicates at each time point. A single reference is available from a mixed tissue sample of male and female individuals. Determining the differential expression and significant gene set enrichment according to chapter 4 by the TS-ABH FDR limits of $\alpha_{genes} = 0.05$ and $\alpha_{sets} = 0.01$ leads to a relatively high number of significant sets for all competing

Table 5.3: TPR and FDR for identifying spiked-in activation profiles in the simulation to compare the five profile algorithms on basis of the OD data.

$|s|_{SIM}$	p_{as}	algorithm	TPR per p_{ag} 0.1	0.2	0.4	0.6	0.8	1	FDR per p_{ag} 0.1	0.2	0.4	0.6	0.8	1
10	0.05	2T-GSA	0.84	5.38	20.84	34.36	45.26	53.16	70.21	26.90	21.18	19.08	20.76	20.01
		1S-GSA	12.70	20.72	52.56	71.46	77.94	81.10	93.63	89.92	77.19	70.57	68.29	66.54
		2S-GSA	1.14	6.36	22.58	35.44	45.82	53.66	82.02	43.72	35.52	34.83	37.01	35.61
		STEM	0.00	0.00	0.00	0.00	0.04	0.20			100.00	100.00	33.33	0.00
		maSigFun	0.00	0.00	0.00	0.00	0.08	2.18					0.00	0.00
	0.1	2T-GSA	0.92	10.14	32.48	49.72	62.78	68.16	70.13	22.71	13.94	10.16	9.33	9.46
		1S-GSA	12.88	18.80	51.22	71.20	77.96	80.84	86.87	81.51	60.82	50.17	46.68	44.28
		2S-GSA	1.70	14.82	37.96	52.66	64.22	69.18	75.57	33.72	25.77	21.07	20.64	19.05
		STEM	0.00	0.00	0.00	0.02	0.04	0.30			100.00	66.67	60.00	0.00
		maSigFun	0.00	0.00	0.00	0.00	0.10	1.68					0.00	0.00
	0.25	2T-GSA	1.82	9.22	34.40	54.76	68.28	72.92	50.00	17.38	6.88	3.35	2.54	2.59
		1S-GSA	16.40	29.08	58.72	71.18	77.14	79.88	66.95	50.93	29.24	23.05	19.53	18.95
		2S-GSA	4.66	19.72	48.48	63.08	72.02	74.44	53.12	20.74	11.18	7.07	5.44	5.51
		STEM	0.02	0.00	0.00	0.08	0.20	0.60	50.00	100.00	100.00	20.00	16.67	9.09
		maSigFun	0.00	0.00	0.00	0.00	0.06	1.94					0.00	0.00
	0.5	2T-GSA	2.40	7.52	24.36	46.18	62.98	71.08	26.38	9.62	2.87	1.16	0.79	0.53
		1S-GSA	15.38	22.54	49.94	63.58	71.76	75.38	38.63	22.54	9.59	8.31	7.86	8.90
		2S-GSA	6.14	17.88	44.02	59.70	70.72	74.04	27.08	11.22	3.25	2.03	1.70	1.28
		STEM	0.00	0.00	0.04	0.08	0.32	0.90	100.00		33.33	55.56	23.81	25.00
		maSigFun	0.00	0.00	0.00	0.00	0.06	2.16					0.00	0.00
20	0.05	2T-GSA	2.22	11.84	35.62	55.62	63.42	68.00	88.77	64.23	40.06	35.42	32.46	31.82
		1S-GSA	18.02	40.00	71.00	79.76	81.68	82.50	91.94	83.37	73.15	69.79	68.59	67.39
		2S-GSA	5.84	18.04	42.74	56.86	64.02	68.76	90.28	77.46	63.19	59.37	56.89	55.21
		STEM	0.00	0.00	0.00	0.04	0.40	1.04	100.00		100.00	60.00	20.00	7.14
		maSigFun	0.00	0.00	0.00	0.00	0.00	2.38					0.00	0.00
	0.1	2T-GSA	4.18	18.38	43.48	60.44	67.62	71.00	23.16	12.89	9.72	9.82	9.70	9.00
		1S-GSA	19.56	39.16	72.88	80.38	82.14	83.14	82.67	69.72	52.01	48.32	46.69	44.71
		2S-GSA	4.90	20.58	46.22	61.48	68.06	71.62	35.36	23.49	19.65	20.30	19.42	19.87
		STEM	0.00	0.00	0.00	0.16	0.60	1.42	100.00	100.00	100.00	38.46	14.29	8.97
		maSigFun	0.00	0.00	0.00	0.00	0.02	2.66					0.00	0.00
	0.25	2T-GSA	6.12	22.52	57.32	70.30	75.32	76.28	16.62	7.70	3.27	2.77	1.67	1.88
		1S-GSA	20.72	42.98	72.98	79.14	81.28	81.90	62.13	40.14	24.79	21.66	19.86	20.04
		2S-GSA	8.80	33.12	64.30	72.40	75.76	76.80	26.42	11.68	6.68	5.36	3.69	4.07
		STEM	0.00	0.00	0.04	0.34	0.90	2.04	100.00	100.00	86.67	22.73	25.00	15.00
		maSigFun	0.00	0.00	0.00	0.00	0.00	3.00					0.00	0.00
	0.5	2T-GSA	5.44	17.48	48.70	65.26	71.98	76.06	11.11	3.74	0.77	0.61	0.61	0.58
		1S-GSA	19.16	34.30	66.02	75.18	78.44	80.68	35.40	21.73	10.23	11.57	12.44	15.25
		2S-GSA	15.50	36.68	63.14	71.82	75.04	77.56	12.23	5.37	2.08	1.70	1.39	1.70
		STEM	0.02	0.02	0.22	0.74	2.18	4.52	75.00	50.00	38.89	39.34	33.13	36.34
		maSigFun	0.00	0.00	0.00	0.00	0.02	3.32					0.00	0.00
50	0.05	2T-GSA	18.34	47.82	75.82	77.92	79.44	81.80	85.41	67.94	55.58	53.55	52.31	49.93
		1S-GSA	33.46	63.18	81.70	83.54	85.18	87.32	88.84	80.40	74.95	73.95	72.65	71.14
		2S-GSA	40.82	65.10	77.84	78.80	80.14	82.52	88.83	82.70	78.58	77.67	76.60	75.13
		STEM	0.02	0.02	0.78	2.70	5.86	9.18	66.67	83.33	9.30	2.88	1.35	1.08
		maSigFun	0.00	0.00	0.00	0.00	0.00	3.24					0.00	0.00
	0.1	2T-GSA	15.86	46.22	74.40	76.96	78.28	79.94	69.14	41.72	28.67	25.08	23.15	18.88
		1S-GSA	33.52	64.78	80.46	81.44	82.78	84.20	76.44	61.76	53.76	51.36	49.10	47.20
		2S-GSA	38.74	63.18	75.52	77.50	78.94	80.88	73.24	60.32	53.28	47.49	43.87	38.15
		STEM	0.02	0.04	0.34	1.32	3.38	5.08	85.71	75.00	22.73	13.16	5.59	5.22
		maSigFun	0.00	0.00	0.00	0.00	0.00	4.08					0.00	0.00
	0.25	2T-GSA	14.30	41.10	69.12	73.50	75.10	77.26	6.17	4.68	2.26	1.71	1.78	1.70
		1S-GSA	42.22	69.82	80.70	81.52	83.32	84.30	46.68	30.64	23.78	22.92	22.62	22.40
		2S-GSA	18.58	47.70	69.90	73.90	75.78	78.00	10.59	8.87	5.26	4.37	3.93	3.85
		STEM	0.04	0.10	0.46	1.78	4.26	6.14	75.00	58.33	28.12	25.83	18.70	24.94
		maSigFun	0.00	0.00	0.00	0.00	0.00	4.20					0.00	0.00
	0.5	2T-GSA	15.90	40.92	68.42	74.56	76.84	77.80	1.49	1.06	0.58	0.40	0.62	0.64
		1S-GSA	38.68	63.62	79.54	81.34	82.76	83.08	24.57	13.68	10.83	13.32	15.27	21.07
		2S-GSA	27.86	55.46	73.36	75.64	77.34	78.38	3.67	2.32	1.42	1.12	1.45	1.68
		STEM	0.10	0.18	1.24	4.34	7.14	11.00	54.55	62.50	34.04	38.35	48.78	57.82
		maSigFun	0.00	0.00	0.00	0.00	0.00	4.20					0.00	0.00
100	0.05	2T-GSA	0.00	0.00	0.00	0.00	0.00	0.00						
		1S-GSA	52.16	76.22	80.68	81.56	83.92	86.62	87.30	81.93	80.18	79.04	77.51	75.90
		2S-GSA	0.00	0.00	0.00	0.00	0.00	0.00						
		STEM	0.12	0.24	6.18	17.80	25.88	32.66	45.45	25.00	1.28	1.00	0.54	0.49
		maSigFun	0.00	0.00	0.00	0.00	0.00	4.24					0.00	0.00
	0.1	2T-GSA	30.68	63.48	77.68	79.78	83.54	85.86	62.10	43.04	36.35	33.39	31.19	30.09
		1S-GSA	49.28	76.68	83.28	85.16	88.62	91.82	73.20	62.94	58.69	56.32	53.82	51.87
		2S-GSA	54.62	73.64	78.54	80.68	84.22	86.52	74.83	67.52	63.75	61.60	59.43	57.04
		STEM	0.06	0.32	2.82	7.84	12.20	16.84	57.14	27.27	2.08	1.51	1.77	1.29
		maSigFun	0.00	0.00	0.00	0.00	0.00	5.36					0.00	0.00
	0.25	2T-GSA	22.64	58.66	72.68	75.06	79.28	82.74	18.68	6.14	3.94	3.17	2.75	1.76
		1S-GSA	55.18	78.04	83.14	86.90	90.00	45.01	32.78	28.81	28.43	26.93	29.69	
		2S-GSA	36.16	63.44	73.18	75.52	80.44	84.44	26.86	13.50	9.99	7.41	6.09	4.54
		STEM	0.04	0.14	1.52	3.94	6.64	8.46	71.43	22.22	21.65	25.66	30.98	47.78
		maSigFun	0.00	0.00	0.00	0.00	0.00	5.74					0.00	0.00
	0.5	2T-GSA	23.84	56.78	72.44	75.06	77.18	79.12	0.83	0.46	0.49	0.69	0.49	0.78
		1S-GSA	51.14	74.62	80.94	82.38	84.38	85.70	17.99	11.10	12.14	15.46	17.31	18.08
		2S-GSA	34.38	64.70	73.36	75.62	77.96	80.38	2.16	1.04	1.19	1.28	1.34	1.50
		STEM	0.10	0.48	3.18	7.96	11.68	16.48	61.54	50.00	39.31	49.94	64.19	66.88
		maSigFun	0.00	0.00	0.00	0.00	0.00	5.78					0.00	0.00

Table 5.4: TPR per algorithm and type of spiked-in profile in the simulation to compare the five profile algorithms on basis of the OD data.

Ovary development data			TPR with $p_{ag} = 0.2$					TPR with $p_{ag} = 0.6$					TPR with $p_{ag} = 1$				
s_{SIM}	p_{as}	algorithm	2T-GSA	1S-GSA	2S-GSA	STEM	maSigFun	2T-GSA	1S-GSA	2S-GSA	STEM	maSigFun	2T-GSA	1S-GSA	2S-GSA	STEM	maSigFun
10	0.05	2T-GSA	8.2	8.9	9.3	0.0	0.5	51.2	53.0	56.2	0.4	11.0	71.0	73.0	76.8	0.3	44.7
		1S-GSA	27.0	27.4	26.5	8.4	14.3	97.7	94.6	96.7	7.4	60.9	99.9	99.8	99.9	9.2	96.7
		2S-GSA	9.8	10.0	10.7	0.2	1.1	52.5	53.7	56.7	0.6	13.7	71.2	73.2	76.9	0.8	46.2
		STEM	0.0	0.0	0.0	0.0	0.0	0.0	0.0	0.0	0.0	0.0	0.0	0.0	0.0	0.0	1.0
		maSigFun	0.0	0.0	0.0	0.0	0.0	0.0	0.0	0.0	0.0	0.0	0.0	0.0	0.1	0.0	10.8
	0.1	2T-GSA	14.6	16.0	17.8	0.2	2.1	70.2	72.1	74.1	0.3	31.9	88.0	90.5	91.4	1.1	69.8
		1S-GSA	25.0	24.4	24.5	7.4	12.7	96.1	95.8	94.8	6.9	62.4	99.7	100.0	99.9	7.5	97.1
		2S-GSA	21.3	23.1	25.0	1.0	3.7	73.1	74.6	76.9	0.6	38.1	88.4	90.9	91.6	1.8	73.2
		STEM	0.0	0.0	0.0	0.0	0.0	0.1	0.0	0.0	0.0	0.0	0.0	0.1	0.0	0.1	1.3
		maSigFun	0.0	0.0	0.0	0.0	0.0	0.0	0.0	0.0	0.0	0.0	0.1	0.0	0.1	0.0	8.2
	0.25	2T-GSA	13.1	15.0	15.2	0.7	2.1	79.7	80.5	81.8	0.6	31.2	97.2	95.9	96.4	0.6	74.5
		1S-GSA	41.3	38.4	38.7	9.1	17.9	95.6	96.0	95.7	5.2	63.4	100.0	100.0	99.7	5.9	94.8
		2S-GSA	28.9	30.9	31.2	1.5	6.1	87.2	88.3	90.3	1.1	48.5	97.8	96.9	97.4	1.0	79.1
		STEM	0.0	0.0	0.0	0.0	0.0	0.0	0.0	0.0	0.1	0.3	0.0	0.1	0.1	0.5	2.3
		maSigFun	0.0	0.0	0.0	0.0	0.0	0.0	0.0	0.0	0.0	0.0	0.0	0.1	0.0	0.0	9.6
	0.5	2T-GSA	11.1	12.4	9.8	0.2	4.1	71.3	67.6	69.7	0.6	21.7	94.8	96.0	94.3	0.7	69.6
		1S-GSA	29.4	31.7	29.3	6.5	15.8	88.2	86.5	86.4	4.6	52.2	94.8	95.8	96.4	5.4	84.5
		2S-GSA	25.7	27.5	26.3	1.4	8.5	87.1	85.3	86.0	1.0	39.1	97.2	98.5	97.8	0.9	75.8
		STEM	0.0	0.0	0.0	0.0	0.0	0.0	0.0	0.0	0.1	0.3	0.0	0.3	0.2	1.8	2.2
		maSigFun	0.0	0.0	0.0	0.0	0.0	0.0	0.0	0.0	0.0	0.0	0.1	0.0	0.1	0.0	10.6
20	0.05	2T-GSA	19.3	17.8	17.9	0.9	3.3	74.1	73.3	74.8	0.5	55.4	84.7	88.3	87.6	3.2	76.2
		1S-GSA	57.7	54.6	55.4	8.2	24.1	99.8	99.5	99.6	5.7	94.2	99.8	100.0	100.0	12.7	100.0
		2S-GSA	28.5	25.3	25.8	2.5	8.1	74.8	73.9	75.5	1.8	58.3	84.9	88.3	87.6	5.4	77.6
		STEM	0.0	0.0	0.0	0.0	0.0	0.0	0.1	0.0	0.0	0.1	0.9	0.2	0.1	1.6	2.4
		maSigFun	0.0	0.0	0.0	0.0	0.0	0.0	0.0	0.0	0.0	0.0	0.0	0.0	0.0	0.0	11.9
	0.1	2T-GSA	28.7	28.8	29.6	0.5	4.3	76.6	76.8	78.8	0.7	69.3	88.1	86.7	90.4	4.1	85.7
		1S-GSA	57.0	52.2	54.4	7.4	24.8	99.7	99.4	100.0	7.3	95.5	99.9	100.0	100.0	15.8	100.0
		2S-GSA	31.1	31.6	32.8	1.1	6.3	77.5	77.3	79.3	1.8	71.5	88.1	86.7	90.4	5.4	87.5
		STEM	0.0	0.0	0.0	0.0	0.0	0.3	0.0	0.0	0.0	0.3	0.2	0.6	0.5	1.9	3.9
		maSigFun	0.0	0.0	0.0	0.0	0.0	0.0	0.0	0.0	0.0	0.0	0.0	0.0	0.0	0.0	13.3
	0.25	2T-GSA	31.7	34.0	37.2	0.3	9.4	91.5	93.1	92.4	0.4	74.1	95.4	95.6	96.8	2.6	91.0
		1S-GSA	61.2	57.8	59.6	7.4	28.9	99.7	99.5	99.6	4.5	92.4	100.0	100.0	100.0	9.8	99.7
		2S-GSA	46.7	49.8	53.5	0.8	14.8	92.6	94.0	93.8	1.0	80.6	95.4	95.6	96.8	4.0	92.2
		STEM	0.0	0.0	0.0	0.0	0.0	0.1	0.1	0.2	0.7	0.6	0.9	1.3	0.9	3.9	3.2
		maSigFun	0.0	0.0	0.0	0.0	0.0	0.0	0.0	0.0	0.0	0.0	0.0	0.0	0.0	0.0	15.0
	0.5	2T-GSA	24.5	27.4	26.9	0.5	8.1	89.6	89.4	89.6	0.5	57.2	96.4	98.9	98.7	0.9	85.4
		1S-GSA	45.7	46.1	47.6	6.8	25.3	98.4	96.8	97.3	5.6	77.8	99.1	98.9	99.1	10.1	96.2
		2S-GSA	50.4	54.2	58.1	1.4	19.3	96.2	96.2	95.9	1.0	69.8	97.9	99.1	99.2	1.4	90.2
		STEM	0.0	0.0	0.0	0.1	0.0	0.2	0.0	0.2	1.4	1.9	1.3	1.5	1.2	13.8	4.8
		maSigFun	0.0	0.0	0.0	0.0	0.0	0.0	0.0	0.0	0.0	0.0	0.0	0.0	0.0	0.0	16.6
50	0.05	2T-GSA	64.4	66.0	67.9	2.4	38.4	99.6	100.0	99.9	2.9	87.2	99.9	100.0	100.0	14.0	95.1
		1S-GSA	82.9	83.4	83.8	8.8	57.0	99.9	100.0	100.0	17.8	100.0	100.0	100.0	100.0	36.6	100.0
		2S-GSA	83.4	84.8	86.5	8.2	62.6	100.0	100.0	100.0	4.5	89.5	100.0	100.0	100.0	15.1	97.5
		STEM	0.0	0.0	0.0	0.1	0.0	0.5	0.3	0.3	7.4	5.0	1.8	3.3	2.6	23.0	15.2
		maSigFun	0.0	0.0	0.0	0.0	0.0	0.0	0.0	0.0	0.0	0.0	0.0	0.0	0.0	0.0	16.2
	0.1	2T-GSA	66.4	63.1	68.4	1.6	31.6	97.0	97.3	97.7	2.0	90.8	96.6	97.8	98.2	11.0	96.1
		1S-GSA	86.7	86.2	88.5	6.2	56.3	99.9	100.0	100.0	7.4	99.9	100.0	100.0	100.0	21.0	100.0
		2S-GSA	81.9	82.5	83.8	6.4	61.3	97.3	97.4	97.7	3.4	91.7	96.7	97.8	98.2	14.0	97.7
		STEM	0.0	0.0	0.1	0.1	0.0	0.6	0.3	0.4	3.0	2.3	1.7	2.8	2.6	14.3	4.0
		maSigFun	0.0	0.0	0.0	0.0	0.0	0.0	0.0	0.0	0.0	0.0	0.0	0.0	0.0	0.0	20.4
	0.25	2T-GSA	59.1	59.1	58.5	0.4	28.4	91.6	91.5	92.9	0.7	90.8	91.5	95.7	96.6	9.3	93.2
		1S-GSA	92.7	92.2	93.4	7.1	63.7	99.9	99.9	100.0	7.8	100.0	100.0	100.0	100.0	21.5	100.0
		2S-GSA	65.4	65.5	65.3	1.0	41.3	91.9	91.8	92.9	1.1	91.8	91.7	95.7	96.6	11.4	94.6
		STEM	0.0	0.1	0.0	0.4	0.0	0.8	0.8	0.8	4.3	2.2	2.1	3.0	3.2	16.5	5.9
		maSigFun	0.0	0.0	0.0	0.0	0.0	0.0	0.0	0.0	0.0	0.0	0.0	0.0	0.0	0.0	21.0
	0.5	2T-GSA	55.9	58.4	57.4	0.3	32.6	93.1	95.6	97.6	0.5	86.0	95.3	99.4	98.9	2.5	92.9
		1S-GSA	86.1	85.1	85.7	6.9	54.3	99.1	99.7	99.7	8.8	99.4	99.1	100.0	100.0	16.5	99.8
		2S-GSA	73.0	75.9	76.1	0.8	51.5	94.9	96.4	98.1	0.7	88.1	97.0	99.5	98.9	3.0	93.5
		STEM	0.0	0.0	0.1	0.6	0.2	0.8	1.7	1.8	13.3	4.1	4.3	5.8	4.1	32.5	8.3
		maSigFun	0.0	0.0	0.0	0.0	0.0	0.0	0.0	0.0	0.0	0.0	0.0	0.0	0.0	0.0	21.0
100	0.05	2T-GSA	0.0	0.0	0.0	0.0	0.0	0.0	0.0	0.0	0.0	0.0	0.0	0.0	0.0	0.0	0.0
		1S-GSA	96.3	95.6	95.8	4.8	88.6	99.9	100.0	100.0	7.9	100.0	100.0	100.0	100.0	33.1	100.0
		2S-GSA	0.0	0.0	0.0	0.0	0.0	0.0	0.0	0.0	0.0	0.0	0.0	0.0	0.0	0.0	0.0
		STEM	0.0	0.2	0.0	0.7	0.3	8.4	9.0	9.4	32.7	29.5	22.3	24.3	24.2	52.7	39.8
		maSigFun	0.0	0.0	0.0	0.0	0.0	0.0	0.0	0.0	0.0	0.0	0.0	0.0	0.0	0.0	21.2
	0.1	2T-GSA	83.9	83.2	84.5	2.3	63.5	99.5	100.0	100.0	7.5	91.9	100.0	100.0	100.0	30.4	98.9
		1S-GSA	94.8	94.6	95.5	10.6	87.9	99.9	100.0	100.0	25.9	100.0	100.0	100.0	100.0	59.1	100.0
		2S-GSA	93.9	92.5	94.0	7.2	80.6	99.8	100.0	100.0	9.3	94.3	100.0	100.0	100.0	33.0	99.6
		STEM	0.1	0.1	0.1	0.5	0.8	1.0	2.7	2.2	17.4	15.9	4.3	10.3	9.3	34.2	26.1
		maSigFun	0.0	0.0	0.0	0.0	0.0	0.0	0.0	0.0	0.0	0.0	0.0	0.0	0.0	0.0	26.8
	0.25	2T-GSA	76.5	76.8	77.6	0.3	62.1	91.6	93.4	93.6	3.9	92.8	92.9	97.2	97.3	31.5	94.8
		1S-GSA	98.2	97.9	98.0	6.6	89.5	99.9	100.0	100.0	15.8	100.0	100.0	100.0	100.0	50.0	100.0
		2S-GSA	81.3	81.3	82.7	1.2	70.7	91.6	93.4	93.4	4.7	94.3	93.1	97.2	97.4	37.4	97.1
		STEM	0.1	0.1	0.0	0.3	0.2	1.7	2.5	1.5	9.3	4.7	2.7	5.9	5.6	18.7	9.4
		maSigFun	0.0	0.0	0.0	0.0	0.0	0.0	0.0	0.0	0.0	0.0	0.0	0.0	0.0	0.0	28.7
	0.5	2T-GSA	73.8	73.4	74.8	0.4	61.5	90.5	95.7	96.8	2.5	89.8	89.7	99.0	99.1	13.9	94.0
		1S-GSA	94.6	94.3	95.9	6.1	82.2	99.3	99.9	100.0	12.8	99.9	99.4	99.9	100.0	29.2	100.0
		2S-GSA	81.9	81.2	82.9	1.0	76.5	91.6	96.0	97.0	3.2	90.3	92.1	99.0	99.1	17.1	94.6
		STEM	0.1	0.5	0.3	1.0	0.5	3.3	4.4	3.5	22.4	6.2	7.0	10.3	10.9	43.0	11.2
		maSigFun	0.0	0.0	0.0	0.0	0.0	0.0	0.0	0.0	0.0	0.0	0.0	0.0	0.0	0.0	28.9

5.3 Simulation Study to Compare the Methods

algorithms. The number of prototype sets yielded by filtering these results and used in the simulation is given in Table 5.1. All procedures provide prototype sets in contrast to the Aldosterone heart experiment, whereas the small number of three sets for the (regression) functional microarray significant profiles (maSigFun) algorithm seems to be characteristic across all four gene expression time series experiments. Applying the five profile algorithms in the simulation study described in section 5.2 yields results, which are elaborated in the following.

The overall true positive rate (TPR) and false discovery rate (FDR) are reported in Table 5.3. All identified gene sets with spiked-in information are considered as true positive (TP) and all gene sets found with a significant activation profile without spiked-in differentially expressed genes are regarded as false positive (FP). In general, the three GSA-type profile algorithms show significantly higher TPR and FDR values than the STEM and maSigFun procedures except from some occasional STEM misdetections, which lead to FDR values up to 100 %. The identification of gene sets with spiked-in genes from significant STEM sets hardly succeeds for any algorithm even with $p_{ag} = 1$ as Table 5.4 reveals. The same table shows, that the GSA-type procedures usually outperform the maSigFun algorithm in identifying sets with spiked-in genes from significant maSigFun sets. The maSigFun procedure seems to be limited to identify gene sets spiked-in with genes from significant maSigFun sets. The three GSA-types detect spiked-in sets from the same types with a high sensitivity even for lower values of p_{ag}, i.e. even for a smaller proportion of spiked-in genes in the *active* set. There is a sensitivity order within the GSA-type algorithms, in which the 1S-GSA procedure clearly outperforms the two competitors and 2S-GSA shows in general a slightly better TPR than 2T-GSA.

There is an obvious anomaly for the algorithms 2T-GSA and 2S-GSA if the simulation parameters are chosen as $|s|_{\text{SIM}} = 100$ and $p_{\text{as}} = 0.05$. No set with a significant profile is identified throughout all simulation turns, which cannot be observed in this extent for the simulations on basis of the other time series experiments. This occurs most likely due to the combination of many true differentially expressed genes in the basis data set and the fact that the proportion of 5 % spiked-in sets is very close to the chosen FDR limit for significant sets. This anomaly is ignored for the following considerations.

The TPR amplifies with increasing $|s|_{\text{SIG}}$ and rising p_{ag} in concordance with the simulation results from the Aldosterone heart data. The relation between TPR and the parameter p_{as} cannot be described uniformly for all algorithms. While the sensitivity for the STEM and maSigFun procedure is highest for the extreme values of the parameter the opposite seems to hold for the profile algorithms of the GSA-type. The differences for varied p_{as} are small by contrast with alterations of p_{ag} or $|s|_{\text{SIM}}$.

Table 5.3 shows a constant FDR of 0 % for the maSigFun algorithm. The FDR of the GSA-algorithms increases for a rising number of genes in the set in contrast to the findings of the Aldosterone simulation. Except for the STEM algorithm, an increasing proportion of active genes (p_{ag}) abates the FDR. The rather high FDR values in comparison with the other simulations can be explained by the high number of differentially expressed genes found in the ovary development data, which may lead to significant sets without spiked-in genes classified as false positives. The reported FDR for the STEM procedure is very high if the proportion of true significant sets is high ($p_{\text{as}} \geq 0.25$) or the proportion of spiked-in genes in active sets is small.

The accuracy (ACC) and negative predictive value (NPV) are reported in Table B.3 in the appendix. Only the NPV of the GSA-type algorithms stands out distinctly from the trivial limit of $1-p_{\text{as}}$, analogously to the Aldosterone simulation. Increasing values of p_{ag} and $|s|_{\text{SIM}}$ enhance the NPV, while a rising p_{as} have a reducing effect due to the lower number of TNs in the gene set universe. The overall ACC for all methods exceeds the trivial border of $1 - p_{\text{as}}$ if the proportion of spiked-in genes is sufficient ($p_{\text{ag}} > 0.5$) except for the GSA methods with $|s|_{\text{SIM}} \geq 50$ and $p_{\text{as}} = 0.05$. The accuracy order within the GSA-type methods is 2T-GSA before 2S-GSA and 1S-GSA, mainly due to the differences in the detection of FPs. The 1S-GSA method regularly fails the trivial ACC limit in cases of a small proportion of active sets ($p_{\text{as}} \leq 0.10$). This occurs as a trade-off in relation to the outstanding TPR performance and reveals a high number of false and true positives.

The ovary development time course experiment provides a high number of differentially expressed genes. Due to the recycling characteristic of the simulation a high

5.3 Simulation Study to Compare the Methods

number of false positives and in consequence a high FDR value can be observed in the simulation results. Nevertheless, the sensitivity of STEM and maSigFun is far smaller than for the GSA-type algorithms. This fact is not compensated by the smaller FDR values, since in general from an exploratory analysis of an expensive study significant findings are appreciated despite the risk of (a controlled number of) false findings. From this point of view, the algorithms 2T-GSA or 2S-GSA are the instruments of choice.

Skin Healing Time Series

The gene expression time series study conducted to analyze the molecular genetic process involved during wound healing is divided into two separate analyses in this thesis. The first data set includes only individuals with injured skin (SH) and the second analysis focuses on the experiments with wounded mucosa on mice tongues (TH). The expression data from three microarrays is available for each of eight time points. The time point before the surgical procedure is used as reference for all compared methods. Analogously to the ovary development data set the TS-ABH FDR limits determining differential gene expression and gene set enrichment are chosen as $\alpha_{\text{genes}} = 0.05$ and $\alpha_{\text{sets}} = 0.01$. Applying the five competing algorithms and filtering the results as described in section 5.2 leads to the number of prototypes per method given in Table 5.1. There are only three prototypes arising from the 2T-GSA algorithm and four maSigFun sets. Generally, from this experiment less significant activation profiles result compared with the OD or TH data set. This corresponds to the lower number of identified differentially expressed genes per time point as reported in Figure 6.7. This fact leads to a lack of suitable genes to spike in differentially expressed genes into *active* gene sets in the simulation, due to the fact that only full gene trajectories are recycled in order to reproduce significant activation profiles. This restriction arising from the data should be kept in mind while examining the simulation results from comparing the five competing algorithms in the following paragraphs.

Table 5.5 reports the overall true positive rate (TPR) and false discovery rate (FDR), whereas all identified gene sets with spiked-in information are considered as true positive (TP) and all gene sets found with a significant activation profile

Table 5.5: TPR and FDR for identifying spiked-in activation profiles in the simulation to compare the five profile algorithms on basis of the SH data.

Skin healing data			TPR per p_{ag}						FDR per p_{ag}				
$|s|_{SIM}$	p_{as}	algorithm	0.1	0.2	0.4	0.6	0.8	1	0.1	0.2	0.4	0.6	0.8	1
10	0.05	2T-GSA	10.24	23.30	39.08	45.22	49.96	53.82	52.59	33.12	20.02	15.89	13.02	10.72
		1S-GSA	18.52	30.46	65.58	81.58	86.28	90.22	93.67	89.77	79.70	75.49	73.81	72.29
		2S-GSA	10.40	24.44	41.38	47.44	52.48	56.56	53.90	34.30	23.09	18.38	15.08	12.34
		STEM	0.00	0.00	0.00	0.04	0.10	0.18	100.00	100.00		0.00	16.67	10.00
		maSigFun	0.00	0.00	0.00	0.02	0.22	15.78				0.00	0.00	0.00
	0.1	2T-GSA	2.72	16.54	34.56	41.66	46.06	50.02	51.25	15.96	7.35	4.10	2.70	1.65
		1S-GSA	19.40	26.90	62.40	80.42	85.60	88.10	86.30	81.50	64.40	56.55	53.82	52.17
		2S-GSA	3.52	20.10	39.52	46.40	50.00	53.56	55.89	19.86	9.52	6.26	3.66	2.48
		STEM	0.00	0.00	0.00	0.00	0.14	0.34	100.00	100.00			0.00	10.53
		maSigFun	0.00	0.00	0.00	0.00	0.26	15.02					0.00	0.00
	0.25	2T-GSA	1.50	8.88	24.56	31.74	36.66	42.28	43.61	8.07	1.37	0.81	0.27	0.09
		1S-GSA	24.12	35.82	64.80	78.28	84.34	86.86	66.72	54.58	37.21	32.56	29.36	29.79
		2S-GSA	2.64	15.74	31.24	38.90	43.92	48.18	44.77	6.75	2.13	1.22	0.36	0.17
		STEM	0.00	0.00	0.00	0.02	0.14	0.48	100.00	100.00	100.00	0.00	0.00	0.00
		maSigFun	0.00	0.00	0.00	0.00	0.34	15.48					0.00	0.00
	0.5	2T-GSA	1.28	4.26	11.96	19.78	28.18	36.94	27.27	4.05	0.33	0.00	0.00	0.05
		1S-GSA	19.94	26.58	47.18	62.02	69.84	73.94	43.70	32.02	18.26	16.75	16.46	18.32
		2S-GSA	2.44	9.52	20.40	28.96	37.28	44.30	29.48	3.84	0.68	0.07	0.00	0.09
		STEM	0.00	0.00	0.04	0.04	0.38	0.76			100.00	0.00	33.33	13.64	17.39
		maSigFun	0.00	0.00	0.00	0.00	0.32	14.74					0.00	0.00
20	0.05	2T-GSA	11.52	29.64	42.46	51.16	54.90	57.90	21.20	13.23	10.27	6.57	5.83	4.64
		1S-GSA	28.08	51.46	82.06	88.20	91.56	93.04	91.32	85.21	77.28	75.16	73.79	73.06
		2S-GSA	11.72	30.28	45.30	53.66	57.04	59.46	21.24	12.99	9.72	7.07	5.78	4.44
		STEM	0.00	0.00	0.02	0.24	0.44	0.96	100.00			66.67	7.69	0.00	2.04
		maSigFun	0.00	0.00	0.00	0.00	0.14	18.48					0.00	0.00
	0.1	2T-GSA	14.24	31.56	45.62	51.04	55.12	57.56	19.37	9.21	4.64	2.93	2.30	1.13
		1S-GSA	25.60	50.74	80.90	88.18	90.44	93.10	83.70	71.63	59.06	56.02	53.51	52.02
		2S-GSA	14.66	33.42	48.72	53.48	56.96	59.40	19.27	9.53	5.03	2.87	2.40	0.74
		STEM	0.00	0.00	0.00	0.12	0.46	0.96	100.00	100.00	100.00	0.00	8.00	2.04
		maSigFun	0.00	0.00	0.00	0.00	0.04	18.52					0.00	0.00
	0.25	2T-GSA	7.48	21.46	33.32	42.92	49.22	54.12	9.22	2.72	0.48	0.19	0.04	0.15
		1S-GSA	28.88	49.10	79.40	86.20	89.88	92.08	61.74	46.37	33.13	30.56	30.28	30.54
		2S-GSA	8.94	26.96	39.42	48.16	52.94	55.94	9.70	2.74	0.40	0.12	0.08	0.11
		STEM	0.00	0.00	0.06	0.32	0.44	1.00	100.00	100.00	50.00	0.00	8.33	9.09
		maSigFun	0.00	0.00	0.00	0.00	0.10	18.38					0.00	0.00
	0.5	2T-GSA	4.54	11.40	23.52	32.56	42.06	50.80	7.35	2.06	0.51	0.18	0.00	0.08
		1S-GSA	24.12	37.70	67.80	79.88	86.56	89.84	39.37	28.27	17.44	17.97	19.57	23.99
		2S-GSA	10.24	20.10	33.00	41.82	48.40	54.00	8.41	2.14	0.54	0.14	0.04	0.07
		STEM	0.02	0.02	0.14	0.56	1.34	3.96	50.00	80.00	36.36	41.67	56.21	71.22
		maSigFun	0.00	0.00	0.00	0.00	0.10	18.18					0.00	0.00
50	0.05	2T-GSA	8.98	24.78	43.56	48.20	50.74	52.12	4.26	2.52	2.36	2.35	2.39	1.92
		1S-GSA	44.58	74.62	86.20	89.52	92.20	94.24	88.95	82.48	79.54	78.02	77.33	76.14
		2S-GSA	9.38	25.26	45.26	50.28	53.20	55.00	4.29	2.17	1.74	1.22	0.86	0.79
		STEM	0.00	0.04	0.46	2.16	3.62	5.76	100.00	50.00	14.81	3.57	2.69	2.37
		maSigFun	0.00	0.00	0.00	0.00	0.00	19.76						0.00
	0.1	2T-GSA	26.76	39.26	51.18	56.04	57.74	59.52	6.82	3.87	1.92	0.78	0.76	0.47
		1S-GSA	47.42	78.24	88.20	91.92	94.72	95.58	77.13	66.14	60.14	57.97	56.18	55.06
		2S-GSA	27.32	41.74	53.30	57.86	59.32	60.64	6.82	3.74	2.06	0.75	0.70	0.43
		STEM	0.02	0.00	0.32	1.18	2.44	3.96	75.00	100.00	11.11	6.35	5.43	5.71
		maSigFun	0.00	0.00	0.00	0.00	0.00	19.80						0.00
	0.25	2T-GSA	24.90	38.96	49.14	53.96	57.00	60.30	3.64	1.27	0.16	0.11	0.11	0.00
		1S-GSA	51.74	78.62	87.60	92.32	94.42	95.80	50.05	37.06	32.20	32.42	31.66	32.46
		2S-GSA	28.80	42.72	52.58	56.48	58.68	60.04	3.68	1.48	0.15	0.14	0.10	0.00
		STEM	0.00	0.04	0.28	0.66	2.84	6.40	100.00	66.67	30.00	45.90	63.68	76.81
		maSigFun	0.00	0.00	0.00	0.00	0.00	19.82						0.00
	0.5	2T-GSA	14.74	25.94	38.40	45.10	53.20	57.76	1.34	0.08	0.05	0.09	0.00	0.03
		1S-GSA	45.16	68.78	85.06	91.34	93.88	95.54	29.02	19.86	18.16	20.01	23.74	25.97
		2S-GSA	22.36	34.30	45.42	50.64	55.22	58.46	1.93	0.35	0.04	0.04	0.00	0.03
		STEM	0.00	0.16	1.18	7.96	15.10	21.48	100.00	55.56	75.52	69.34	70.02	72.19
		maSigFun	0.00	0.00	0.00	0.00	0.00	19.24						0.00
100	0.05	2T-GSA	8.42	17.90	28.68	32.62	38.54	42.68	42.49	45.19	45.43	43.41	38.83	35.88
		1S-GSA	58.60	83.92	92.92	95.06	96.54	96.92	87.75	82.97	80.71	79.98	79.10	78.59
		2S-GSA	9.02	19.92	30.36	34.76	39.40	43.54	40.66	42.23	43.76	41.00	37.56	34.17
		STEM	0.02	0.66	5.82	13.82	18.18	21.34	90.91	17.50	2.68	1.29	1.41	1.57
		maSigFun	0.00	0.00	0.00	0.00	0.00	19.24						0.00
	0.1	2T-GSA	17.82	34.10	46.00	51.36	55.18	57.24	1.87	0.87	1.37	0.47	0.68	0.66
		1S-GSA	61.60	82.54	89.30	91.64	94.82	95.60	75.15	67.90	64.90	62.79	60.66	59.59
		2S-GSA	18.70	36.52	48.74	54.12	56.96	58.36	1.48	0.54	0.33	0.29	0.35	0.27
		STEM	0.02	0.36	2.10	3.94	5.46	6.64	80.00	14.29	7.89	9.22	12.78	20.95
		maSigFun	0.00	0.00	0.00	0.00	0.00	19.98						0.00
	0.25	2T-GSA	32.14	44.12	52.58	55.40	58.98	63.10	1.59	0.27	0.30	0.04	0.10	0.09
		1S-GSA	68.70	84.84	92.14	94.34	96.00	96.72	47.99	39.43	36.37	34.81	34.55	35.37
		2S-GSA	35.38	47.78	54.88	56.52	59.28	61.32	1.50	0.25	0.25	0.04	0.07	0.03
		STEM	0.00	0.14	1.48	4.68	10.46	16.22	100.00	68.18	54.88	78.13	84.84	88.26
		maSigFun	0.00	0.00	0.00	0.00	0.00	19.90						0.00
	0.5	2T-GSA	22.98	32.98	43.16	48.28	54.22	60.28	0.78	0.18	0.14	0.00	0.04	0.00
		1S-GSA	56.60	79.90	89.60	93.74	95.46	97.24	23.37	17.29	17.75	18.59	22.19	24.73
		2S-GSA	28.02	39.50	47.46	51.06	55.00	60.38	0.71	0.15	0.13	0.00	0.00	0.00
		STEM	0.10	0.62	8.38	19.42	25.40	28.26	76.19	73.04	70.35	72.98	75.35	75.67
		maSigFun	0.00	0.00	0.00	0.00	0.00	19.46						0.00

Table 5.6: TPR per algorithm and type of spiked-in profile in the simulation to compare the five profile algorithms on basis of the SH data.

s_{SIM}	p_{as}	Skin healing data algorithm	TPR with $p_{ag} = 0.2$ 2T-GSA	1S-GSA	2S-GSA	STEM	maSigFun	TPR with $p_{ag} = 0.6$ 2T-GSA	1S-GSA	2S-GSA	STEM	maSigFun	TPR with $p_{ag} = 1$ 2T-GSA	1S-GSA	2S-GSA	STEM	maSigFun
10	0.05	2T-GSA	43.9	25.1	35.3	0.6	11.6	77.6	53.2	70.3	0.7	24.3	87.4	64.8	83.0	1.0	32.9
		1S-GSA	36.2	31.0	33.6	14.5	37.0	97.4	90.2	91.5	28.8	100.0	100.0	99.1	99.1	52.9	100.0
		2S-GSA	45.8	26.4	37.3	0.6	12.1	81.1	56.4	73.8	0.8	25.1	91.1	67.2	88.3	1.1	35.1
		STEM	0.0	0.0	0.0	0.0	0.0	0.0	0.0	0.0	0.0	0.2	0.0	0.0	0.0	0.0	0.9
		maSigFun	0.0	0.0	0.0	0.0	0.0	0.0	0.0	0.0	0.0	0.1	1.7	0.0	2.0	0.1	75.1
	0.1	2T-GSA	31.1	17.1	23.5	0.8	10.2	69.9	48.1	65.5	0.2	24.6	84.8	58.8	77.0	0.7	28.8
		1S-GSA	31.6	27.2	29.1	14.3	32.3	96.1	88.1	90.5	27.6	99.8	99.9	97.9	98.6	44.1	100.0
		2S-GSA	36.7	21.9	28.8	0.9	12.2	78.2	54.4	71.3	0.4	27.7	90.3	62.4	83.1	0.9	31.1
		STEM	0.0	0.0	0.0	0.0	0.0	0.0	0.0	0.0	0.0	0.0	0.0	0.1	0.0	0.1	1.5
		maSigFun	0.0	0.0	0.0	0.0	0.0	0.0	0.0	0.0	0.0	0.0	2.7	0.0	1.1	0.1	71.2
	0.25	2T-GSA	16.7	7.8	13.5	0.2	6.2	56.1	34.5	48.8	0.2	19.1	73.7	47.1	66.3	0.3	24.0
		1S-GSA	45.9	36.1	41.9	15.0	40.2	95.9	85.2	88.3	22.4	99.6	99.0	97.0	97.4	40.9	100.0
		2S-GSA	28.6	15.7	24.2	0.3	9.9	68.4	42.5	59.9	0.2	23.5	84.2	54.0	76.8	0.3	25.6
		STEM	0.0	0.0	0.0	0.0	0.0	0.0	0.0	0.0	0.0	0.1	0.1	0.1	0.0	0.0	2.2
		maSigFun	0.0	0.0	0.0	0.0	0.0	0.0	0.0	0.0	0.0	0.0	1.5	0.1	2.2	0.1	73.5
	0.5	2T-GSA	8.7	4.5	4.4	0.4	3.3	35.2	19.6	27.0	0.1	17.0	62.1	34.9	51.7	0.1	35.9
		1S-GSA	33.4	28.4	28.0	13.9	29.2	77.5	64.1	67.4	24.5	76.6	91.8	77.9	82.1	37.2	80.7
		2S-GSA	19.0	10.3	11.4	0.4	6.5	53.4	26.3	43.6	0.3	21.2	78.2	43.6	65.8	0.1	33.8
		STEM	0.0	0.0	0.0	0.0	0.0	0.0	0.0	0.0	0.0	0.2	0.0	0.1	0.0	1.2	2.5
		maSigFun	0.0	0.0	0.0	0.0	0.0	0.0	0.0	0.0	0.0	0.0	1.6	0.2	1.5	0.0	70.4
20	0.05	2T-GSA	53.9	28.3	45.8	0.3	19.9	87.0	64.9	82.1	0.3	21.5	96.2	72.7	93.5	0.4	26.7
		1S-GSA	64.7	51.2	56.5	19.2	65.7	100.0	98.4	99.3	43.3	100.0	100.0	100.0	100.0	65.2	100.0
		2S-GSA	54.9	28.7	47.6	0.3	19.9	90.4	67.6	86.3	0.3	23.7	97.9	74.5	95.3	0.5	29.1
		STEM	0.0	0.0	0.0	0.0	0.0	0.0	0.0	0.0	0.0	1.2	0.0	0.1	0.1	0.3	4.3
		maSigFun	0.0	0.0	0.0	0.0	0.0	0.0	0.0	0.0	0.0	0.0	0.0	0.0	0.1	0.0	92.4
	0.1	2T-GSA	55.9	30.1	49.5	0.3	22.0	88.1	60.4	81.8	0.3	24.6	96.2	71.1	92.1	0.6	27.8
		1S-GSA	62.4	51.1	56.9	17.9	65.4	99.7	98.4	98.4	44.4	100.0	100.0	99.9	100.0	65.6	100.0
		2S-GSA	59.4	32.4	52.7	0.3	22.3	92.9	62.0	85.5	0.3	26.7	98.6	73.7	94.4	0.7	29.6
		STEM	0.0	0.0	0.0	0.0	0.0	0.0	0.0	0.0	0.0	0.6	0.1	0.1	0.1	0.3	4.2
		maSigFun	0.0	0.0	0.0	0.0	0.0	0.0	0.0	0.0	0.0	0.0	0.0	0.0	0.1	0.0	92.5
	0.25	2T-GSA	38.9	17.5	32.1	0.6	18.2	74.2	46.6	70.5	0.3	23.0	91.5	56.8	87.1	0.2	35.0
		1S-GSA	57.9	51.0	54.0	20.8	61.8	99.3	97.4	97.4	36.9	100.0	100.0	99.2	99.8	61.4	100.0
		2S-GSA	49.4	23.1	40.4	0.6	21.3	86.2	50.2	78.8	0.3	25.3	96.1	62.5	92.1	0.2	28.8
		STEM	0.0	0.0	0.0	0.0	0.0	0.0	0.0	0.0	0.0	1.5	0.0	0.1	0.0	0.7	4.2
		maSigFun	0.0	0.0	0.0	0.0	0.0	0.0	0.0	0.0	0.0	0.0	0.0	0.0	0.1	0.0	91.9
	0.5	2T-GSA	21.0	10.5	14.9	0.3	10.3	60.8	24.5	52.2	0.3	25.0	81.8	45.8	76.4	0.1	49.9
		1S-GSA	45.2	39.7	38.1	17.1	48.4	93.1	83.7	86.1	36.7	99.8	98.1	92.3	94.1	65.5	99.2
		2S-GSA	38.7	17.9	29.0	0.4	14.5	78.5	34.4	67.4	0.4	28.4	92.5	55.0	86.4	0.1	36.0
		STEM	0.0	0.0	0.0	0.1	0.0	0.0	0.1	0.0	0.7	2.0	0.0	0.3	0.0	8.0	11.5
		maSigFun	0.0	0.0	0.0	0.0	0.0	0.0	0.0	0.0	0.0	0.0	0.0	0.0	0.1	0.0	90.8
50	0.05	2T-GSA	54.4	22.3	47.1	0.0	0.1	91.7	56.2	92.1	0.0	1.0	88.7	68.2	98.8	0.0	4.9
		1S-GSA	93.7	81.0	87.1	17.3	94.0	100.0	100.0	100.0	47.6	100.0	100.0	100.0	100.0	71.2	100.0
		2S-GSA	54.9	23.0	48.2	0.0	0.2	95.4	59.0	94.2	0.0	2.8	94.4	71.0	99.6	0.0	10.0
		STEM	0.0	0.0	0.0	0.0	0.2	0.0	0.3	0.0	0.0	0.6	0.0	1.2	0.0	3.7	23.9
		maSigFun	0.0	0.0	0.0	0.0	0.0	0.0	0.0	0.0	0.0	0.0	0.0	0.0	0.0	0.0	98.8
	0.1	2T-GSA	72.2	41.9	67.7	0.3	14.2	93.8	67.9	96.1	0.2	22.2	95.8	74.8	99.5	0.8	26.7
		1S-GSA	95.3	85.8	89.0	23.2	97.9	100.0	100.0	100.0	59.6	100.0	100.0	99.7	100.0	78.2	100.0
		2S-GSA	76.8	44.6	72.0	0.3	15.0	96.3	71.5	97.5	0.2	23.8	97.7	76.4	100.0	0.8	28.3
		STEM	0.0	0.0	0.0	0.0	0.0	0.0	0.2	0.0	0.2	5.5	0.0	1.1	0.0	2.8	15.9
		maSigFun	0.0	0.0	0.0	0.0	0.0	0.0	0.0	0.0	0.0	0.0	0.0	0.0	0.0	0.0	99.0
	0.25	2T-GSA	70.6	39.7	64.0	0.2	20.3	89.5	62.5	91.6	0.1	26.1	89.4	73.1	98.0	0.0	41.0
		1S-GSA	95.9	84.9	91.3	25.5	95.5	100.0	99.7	99.9	62.0	100.0	100.0	99.9	100.0	79.1	100.0
		2S-GSA	77.6	45.1	68.6	0.2	22.1	95.0	67.0	94.8	0.1	25.5	94.5	76.2	99.4	0.0	30.1
		STEM	0.0	0.0	0.0	0.0	0.1	0.0	0.0	0.0	0.0	0.9	0.0	0.5	0.0	10.9	20.6
		maSigFun	0.0	0.0	0.0	0.0	0.0	0.0	0.0	0.0	0.0	2.4	0.0	0.0	0.0	0.0	99.1
	0.5	2T-GSA	48.3	21.6	42.4	0.1	17.3	74.5	41.0	82.6	0.0	27.4	76.8	58.2	96.4	0.2	57.2
		1S-GSA	86.0	70.8	76.1	25.8	85.2	97.7	96.1	98.9	64.0	100.0	99.4	98.8	99.8	79.7	100.0
		2S-GSA	65.4	30.2	55.7	0.1	20.1	88.1	49.1	90.9	0.0	25.1	86.9	65.4	98.4	0.2	41.4
		STEM	0.0	0.0	0.0	0.0	0.3	0.0	0.2	0.5	0.0	18.5	20.6	0.1	3.5	45.8	58.0
		maSigFun	0.0	0.0	0.0	0.0	0.0	0.0	0.0	0.0	0.0	0.0	0.0	0.0	0.0	0.0	96.2
100	0.05	2T-GSA	41.2	7.0	39.7	0.8	0.8	59.5	16.8	85.7	0.3	0.8	82.4	35.3	95.7	0.0	0.0
		1S-GSA	97.4	91.5	97.1	37.2	96.4	98.2	99.7	100.0	77.4	100.0	100.0	100.0	100.0	84.6	100.0
		2S-GSA	44.1	9.7	44.1	0.9	0.8	62.6	21.3	88.9	0.2	0.8	88.2	32.9	96.6	0.0	0.0
		STEM	0.0	0.0	0.0	0.0	2.4	0.0	3.1	0.0	40.3	25.7	0.0	10.7	0.9	59.4	35.7
		maSigFun	0.0	0.0	0.0	0.0	0.0	0.0	0.0	0.0	0.0	0.0	0.0	0.0	0.0	0.0	96.2
	0.1	2T-GSA	70.7	27.0	72.3	0.1	0.4	85.6	64.0	98.7	0.0	8.5	99.2	72.5	100.0	0.0	14.5
		1S-GSA	98.6	93.5	96.9	23.7	100.0	100.0	100.0	100.0	58.2	100.0	100.0	100.0	100.0	78.0	100.0
		2S-GSA	75.5	30.4	75.5	0.0	1.2	91.4	67.5	99.4	0.0	12.3	99.9	73.9	100.0	0.0	18.0
		STEM	0.0	0.0	0.0	0.2	1.6	0.0	1.0	0.1	1.9	16.7	0.0	3.8	0.1	3.7	25.6
		maSigFun	0.0	0.0	0.0	0.0	0.0	0.0	0.0	0.0	0.0	0.0	0.0	0.0	0.0	0.0	99.9
	0.25	2T-GSA	78.7	44.4	78.7	0.1	18.7	82.6	67.7	98.5	0.4	27.8	97.6	72.4	100.0	2.0	43.5
		1S-GSA	99.2	95.0	98.1	32.1	99.8	100.0	99.8	100.0	71.9	100.0	100.0	100.0	100.0	83.6	100.0
		2S-GSA	86.4	47.9	83.9	0.2	20.5	87.5	68.7	99.6	0.2	26.6	99.5	73.7	100.0	0.4	33.0
		STEM	0.0	0.1	0.0	0.1	0.5	0.1	0.9	0.2	7.9	14.3	0.0	4.8	0.0	29.7	46.6
		maSigFun	0.0	0.0	0.0	0.0	0.0	0.0	0.0	0.0	0.0	0.0	0.0	0.0	0.0	0.0	99.5
	0.5	2T-GSA	62.3	21.4	63.1	0.1	18.0	61.1	48.4	95.4	0.0	36.5	79.2	58.8	99.0	1.5	62.9
		1S-GSA	94.9	83.6	91.1	30.5	99.4	98.9	96.5	99.8	71.5	100.0	100.0	98.1	99.9	88.2	100.0
		2S-GSA	75.3	29.1	72.6	0.2	20.3	74.9	53.0	98.8	0.0	28.6	92.8	62.4	100.0	1.1	45.6
		STEM	0.0	0.1	0.0	1.7	1.3	0.0	3.0	0.0	47.0	47.1	0.2	6.6	0.6	72.9	61.0
		maSigFun	0.0	0.0	0.0	0.0	0.0	0.0	0.0	0.0	0.0	0.0	0.0	0.0	0.0	0.0	97.3

without spiked-in differentially expressed genes are treated as false positive (FP). In contrast to the simulation based on the AH data set, prototype sets derived from all methods are used to spike in original differential genes expression trajectories in selected artificial gene sets. The 1S-GSA profile algorithm has both an outstanding sensitivity and a very high FDR as shown in Table 5.5. The FDR equals 0 % for the maSigFun method, but the method's sensitivity is restricted to the identification of maSigFun generated sets including exclusively *active* genes ($p_{\text{ag}} = 1$) as can be concluded from Table 5.6. The competing methods are for $p_{\text{ag}} < 1$ more sensitive in detecting maSigFun sets. This holds also for the STEM algorithm in simulation settings with small gene set sizes. The STEM procedure in general shows very low sensitivity for the identification of the *active* sets build from significant GSA-type prototypes. The 1S-GSA algorithm has a higher TPR performance than the STEM procedure itself in identifying STEM sets across all simulation parameter settings (see Table 5.6). The anomaly for the 2T-GSA and 2S-GSA algorithm, which occured in the OD simulation, is found in this simulation only attenuated in form of a ten percent loss in TPR and an exceptionally high FDR (compare Table 5.5 with parameters $|s|_{\text{SIM}} = 100$, $p_{\text{as}} = 0.05$). Apart from this inconsistency, the general effect of the simulation parameters is discussed in the following.

The sensitivity enhances with a growing proportion of spiked-in genes (increasing p_{ag}) as expected and analogously to the simulations based on the AH and OD data sets. Only the GSA-type algorithms show a meaningful TPR for small p_{ag} values. Apart from the anomaly mentioned above, a rising share of active genes (larger p_{as}) dilutes the TPR for the GSA-type algorithms. For the STEM procedure, the TPR seems to be maximal if p_{as} is assigned to its maximal value. In this case, there are only ten sets in the gene set universe, which facilitates the identification of the single *active* set with spiked-in genes from an originally significant STEM gene set by clustering. All methods have the tendency to yield a higher TPR for larger gene set size $|s|_{\text{SIM}}$. There is a characteristic order in the TPR values of the five competing algorithms per specific constellation of simulation parameters. 1S-GSA performs with the highest sensitivity followed with a significant gap by 2S-GSA and 2T-GSA. All GSA-type methods clearly outperform the maSigFun and STEM algorithms.

5.3 Simulation Study to Compare the Methods

The FDR decreases in general for an increasing p_{ag}, an rising $|s|_{SIM}$ or an incremented p_{as}, but not for all instances. 1S-GSA and STEM turn out to show extraordinary high FDR values for some parameter settings in contrast to the three competing algorithms. Apart from the always 0 % FDR of the maSigFun procedure, the smallest false discovery rate is achieved by the 2S-GSA and 2T-GSA methods, typically in this order except for the smallest examined gene set size ($|s|_{SIM} = 10$).

The accuracies (ACCs) and negative predictive values (NPVs) are given in Table B.4 in the appendix. The NPV is only for the GSA-type algorithms clearly above the trivial limit of $1 - p_{as}$, whereas for STEM the NPV it stays near to this bound but falls clearly below it for $|s|_{SIM} \geq 20$ and $p_{as} = 0.5$. The maSigFun procedure rises only slightly above the trivial NPV limit and this exclusively for high p_{ag} values. The ACC is only for some instances of the 1S-GSA and STEM algorithms below the trivial limit of $1 - p_{as}$. The maximal ACC value is commonly achieved by the 2S-GSA method with the exception of $p_{as} = 0.5$ where 1S-GSA performs best. The two algorithms 2T-GSA and 2S-GSA have very similar and high ACC values in relation to STEM and maSigFun.

Despite of the small number of differentially expressed genes identified in the skin healing data set, the simulation based on this data set finds clear differences between the performance of the five examined algorithms. While 1S-GSA shows in general the highest sensitivity its specificity and accuracy are clearly below the two other GSA algorithms. The STEM and maSigFun procedures show in most settings only very poor sensitivity for the detection of gene sets spiked-in with a moderate or even high proportion of genes from significant sets found by GSA-type algorithms. This result further strengthens the impression that the significant findings of the maSigFun and STEM algorithms are limited to very specific types of signal while the GSA-type algorithms are at least sensitive for the maSigFun signal type in addition to their own specific profiles.

Tongue Healing Data Set

The gene expression time series study conducted to analyze the wounded mucosa on mice tongue (TH) results in much larger numbers of significant genes per time point than its sister data set (SH) as shown in Figure 6.8, but still by a magnitude smaller than in the ovary development experiment (compare Figure 6.6). The application of the five competing algorithms and the subsequent filtering according to the procedure given in section 5.2 lead to a moderate number of prototypes for each method except the relative small number of significant gene sets identified by the STEM method. The prototype numbers per algorithm are reported in Table 5.1. The middle position between the case of high quantity of significant genes in the OD data set simulation and the case of a very small quantity in the SH data set simulation supports the validity of the results obtained by the simulation based on the TH gene expression time series experiment. The simulation results are described in the following.

In comparison with the simulation results based on the other three data sets, the TH simulation study yields higher TPR values for small sets and generally lower FDR values particularly for small gene set sizes and with the exception of STEM and 1S-GSA procedures. This finding can be understood by comparing Table 5.7 with Tables 5.2, 5.3 and 5.5. One reason why the 1S-GSA procedure stands out from this general finding is that the method identifies differentially expressed genes under each data condition due to a quantile criteria regardless of whether this is really a significant difference. The STEM procedure has a similar drawback due to its characteristic of obligatory allocation of each gene to a model profile, which results in clusters independently of the differential expression of the included genes. The sensitivity of the maSigFun is limited to gene sets including exclusively genes from a former significant maSigFun gene set (i.e. $p_{ag} = 1$). This becomes even more obvious in Table 5.8, which shows the TPR for the cross table of all algorithms and origins of spiked-in genes. The effect of the simulation parameters on the different accuracy measures is elaborated in the following paragraphs.

For all five competing algorithms the sensitivity (TPR) enhances with a rising proportion of spiked-in differentially expressed genes in the *active* sets as reported in Table 5.7. Excluding the anomaly known from the previous simulation studies with parameters $|s|_{\text{SIM}} = 100$ and $p_{as} = 0.05$, the TPR for all algorithms except maSigFun shows a tendency to increase with larger gene set size ($|s|_{\text{SIM}}$). The differences in TPR with respect to the proportion of *active* sets in the gene universe are inconsistent, but at least for the GSA-type algorithm a moderate value ($0.05 < p_{as} < 0.5$) seems

Table 5.7: TPR and FDR for identifying spiked-in activation profiles in the simulation to compare the five profile algorithms on basis of the TH data.

	Tongue healing data				TPR per p_{ag}					FDR per p_{ag}				
$|s|_{SIM}$	p_{as}	algorithm	0.1	0.2	0.4	0.6	0.8	1	0.1	0.2	0.4	0.6	0.8	1
10	0.05	2T-GSA	13.28	29.72	51.82	61.28	68.76	72.22	24.20	16.66	12.20	7.38	6.17	5.20
		1S-GSA	20.50	29.36	66.10	87.48	94.54	97.44	93.13	90.27	80.10	74.25	72.41	70.83
		2S-GSA	13.28	29.70	52.12	62.62	69.84	72.94	24.29	16.71	12.64	8.53	7.18	5.96
		STEM	0.00	0.00	0.00	0.00	0.00	0.06	100.00	100.00			0.00	0.00
		maSigFun	0.00	0.00	0.00	0.00	0.00	20.54					0.00	0.00
	0.1	2T-GSA	1.20	22.98	48.66	61.20	69.50	73.64	27.71	3.04	1.50	1.54	1.64	1.39
		1S-GSA	18.96	28.72	63.80	86.54	94.60	97.10	86.82	80.81	63.37	54.80	51.10	49.02
		2S-GSA	1.26	23.60	52.14	64.00	71.32	74.82	32.26	4.30	2.10	2.53	2.81	2.45
		STEM	0.00	0.00	0.00	0.00	0.02	0.02	100.00	100.00			0.00	50.00
		maSigFun	0.00	0.00	0.00	0.00	0.26	20.32					0.00	0.00
	0.25	2T-GSA	1.20	12.96	40.46	55.62	65.70	71.50	11.76	3.57	0.78	0.22	0.30	0.14
		1S-GSA	24.74	38.94	72.78	86.66	93.82	96.98	67.23	52.51	33.22	25.89	24.03	22.76
		2S-GSA	1.76	21.40	48.78	60.12	69.58	73.72	14.56	2.73	1.09	0.53	0.49	0.27
		STEM	0.00	0.00	0.00	0.02	0.02	0.04		100.00		66.67	66.67	33.33
		maSigFun	0.00	0.00	0.00	0.00	0.24	19.86					0.00	0.00
	0.5	2T-GSA	1.46	6.40	20.70	40.90	53.02	62.38	12.05	1.54	0.29	0.00	0.04	0.03
		1S-GSA	19.96	28.96	56.86	73.80	84.00	88.58	41.29	28.60	13.51	10.28	9.60	10.87
		2S-GSA	2.18	14.56	36.48	51.44	60.34	67.44	8.40	1.09	0.33	0.08	0.10	0.03
		STEM	0.02	0.00	0.04	0.06	0.16	0.34	50.00		0.00	40.00	57.89	15.00
		maSigFun	0.00	0.00	0.00	0.02	0.08	19.82					0.00	0.00
20	0.05	2T-GSA	12.76	39.16	59.84	68.54	72.60	74.26	4.20	4.02	2.92	4.17	4.75	4.40
		1S-GSA	27.76	52.78	88.30	96.82	98.40	99.02	91.61	84.94	76.11	73.78	72.41	71.48
		2S-GSA	12.76	39.16	59.84	68.56	72.74	74.34	4.20	4.02	2.92	4.25	5.34	4.81
		STEM	0.00	0.00	0.00	0.02	0.12	0.22	100.00		100.00	0.00	0.00	0.00
		maSigFun	0.00	0.00	0.00	0.00	0.08	20.08					0.00	0.00
	0.1	2T-GSA	19.20	43.18	62.70	71.86	76.52	77.64	3.23	1.82	2.18	1.91	1.77	1.52
		1S-GSA	25.94	50.36	87.46	95.90	98.28	99.28	84.18	72.07	57.19	52.68	51.33	49.92
		2S-GSA	19.20	43.20	63.10	72.50	76.88	77.78	3.23	1.86	2.29	1.95	1.91	1.84
		STEM	0.00	0.00	0.00	0.02	0.06	0.32	100.00	100.00	100.00	0.00	25.00	5.88
		maSigFun	0.00	0.00	0.00	0.00	0.12	20.04					0.00	0.00
	0.25	2T-GSA	11.68	35.16	59.28	72.08	76.42	79.08	2.18	0.51	0.27	0.08	0.13	0.03
		1S-GSA	30.08	51.04	87.18	96.18	98.54	99.48	61.49	45.32	29.12	24.92	24.76	23.81
		2S-GSA	12.12	39.58	62.36	73.78	77.40	79.46	2.26	0.55	0.42	0.14	0.23	0.05
		STEM	0.00	0.00	0.00	0.02	0.14	0.54	100.00		100.00	0.00	50.00	6.90
		maSigFun	0.00	0.00	0.00	0.00	0.06	20.12					0.00	0.00
	0.5	2T-GSA	6.38	19.88	47.14	59.84	70.52	75.22	1.54	0.60	0.04	0.07	0.00	0.00
		1S-GSA	25.30	41.38	78.74	91.70	96.92	98.60	37.68	24.27	13.45	12.73	13.42	15.70
		2S-GSA	13.68	35.58	55.52	66.84	74.24	77.38	1.30	0.67	0.18	0.15	0.00	0.00
		STEM	0.02	0.04	0.00	0.26	0.88	2.02	66.67	50.00	100.00	45.83	61.06	74.30
		maSigFun	0.00	0.00	0.00	0.00	0.08	19.96					0.00	0.00
50	0.05	2T-GSA	36.64	55.90	68.04	71.60	73.68	74.52	6.91	9.46	11.27	8.91	7.85	7.36
		1S-GSA	47.28	79.12	95.36	98.24	98.58	99.12	88.26	81.42	77.34	76.03	75.05	74.12
		2S-GSA	36.64	55.90	68.24	71.78	73.68	74.68	6.91	9.46	12.47	9.98	8.70	8.41
		STEM	0.00	0.00	0.00	0.14	0.28	0.84	1.22	100.00	100.00	12.50	0.00	2.33	4.69
		maSigFun	0.00	0.00	0.00	0.00	0.00	20.08					0.00	0.00
	0.1	2T-GSA	38.12	57.68	70.92	74.34	76.84	78.20	1.50	1.70	1.42	1.38	0.83	0.89
		1S-GSA	48.18	83.30	97.16	98.86	99.62	99.90	76.91	64.01	57.49	54.97	53.02	51.73
		2S-GSA	38.12	57.80	71.08	74.46	76.96	78.34	1.50	1.80	1.52	1.48	1.21	1.31
		STEM	0.00	0.00	0.02	0.32	0.70	0.98	100.00	100.00	85.71	11.11	7.89	9.26
		maSigFun	0.00	0.00	0.00	0.00	0.00	20.14					0.00	0.00
	0.25	2T-GSA	36.20	55.66	73.70	78.48	79.98	80.62	0.39	0.43	0.27	0.20	0.02	0.02
		1S-GSA	53.92	85.14	97.72	99.42	99.80	99.94	49.62	34.45	28.29	25.62	25.85	26.45
		2S-GSA	36.80	58.16	74.86	78.72	80.10	80.70	0.54	0.45	0.32	0.23	0.07	0.02
		STEM	0.00	0.02	0.16	0.52	1.26	2.48	100.00	66.67	33.33	49.02	71.36	90.47
		maSigFun	0.00	0.00	0.00	0.00	0.00	20.12					0.00	0.00
	0.5	2T-GSA	23.48	45.56	67.24	74.56	78.56	79.68	0.25	0.04	0.03	0.00	0.00	0.00
		1S-GSA	45.88	78.38	96.12	99.02	99.48	99.66	27.91	15.65	14.03	15.18	17.04	19.87
		2S-GSA	33.52	53.14	71.78	77.06	79.40	80.54	0.24	0.15	0.03	0.00	0.03	0.00
		STEM	0.02	0.08	0.68	4.28	8.32	11.78	87.50	73.33	77.48	82.34	83.41	83.67
		maSigFun	0.00	0.00	0.00	0.00	0.02	20.12					0.00	0.00
100	0.05	2T-GSA	0.66	2.18	10.08	23.22	35.58	45.28	80.00	64.84	45.40	32.66	29.07	29.07
		1S-GSA	67.06	92.02	98.88	99.82	99.90	100.00	86.14	81.38	79.09	78.17	77.32	76.93
		2S-GSA	0.60	2.28	10.88	25.74	38.66	47.62	81.48	60.28	35.85	23.39	23.42	25.66
		STEM	0.04	0.50	6.82	15.04	21.86	29.62	80.00	24.24	2.85	1.70	1.26	0.67
		maSigFun	0.00	0.00	0.00	0.00	0.00	20.18					0.00	0.00
	0.1	2T-GSA	47.24	63.68	72.12	74.80	76.04	78.44	5.52	4.07	2.91	2.25	1.50	1.18
		1S-GSA	63.64	87.64	96.94	97.84	97.56	97.86	74.53	66.21	61.97	59.95	59.03	58.47
		2S-GSA	47.28	64.22	72.36	74.90	76.56	79.82	5.55	4.69	3.31	3.20	2.92	2.21
		STEM	0.02	0.06	0.32	1.20	1.72	2.70	87.50	25.00	23.81	17.81	25.22	43.51
		maSigFun	0.00	0.00	0.00	0.00	0.00	20.06					0.00	0.00
	0.25	2T-GSA	44.64	64.88	76.52	78.58	79.36	79.82	0.62	0.46	0.03	0.10	0.08	0.10
		1S-GSA	72.34	92.28	98.74	99.50	99.90	99.94	46.48	37.47	33.74	33.33	34.61	35.97
		2S-GSA	45.38	66.44	76.76	78.76	79.48	80.16	0.61	0.48	0.08	0.13	0.13	0.10
		STEM	0.02	0.04	0.46	3.10	6.72	9.88	80.00	84.62	79.46	89.46	93.40	93.86
		maSigFun	0.00	0.00	0.00	0.00	0.00	20.10					0.00	0.00
	0.5	2T-GSA	37.22	56.52	74.56	77.44	77.70	78.78	0.11	0.11	0.00	0.00	0.00	0.00
		1S-GSA	61.92	87.68	98.12	99.30	99.64	99.80	21.30	14.74	15.05	17.68	22.87	26.70
		2S-GSA	40.52	61.50	76.42	78.48	78.28	79.34	0.10	0.10	0.00	0.03	0.00	0.00
		STEM	0.06	0.40	6.08	12.22	16.10	18.64	76.92	76.47	80.94	82.21	81.70	81.80
		maSigFun	0.00	0.00	0.00	0.00	0.00	20.10					0.00	0.00

Table 5.8: TPR per algorithm and type of spiked-in profile in the simulation to compare the five profile algorithms on basis of the TH data.

Tongue healing data s|SIM p_{as}	algorithm	TPR with $p_{ag} = 0.2$ 2T-GSA 1S-GSA 2S-GSA STEM maSigFun					TPR with $p_{ag} = 0.6$ 2T-GSA 1S-GSA 2S-GSA STEM maSigFun					TPR with $p_{ag} = 1$ 2T-GSA 1S-GSA 2S-GSA STEM maSigFun				
10	0.05 2T-GSA	32.3	27.2	33.2	0.6	55.3	74.1	70.8	75.3	0.1	86.1	87.8	87.1	90.9	0.5	94.8
	1S-GSA	32.8	31.7	33.0	19.7	29.6	91.5	91.8	91.8	63.7	98.6	99.7	99.4	99.7	88.4	100.0
	2S-GSA	32.3	27.2	33.1	0.6	55.3	75.8	73.0	76.9	0.1	87.3	88.3	88.5	91.6	0.5	95.8
	STEM	0.0	0.0	0.0	0.0	0.0	0.0	0.0	0.0	0.0	0.0	0.0	0.0	0.0	0.0	0.3
	maSigFun	0.0	0.0	0.0	0.0	0.0	0.0	0.0	0.0	0.0	0.0	3.1	1.2	1.7	0.0	96.7
	0.1 2T-GSA	20.4	19.0	22.8	0.0	52.7	76.2	67.7	75.5	0.7	85.9	89.7	89.7	91.4	1.4	96.0
	1S-GSA	32.2	31.3	30.7	20.1	29.3	89.7	88.0	90.2	66.4	98.4	99.1	98.9	99.0	88.5	100.0
	2S-GSA	21.2	20.1	23.8	0.0	52.9	79.8	72.1	80.1	0.9	87.1	91.3	91.6	93.5	1.9	95.8
	STEM	0.0	0.0	0.0	0.0	0.0	0.0	0.0	0.0	0.0	0.0	0.0	0.0	0.0	0.0	0.1
	maSigFun	0.0	0.0	0.0	0.0	0.0	0.0	0.0	0.0	0.0	0.0	2.8	1.4	1.5	0.0	95.9
	0.25 2T-GSA	14.4	11.2	13.5	0.3	25.4	66.4	59.5	64.1	1.1	87.0	85.7	84.4	89.1	1.9	96.4
	1S-GSA	44.1	40.6	42.2	27.1	40.7	90.6	89.4	89.4	66.8	97.1	97.8	97.7	98.4	91.1	99.9
	2S-GSA	22.9	17.6	21.5	0.4	44.6	73.3	67.3	71.9	1.2	86.9	88.7	89.2	92.0	2.2	96.5
	STEM	0.0	0.0	0.0	0.0	0.0	0.0	0.0	0.0	0.1	0.0	0.0	0.0	0.0	0.1	0.1
	maSigFun	0.0	0.0	0.0	0.0	0.0	0.0	0.0	0.0	0.0	0.0	1.8	0.9	2.3	0.0	94.3
	0.5 2T-GSA	9.1	7.9	6.8	0.1	8.1	45.5	38.9	44.4	1.3	74.4	72.0	69.0	75.7	1.6	93.6
	1S-GSA	30.8	32.0	33.4	23.1	25.5	73.1	73.2	72.8	59.2	90.7	87.5	86.2	87.5	85.7	96.0
	2S-GSA	16.5	14.4	15.4	0.3	26.2	58.9	52.7	59.0	1.4	85.2	80.5	78.0	83.4	2.5	92.8
	STEM	0.0	0.0	0.0	0.0	0.0	0.0	0.0	0.0	0.2	0.1	0.1	0.0	0.0	0.9	0.7
	maSigFun	0.0	0.0	0.0	0.0	0.0	0.0	0.1	0.0	0.0	0.0	1.8	1.4	1.8	0.0	94.1
20	0.05 2T-GSA	39.6	30.7	43.8	0.0	81.6	85.5	80.2	84.6	0.7	91.7	93.1	89.4	90.3	0.9	97.6
	1S-GSA	61.7	57.6	59.8	28.1	56.7	99.4	99.0	99.1	86.6	100.0	100.0	100.0	99.9	95.2	100.0
	2S-GSA	39.6	30.7	43.8	0.1	81.6	85.5	80.3	84.6	0.7	91.7	93.1	89.7	90.3	0.9	97.7
	STEM	0.0	0.0	0.0	0.0	0.0	0.0	0.0	0.0	0.0	0.1	0.1	0.2	0.0	0.4	0.4
	maSigFun	0.0	0.0	0.0	0.0	0.0	0.0	0.0	0.0	0.0	0.0	0.0	0.1	0.4	0.0	99.9
	0.1 2T-GSA	45.8	38.9	48.1	0.1	83.0	89.3	85.7	91.3	0.3	92.7	96.4	94.8	95.9	1.8	99.3
	1S-GSA	57.0	56.6	55.2	30.0	53.0	98.7	98.8	99.3	82.8	99.9	100.0	99.9	100.0	96.5	100.0
	2S-GSA	45.9	38.8	48.2	0.1	83.0	89.9	87.3	91.4	0.4	93.5	96.5	95.2	95.9	1.9	99.4
	STEM	0.0	0.0	0.0	0.0	0.0	0.0	0.0	0.0	0.0	0.1	0.2	0.0	0.0	0.8	0.6
	maSigFun	0.0	0.0	0.0	0.0	0.0	0.0	0.0	0.0	0.0	0.0	0.0	0.0	0.3	0.0	99.9
	0.25 2T-GSA	35.0	31.2	35.7	0.2	73.7	87.9	84.0	91.0	1.3	96.2	97.5	96.8	97.9	3.8	99.4
	1S-GSA	57.7	56.2	53.7	31.4	56.2	98.7	97.6	98.1	86.5	100.0	99.9	100.0	100.0	97.5	100.0
	2S-GSA	39.9	34.6	39.8	0.2	83.4	90.6	87.6	92.5	1.9	96.3	98.1	97.5	98.2	4.0	99.5
	STEM	0.0	0.0	0.0	0.0	0.0	0.0	0.0	0.0	0.1	0.0	0.0	0.0	0.2	1.9	0.6
	maSigFun	0.0	0.0	0.0	0.0	0.0	0.0	0.0	0.0	0.0	0.0	0.0	0.0	0.7	0.0	99.9
	0.5 2T-GSA	25.1	21.0	20.5	0.3	32.5	71.1	62.6	71.0	1.6	92.9	89.3	89.0	94.0	4.5	99.3
	1S-GSA	44.5	44.4	40.6	33.6	43.8	90.7	88.4	91.0	89.0	99.4	98.0	98.0	99.1	98.0	99.9
	2S-GSA	38.0	34.5	38.2	0.4	66.8	82.2	75.4	81.8	2.0	92.8	93.7	92.9	96.5	4.7	99.1
	STEM	0.0	0.0	0.0	0.2	0.0	0.0	0.0	0.0	0.7	0.6	0.1	0.2	0.1	8.9	0.8
	maSigFun	0.0	0.0	0.0	0.0	0.0	0.0	0.0	0.0	0.0	0.0	0.0	0.0	0.7	0.0	99.1
50	0.05 2T-GSA	67.1	59.9	68.3	0.4	83.8	86.3	88.0	88.1	0.4	95.2	89.8	89.9	91.6	2.0	99.3
	1S-GSA	87.1	84.6	88.3	39.6	96.0	100.0	100.0	100.0	91.2	100.0	100.0	100.0	100.0	95.6	100.0
	2S-GSA	67.1	59.9	68.3	0.4	83.8	86.4	88.1	88.4	0.4	95.6	90.1	90.0	91.6	2.3	99.4
	STEM	0.0	0.0	0.0	0.0	0.0	0.0	0.0	0.1	0.8	0.5	0.2	0.7	0.3	3.2	1.7
	maSigFun	0.0	0.0	0.0	0.0	0.0	0.0	0.0	0.0	0.0	0.0	0.0	0.0	0.4	0.0	100.0
	0.1 2T-GSA	70.9	61.0	71.2	0.1	85.2	90.6	91.0	91.4	0.1	98.6	96.0	95.9	95.6	3.8	99.7
	1S-GSA	89.4	86.8	90.5	51.0	98.8	100.0	100.0	100.0	94.3	100.0	100.0	100.0	100.0	99.5	100.0
	2S-GSA	71.2	61.1	71.4	0.1	85.2	90.7	91.1	91.6	0.1	98.8	96.1	96.0	95.7	4.2	99.7
	STEM	0.0	0.0	0.0	0.0	0.0	0.1	0.4	0.0	0.7	0.4	0.3	0.4	0.3	2.7	1.2
	maSigFun	0.0	0.0	0.0	0.0	0.0	0.0	0.0	0.0	0.0	0.0	0.0	0.0	0.7	0.0	100.0
	0.25 2T-GSA	66.7	57.3	68.4	0.3	85.6	96.6	96.3	98.0	1.9	99.6	98.9	98.9	99.4	5.9	100.0
	1S-GSA	90.8	88.9	90.7	57.3	98.0	100.0	100.0	100.0	97.1	100.0	100.0	100.0	100.0	99.7	100.0
	2S-GSA	70.4	61.4	72.8	0.3	85.9	97.2	96.7	98.2	2.0	99.5	99.0	98.9	99.4	6.2	100.0
	STEM	0.0	0.0	0.0	0.1	0.0	0.1	0.0	0.1	1.6	0.8	0.4	0.6	0.6	9.7	1.1
	maSigFun	0.0	0.0	0.0	0.0	0.0	0.0	0.0	0.0	0.0	0.0	0.0	0.0	0.6	0.0	100.0
	0.5 2T-GSA	54.0	40.8	49.5	0.2	83.3	91.2	85.0	94.7	2.6	99.3	93.8	97.4	99.5	7.7	100.0
	1S-GSA	79.9	74.7	79.6	61.8	95.9	98.3	99.5	99.5	97.8	100.0	99.1	99.7	100.0	99.5	100.0
	2S-GSA	65.4	53.4	62.5	0.3	84.1	94.3	91.6	97.2	3.0	99.2	97.1	98.0	99.7	7.9	100.0
	STEM	0.0	0.0	0.0	0.3	0.1	0.0	0.1	0.1	19.6	1.6	1.9	1.8	0.6	48.7	5.9
	maSigFun	0.0	0.0	0.0	0.0	0.0	0.0	0.0	0.0	0.0	0.0	0.0	0.0	0.6	0.0	100.0
100	0.05 2T-GSA	3.0	1.7	2.3	0.1	3.8	14.4	16.4	31.3	0.0	54.0	28.1	43.6	66.3	7.8	80.6
	1S-GSA	94.9	93.7	95.1	76.4	100.0	100.0	100.0	100.0	99.1	100.0	100.0	100.0	100.0	100.0	100.0
	2S-GSA	3.1	3.1	3.0	0.0	2.2	20.5	22.6	33.8	0.0	51.8	36.8	52.3	68.8	0.3	79.9
	STEM	0.1	0.0	0.0	1.4	1.0	4.1	3.8	3.5	45.7	18.1	19.7	20.8	15.8	72.3	19.5
	maSigFun	0.0	0.0	0.0	0.0	0.0	0.0	0.0	0.0	0.0	0.0	0.0	0.0	0.9	0.0	100.0
	0.1 2T-GSA	79.3	74.3	78.3	0.7	85.8	90.7	92.4	90.9	0.5	99.5	94.6	97.3	96.8	3.6	99.9
	1S-GSA	97.5	95.4	97.5	47.8	100.0	100.0	100.0	100.0	89.2	100.0	100.0	100.0	100.0	89.3	100.0
	2S-GSA	80.4	74.9	78.7	0.7	86.4	91.1	92.4	90.9	0.6	99.5	96.1	97.9	97.2	8.0	99.9
	STEM	0.0	0.0	0.0	0.1	0.2	0.1	0.3	0.2	3.4	2.0	0.7	1.3	0.4	8.6	2.5
	maSigFun	0.0	0.0	0.0	0.0	0.0	0.0	0.0	0.0	0.0	0.0	0.0	0.0	0.3	0.0	100.0
	0.25 2T-GSA	78.4	75.0	81.0	0.2	89.8	96.1	98.2	97.5	1.1	100.0	98.0	98.0	99.3	3.8	100.0
	1S-GSA	98.4	97.2	98.4	67.5	99.9	100.0	100.0	100.0	97.5	100.0	100.0	100.0	100.0	99.7	100.0
	2S-GSA	81.0	77.4	83.0	0.2	90.6	96.7	98.2	97.5	1.4	100.0	98.7	98.0	99.3	4.8	100.0
	STEM	0.0	0.0	0.0	0.2	0.0	0.1	0.6	0.2	12.3	2.3	0.8	0.6	0.6	42.9	4.5
	maSigFun	0.0	0.0	0.0	0.0	0.0	0.0	0.0	0.0	0.0	0.0	0.0	0.0	0.5	0.0	100.0
	0.5 2T-GSA	67.1	56.7	71.0	0.1	87.7	91.0	94.9	98.9	2.5	99.9	90.5	98.9	99.9	4.6	100.0
	1S-GSA	91.6	87.9	92.5	57.8	98.6	99.2	99.5	100.0	97.8	100.0	99.3	100.0	100.0	99.7	100.0
	2S-GSA	74.2	66.0	78.7	0.2	88.4	93.7	96.4	99.6	2.8	99.9	93.2	99.3	99.9	4.3	100.0
	STEM	0.0	0.1	0.0	1.8	0.1	1.0	1.5	0.7	53.3	4.6	3.3	3.9	0.9	75.9	9.2
	maSigFun	0.0	0.0	0.0	0.0	0.0	0.0	0.0	0.0	0.0	0.0	0.0	0.0	0.5	0.0	100.0

5.3 Simulation Study to Compare the Methods

to yield highest sensitivity. In agreement with the results of the previous simulations usually the 1S-GSA procedure performs with highest over all TPR values followed by 2S-GSA and 2T-GSA algorithms, which achieve both a very similar TPR level. The STEM and maSigFun procedures show a much worse ability in identifying *active* sets with spiked-in information found by other algorithms than their own (see Table 5.8). Hence, their overall TPR lies clearly below the values of the three GSA procedures. Only the sensitivity of the GSA-type methods turn out to be robust against a dilution of the proportion of spiked-in genes in *active* sets to a small value for p_{ag}. This property is particularly true for larger gene sets, because a small number of differentially genes can be detected as enrichment with more power in larger sets. Gene sets with spiked in genes from significant STEM sets are only identified with high confidence by the 1S-GSA algorithm as is apparent from Tables 5.7 and 5.8.

The outstanding sensitivity of the 1S-GSA algorithm has to be paid with extraordinary high FDR values, which are regularly by a magnitude higher than the reasonable values for the other GSA-type methods, which take into account the significance of the differential expression. As Table 5.7 states, there are literally no false discoveries applying the maSigFun procedure. STEM shows a very high proportion of false discoveries if the parameters $|s|_{\text{SIM}}$ and p_{as} become larger. In these simulation settings a significant set is harder to detect due to its enrichment with genes from a particular model cluster, since a smaller number of genes results in significant enrichment and the number of sets becomes smaller. At least for the GSA-type algorithms, an increase of p_{ag} reduces the FDR. A decreasing FDR can also be observed for the GSA-type methods with incremented values of p_{as} (except for the known anomaly with $|s|_{\text{SIM}} = 100$ and $p_{\text{as}} = 0.05$). The 2T-GSA and the 2S-GSA algorithm show very similar FDR values across all simulation settings, whereas the 2T-GSA algorithm has a slight advantage.

Table B.5 in the appendix reports the overall accuracy (ACC) and negative predictive value (NPV) values under each simulation setting. It turns out that the STEM procedure and the 2T-GSA fall below the trivial ACC limit for some parameter constellations, whereas this is observed for STEM in combination with large p_{as} values and for 1S-GSA in settings with small p_{as} values. The maSigFun algorithm surpasses the limit of $1 - p_{as}$ only for a high proportion of active genes ($p_{ag} \geq 0.8$) spiked-in into the corresponding active sets. The two algorithms 2T-GSA and 2S-GSA show very similar ACC performance, but are the only ones that achieve a considerable distance to the trivial limit across all simulation parameter settings. The NPV values of the STEM procedure stated in Table B.5 are very close to the trivial limit of $1 - p_{as}$ for small gene set sizes but fall considerably below this border if the gene set size is incremented and the proportion of active sets becomes high ($p_{as} \geq 0.25$). The maSigFun procedure shows only for large proportions of active genes ($p_{ag} \geq 0.8$) a substantial increase over the trivial ACC limit. The best NPV performance is achieved by the three GSA-type algorithms across all simulation settings, whereas the NPV of the 1S-GSA algorithm clearly exceeds the 2S-GSA method, which achieves similar or slightly better values than the 2T-GSA procedure.

In summary, the simulation based on the tongue healing data set has proven to be appropriate to examine the differences between the five competing profile algorithms in a realistic simulation setting. The 1S-GSA method shows by far highest sensitivity for all types of spiked-in significant gene sets and performs with highest NPV values across all parameter settings. The drawback of the method is the very low specificity and the disappointing accuracy measurements in contrast to the two other GSA methods. Spiked-in genes from significant STEM sets are only detected by the 1S-GSA and STEM procedure, where the latter needs larger set sizes and a higher proportion of active sets (p_{as}) to achieve reasonable TPR values. The maSigFun method is restricted to identify significant sets exclusively spiked-in with informative genes from an originally significant maSigFun gene set ($p_{ag} = 1$). This is only realistic for small gene sets as emphasized in data analysis chapter 6. The best accuracy is observed for the GSA-type methods, which take into account the significance of the differential expression. Commonly, the 2S-GSA method slightly outperforms the 2T-GSA algorithm in terms of accuracy, sensitivity and NPV.

5.3 Simulation Study to Compare the Methods

Summary of Simulation Results for the Comparison of the Five Profile Algorithms
Looking only at the best performing profile algorithm in the simulation results reveals a very heterogeneous picture across the chosen parameters and data sets. Figure 5.3 shows the best performing profile algorithm in terms of accuracy (ACC) and FDR for each data set and simulation parameter setting in a heat map. STEM seems to be the superior profile algorithm for the largest gene group setting and $p_{as} = 0.05$, but only because of the fact that virtually no gene set was assigned with a significant activation profile by this method. The FDR of maSigFun is smallest among all algorithms for the same reason (omitted in case of no significant gene set activation profiles). By disregarding those simulation parameter settings, where the three GSA-type algorithms identified a lot more false positives than true positives, the dominance of these algorithm type becomes clear for the accuracy and FDR (even more clearly for sensitivity and specificity, not shown). While 2T-GSA performs with a substantial advantage for the FDR, the ACC is higher for the 2S-GSA and 1S-GSA. The latter seems to be best for the not very common situation, where one half of the analyzed gene sets is active across the experimental time period ($p_{as} = 0.5$).

Despite the differences of the four data set bases for the simulation study, there are common characteristics in the simulation results. An increasing sensitivity and specificity is observed for an enhanced proportion of differentially expressed genes in the active gene sets across all simulations. This property is true only with restrictions for the modified STEM procedure, which turns out to be not very reliable in the simulation setting. The maSigFun procedure is reliable, but only in the very restricted case that almost all genes in the gene set fit well to a common model, which is not very often the case in the gene expression time series analyses. The claim in the original paper (M. Nueda, Sebastián, et al. 2009) that 70% to 80% changing genes would be sufficient for a high sensitivity cannot be confirmed, although the R^2 value was chosen at a moderate level of 0.5. In general, it turns out that the profile of larger gene sets can be correctly identified easier by all methods if the proportion of spiked-in informative genes is hold fix. The sometimes large deviations in the specific accuracy values with identical parameter constellation can

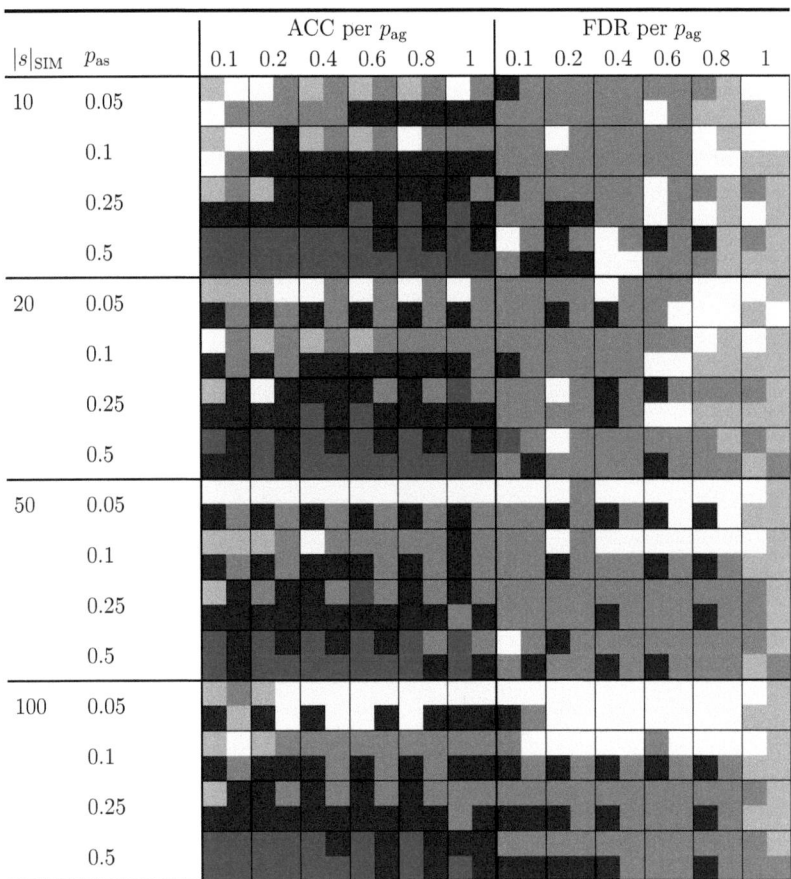

Figure 5.3: Heat Map of the profile algorithm performance in terms of ACC and FDR in the different settings of the simulation to compare the five algorithms (STEM , maSigFun , 2T-GSA , 1S-GSA , 2S-GSA). Each combination of simulation parameters is represented by four colored fields, wherein the upper left corner marks the best performing algorithm on the AH data set. The other three fields represent clockwise the superior profile algorithm for the simulations based on the OD, TH and SH experiments.

5.3 Simulation Study to Compare the Methods

be explained by differences in the data quality of the underlying gene expression time series data sets.

The 1S-GSA procedure has an outstanding sensitivity, which includes the ability to detect genes sets spiked-in with STEM genes, but with the drawback of the by far highest FDR values among all profile algorithms. The two methods which take into account the significance of the differential expression at the different time points perform with an acceptable sensitivity in relation to the three competitors and with best accuracy and relative good specificity values. They suffer from a simulation specific anomaly, which results from the TS-ABH FDR adjustment in a simulation setting with only few significant sets. All in all, the GSA-type algorithms clearly dominate the two competing methods from the literature. The reservation must be made, however, that the right detection of the significant gene set may not be equated with the identification of the true (spiked-in) profile. For instance, the 98.2 % TPR in Table 5.8 ($|s|_{\text{SIM}} = 20$, $p_{\text{as}} = 0.25$, $p_{ag} = 1$, 2S-GSA-2S-GSA) shrinks to 8.1 % when considering only exact matches of the yielded and spiked-in profiles as true positives, as shown in Table 5.9. Table 5.9 reveals that the recall of the exact matching of the spiked-in prototype profile and the application of the profile algorithms in the simulation runs turns out to be very data set dependent. Although the overall TPR is not maximal in the simulation results of the OD data set, the recall for exact matches clearly surpasses the recall values for the other data sets, in general. The 1S-GSA algorithm outperforms the other GSA-type algorithms for the most cases in the simulation based on the OD data at least for the gene set sizes below 100. This does not hold for the other data sets, where 1S-GSA dominates the recall only on those sets, where also a 1S-GSA prototype was used. In relation to the recall rates considering all reported *active* sets with spiked-in genes as TPs, STEM shows by far the highest recall performance stating only exact matches as TPs. This can be explained by the clustering step in the STEM procedure and the simulation algorithm, which spikes-in whole gene trajectories fitting well to the identified model profile. The quality of the resulting profiles with the different GSA-type procedures may be improved by smoothing the profiles as examined in the second simulation study, which follows.

Table 5.9: TPR (or recall) for the exact profile matching of the spiked-in gene set with the identified profile in the simulation to compare the profile algorithms. The method maSigFun is missing due to the continuous character of the resulting model parameters. The columns contain the type of spike-in information and the rows show the recall when applying the current algorithm. The matching of the attached STEM model profiles is only compared with the corresponding STEM profile of the spiked-in gene set information. The other possible algorithm-input combinations of the AH simulation are omitted (all 0 %).

Exact Recall with $p_{ag}=1$			AH	Ovary development data				Skin healing data				Tongue healing data			
$|s|_{SIM}$	p_{as}	algorithm	STEM	2T-GSA	1S-GSA	2S-GSA	STEM	2T-GSA	1S-GSA	2S-GSA	STEM	2T-GSA	1S-GSA	2S-GSA	STEM
10	0.05	2T-GSA		36.4	39.6	38.8		16.8	4.0	14.8		13.2	9.2	11.2	
		1S-GSA		53.2	64.0	56.0		0.4	14.8	4.4		8.4	14.0	8.0	
		2S-GSA		39.2	41.6	40.4		16.0	6.0	14.4		14.0	10.8	12.4	
		STEM	0.4				0.0				0.0				0.0
	0.1	2T-GSA		38.4	54.4	49.6		14.4	9.2	15.2		7.6	6.8	10.4	
		1S-GSA		49.2	64.0	56.0		3.2	18.8	4.0		6.8	13.2	4.8	
		2S-GSA		41.6	56.4	52.8		20.4	10.8	16.8		10.0	10.8	12.8	
		STEM	0.4				0.0				0.0				0.0
	0.25	2T-GSA		50.4	63.2	55.2		10.8	2.4	15.2		9.6	8.0	9.6	
		1S-GSA		53.6	69.2	63.2		6.4	17.6	4.0		4.0	15.6	4.4	
		2S-GSA		51.6	66.0	59.2		18.8	5.2	16.4		9.6	7.2	11.6	
		STEM	0.4				0.0				0.0				0.4
	0.5	2T-GSA		45.2	39.2	40.4		8.8	0.4	7.6		3.2	4.0	2.8	
		1S-GSA		46.0	49.2	45.6		8.4	6.8	6.8		4.4	5.2	3.6	
		2S-GSA		50.8	52.0	50.0		14.8	2.8	14.0		7.6	5.2	4.0	
		STEM	0.8				0.0				0.0				0.0
20	0.05	2T-GSA		47.6	58.4	54.0		28.8	8.4	30.4		14.0	11.6	16.4	
		1S-GSA		56.4	66.0	61.6		0.0	13.2	1.2		6.4	14.0	6.4	
		2S-GSA		47.6	58.8	53.2		29.6	8.0	28.8		14.0	12.0	16.0	
		STEM	2.0				0.8				0.0				0.0
	0.1	2T-GSA		50.8	58.4	53.6		27.6	4.4	23.2		11.2	12.8	10.0	
		1S-GSA		58.0	63.2	58.4		0.0	17.6	0.8		2.8	16.4	4.4	
		2S-GSA		51.2	58.4	53.6		24.8	5.6	24.8		12.0	13.2	10.0	
		STEM	1.6				0.4				0.0				0.8
	0.25	2T-GSA		51.2	60.8	60.0		14.8	3.6	18.0		10.4	12.0	7.2	
		1S-GSA		54.4	63.2	59.6		1.2	12.8	3.2		4.8	19.6	4.8	
		2S-GSA		50.4	61.6	59.6		19.2	7.2	20.0		8.8	13.2	9.2	
		STEM	2.4				0.4				0.0				0.8
	0.5	2T-GSA		49.6	61.2	50.8		8.4	2.4	14.0		6.4	7.2	6.4	
		1S-GSA		49.2	63.2	53.6		2.4	12.8	3.6		4.0	11.6	4.8	
		2S-GSA		53.2	64.0	53.6		16.4	4.4	16.8		8.0	10.4	10.4	
		STEM	4.8				2.0				0.0				8.8
50	0.05	2T-GSA		56.4	65.2	62.4		43.6	4.4	61.6		14.8	17.6	9.6	
		1S-GSA		59.2	63.6	61.6		0.0	12.0	0.0		6.8	22.8	7.6	
		2S-GSA		57.6	65.6	62.4		40.8	4.0	63.6		14.4	17.6	9.6	
		STEM	7.6				4.8				0.8				2.8
	0.1	2T-GSA		52.8	65.2	58.0		15.6	10.8	24.0		11.6	20.4	4.4	
		1S-GSA		53.6	65.2	58.0		0.0	6.4	0.0		6.4	23.6	3.2	
		2S-GSA		50.0	65.2	57.6		18.4	8.8	23.2		11.6	20.4	4.0	
		STEM	6.8				3.6				1.2				2.8
	0.25	2T-GSA		52.0	63.2	62.8		6.8	10.0	17.6		10.0	23.6	8.4	
		1S-GSA		52.0	68.0	64.0		0.0	8.0	0.0		2.8	24.0	4.4	
		2S-GSA		53.2	63.2	62.8		10.4	8.4	20.8		10.4	23.2	8.4	
		STEM	5.6				4.4				0.0				7.6
	0.5	2T-GSA		57.6	64.0	67.2		4.0	4.8	14.8		10.8	14.8	9.2	
		1S-GSA		59.2	64.4	66.4		0.0	8.8	0.4		2.4	18.0	5.2	
		2S-GSA		58.0	63.6	66.0		7.2	3.6	14.0		8.0	16.0	9.2	
		STEM	7.2				10.0				0.8				45.2
100	0.05	2T-GSA		0.0	0.0	0.0		0.0	0.0	0.4		0.4	0.0	10.8	
		1S-GSA		64.4	72.0	64.4		0.0	13.2	0.0		6.4	22.0	7.2	
		2S-GSA		0.0	0.0	0.0		0.0	0.4	5.6		0.0	0.0	6.8	
		STEM	35.6				0.4				0.4				1.6
	0.1	2T-GSA		61.2	73.2	64.0		37.6	9.2	42.0		11.2	20.0	6.8	
		1S-GSA		60.8	72.0	63.6		0.0	5.2	0.0		5.6	20.4	5.2	
		2S-GSA		59.6	72.8	64.0		39.6	7.6	40.8		11.2	19.6	6.4	
		STEM	36.0				4.8				0.4				8.0
	0.25	2T-GSA		61.2	68.0	66.0		4.8	6.4	8.0		5.2	22.0	5.2	
		1S-GSA		60.4	67.6	67.2		0.0	6.4	0.0		2.0	20.0	4.4	
		2S-GSA		58.0	68.0	66.0		8.0	5.6	8.8		4.8	20.8	5.2	
		STEM	43.6				8.4				0.4				45.6
	0.5	2T-GSA		44.8	69.2	66.4		4.0	6.4	14.4		4.4	14.0	10.4	
		1S-GSA		46.8	68.8	67.2		0.0	7.2	0.0		2.0	17.2	6.4	
		2S-GSA		48.4	69.2	66.4		10.8	7.6	18.8		5.6	15.6	10.0	
		STEM	58.4				11.2				1.2				73.2

5.4 Simulation Study to Evaluate the Smoothing Algorithms

The second simulation study in this thesis is conducted to validate the smoothing algorithms in the three proposed GSA-type gene set activation profile algorithms. The construction of the simulation study is described in section 5.2. The simulation study uses all measured gene expression values from one of the four application data sets briefly described in section 5.1 and the corresponding gene set definitions given in chapter 6 (e.g. see Table 6.1 for an overview). The used three activation profile algorithms (2T-GSA, 1S-GSA and 2S-GSA) are defined in chapter 4. This simulation study is neither conducted to compare the data sets nor to compare the profile algorithms, but it shows the benefit of smoothing for the accuracy of the resulting gene set activation profiles. The compared smoothing methods are explained in section 4.3. There are different smoothing parameters examined for each data set and smoothing method in the simulation scenario. On each data set 2500 simulation runs for each combination of data set and pair of smoothing parameters within a smoothing method are used to calculate the sensitivity and specificity regarding the identification of preset activation profile prototypes. The preset profiles are selected in consideration of the number of identified up and down regulated genes at the examined time points in the data sets and are shown in Table 5.10. On the one hand continuous profiles are chosen, which can be re-created by the smoothing approaches. On the other hand at least one preset profile per data set cannot be achieved by the proposed smoothing techniques (e.g. ++oo-- for OD experiment data). The factors in Table 5.10 are used to multiply the use of each preset profile in the each simulation setting. They are chosen in consideration of the overall number of differentially expressed genes in the four experiments in order to obtain a maximal number of sets with preset profile, but without exceeding the number of originally available differentially expressed genes. The determination of the used smoothing parameters is explained in the next subsection.

Table 5.10: Preset activation profile per data set used in the simulation study to validate the smoothing algorithms. Each profile is the prototype activation profile for a number of *factor* gene sets, which leads to the given total number of preset gene sets.

	Aldosterone effect	Ovary development	Skin healing	Tongue healing
Preset Profiles	---o+++	---+++	++++ooo	--o-+++
	++++ooo	+++ooo	+++oooo	++++ooo
	ooo----	ooo---	++oooooo	ooo----
	+++++++	++++++	+oooooo	+++++++
	-------	------	o--oooo	-------
	++ooo--	++oo--		++ooo--
Factor	5	15	2	8
Total	30	90	10	48

Parameter Selection for Smoothing Methods in Simulation Study

The smoothing algorithm parameters λ_{fill} and λ_{wipe} determine the extent of smoothing for each method (except the parameterless smoothing by forcing continous differential expression status for all genes in the gene set (GE)). The smoothing decision of a smoothing algorithm with fixed smoothing parameters depends on three numbers: the total of differentially expressed genes in the gene set, the overall number of differentially expressed genes and the gene set size. Only the last one is constant over the three time points, which are regarded in the smoothing decision. An adaptive choice of smoothing parameters in dependence on the gene set size and the total numbers of differential expressions would be desirable, but increases the complexity and the computational effort especially for the extensive simulation study. However, the parameters are supposed to be constant for all gene sets and time points within one algorithm and data set. The smoothing is applied in order to increase the reliability of the gene set activation profiles resulting from the proposed GSA-type profile algorithms in section 4.2. The first simulation study is conducted to evaluate the smoothing algorithms and to find suitable fixed smoothing parameters for each data set. The starting parameters chosen for this simulation study are determined

5.4 Simulation Study to Evaluate the Smoothing Algorithms

in a random based procedure below. The final parameters for each simulation study, data set and smoothing method are listed in Table B.6 (pp. 222 in the Appendix).

Determining Smoothing Parameters via GSP

The selection criterion for determining the smoothing parameter is the general smoothing proportion (GSP) value aggregated across the whole gene set universe defined for the underlying data set. The GSP is calculated separately for each combination of data set, smoothing algorithm, smoothing direction (i.e. distinguishing between the effect of λ_{fill} and λ_{wipe}) and smoothing parameter. The assumed number of differential expressed genes is hold constant as median of differentially expressed genes at all time points (3194 in the ovary development data set, 456 in the aldosterone heart data set, 16 in the skin healing data set, 711 in the tongue healing data set) for the calculation of the GSP. The significance limit is chosen as the p-value analogue to the TS-ABH FDR limit in the application (AH: $1.031\,10^{-5}$, OD: $5.4872\,10^{-4}$, SH: $9.22\,10^{-6}$, TH: $2.3441\,10^{-4}$). A total of 8000 enrichment test p-value triples possible regarding the given gene set size and suitable for smoothing (e.g. one p-value below significance border and two above or vice versa) are randomly generated for each given gene set size (restricted to those gene set sizes, where smoothing may occur with respect to the fixed number of differentially expressed genes). That means, first all possible p-values in the enrichment test for the fixed gene set size and total number of differentially expressed genes are determined. If there are values below the data set dependent significance border, 8000 triples are generated, in which each p-value from below and above the border separately had the same probability to be selected. This is done twice, once for the case of smoothing by adding a significant position (*fill*: two p-values below α_{sets}, one above) and once for the case of smoothing by erasing a significant position (*wipe*: one p-value below α_{sets}, two above). The smoothing algorithm with a fixed smoothing parameter is applied on all triplets across all gene set sizes and the proportion of smoothing events is determined for a given smoothing parameter (λ_{fill}, λ_{wipe}). The whole procedure is repeated three times and the resulting mean proportion of smoothing events is thereafter weighted with the set size frequency. The aggregated weighted sum of smoothing proportions across all gene set sizes is denoted as general smoothing proportion and lies between 0 and 1. In the following, percentage values are used for displaying the GSP. Those methods that need the number of differentially expressed genes in the group or the full contingency table respectively, restore the corresponding values from the p-values in the triplet. The directed and undirected sequential difference smoothing methods are disadvantaged, because their smoothing decision is based on another sampling according to the original time point to time point differences in the data set.

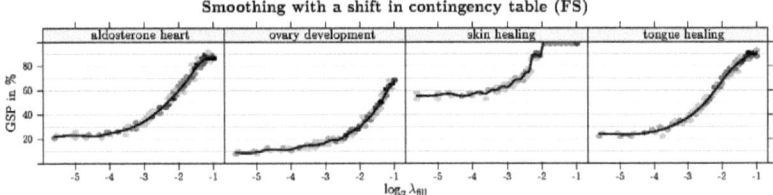

Figure 5.4: Hexbin Plot of the cumulated smoothing proportion in dependence of the smoothing parameter λ_{fill} for the smoothing due to a shift in contingency table and the four used data sets. Hexbin Plots use shaded hexagons to represent the number of observations on the covered area (darker areas symbolize more points). This form of visualization is chosen in order to emphasize the functional relationship and mitigate the impression of variation in a normal scatter plot. The smoothing parameter values, which correspond with the objective GSPs (horizontal lines) in the solid spline curve are used in the simulation setting. There could not be found any λ_{fill} in the greedy search as suitable smoothing parameter for the skin healing data set, which leads to a general smoothing proportion of less than 50%. There are also missing parameter ranges for the other data sets. Hence, there are less parameter combinations possible for the simulation to validate this smoothing method than for the other smoothing algorithms.

A greedy search algorithms varies the smoothing parameter and calculates the resulting GSP for each smoothing algorithm and data set. The calculated GSP values show a variance due to the random character of the creating algorithm, which is variable in dependence of parameter region, data set and smoothing algorithm. The objective values for the GSP are 2.5, 5, 7.5, 10, 15, 20, 25, 30, 35, 40, 45, 50, 60, 70, 80 and 90 % for each smoothing algorithm, direction and data set. A smoothing spline model is used to obtain a certain smoothing parameter for each GSP objective value from the greedy search results. These objective values cannot be fulfilled for all smoothing methods and data sets as demonstrated for instance in Figure 5.4 or Figures B.1 to B.16 in the appendix. The method failed to identify suitable parameters in case of the SU and SD smoothing for the tongue healing data and in the case of the SD algorithm also for the skin healing data. Therefore, the determined smoothing parameters from the aldosterone heart data are recycled in the given cases.

The segmentation algorithms (1S-GSA and 2S-GSA) can only be combined with the averaging smoothing methods (AM, GM, IN and IX), since the other smoothing procedures assume a fixed number of differentially expressed genes.

Each possible combination from pairs of λ_{fill} and λ_{wipe} in Table B.6 for each of the three profile algorithms is used in the simulation to validate the impact of smoothing in terms of sensitivity and specificity, predictive value and accuracy.

Simulation Results

The four data sets used as basis for the simulation study to validate the smoothing methods differ in numbers of time points, numbers of differentially expressed genes and slightly in the defined gene sets (see chapter 6). The smoothing parameters, the number and patterns of preset activation profiles are adapted to the specific data set situation. Hence, the validation of the smoothing algorithm is considered separately on each data set.

As remarked earlier, the disclosure of the surprisingly high FDR values would be misleading. The high number of false positives (FPs), i.e. the number of not preset gene sets with a non-constant o-profile, occur due to the construction of the simulation study. The gene sets are not independent, for instance all genes from a gene set defined by a GO term are included in all parent terms, too. Hence, by fulfilling the preset profile in a gene set, it is likely that related terms inherit the profile (or at least some significant positions) and are counted as FPs. The second reason is the recycling of all gene values from the underlying data set, i.e. not all differentially expressed genes belong to the preset gene sets. Hence, an occasional enrichment of not preset genes is possible. Nevertheless, this construction of a simulation study is more realistic than spiking in signal for only a few gene sets and adding artificial random noise.

In the following, four accuracy measures are discussed on the basis of the corresponding plots, which show the simulation results. The sensitivity or true positive rate (TPR) presents the proportion of preset gene sets, whose preset profile was identified correctly by the profile algorithm and the subsequent smoothing. The specificity or one minus false positive rate (1-FPR) states the proportion of constant-o-profiles (i.e. true negative (TN) sets) among the gene sets without a preset profile. The accuracy (ACC) combines both ratios and denotes the proportion of all correctly identified profiles (TP + TN) among all profiles. The fourth measure is the positive predictive value restricted to the preset profile types and a maximum similarity of 90% with preset gene sets (PPV_{p90}). This value gives the proportion of correctly identified gene sets annotated to any of the preset profile prototypes considering only those findings as FP, which do share a maximum of 90 % of genes with any preset gene set. The focus does not lie on the absolute values of these four measures but on the way and extent they are affected by the different combination of data set, profile algorithm and smoothing procedure.

Aldosterone Heart Data Set

The particular characteristic of this data set is the missing of replicated measurements ($M = 1$) and hence the use of a fold change (1.5) to identify up and down regulated genes. The number of differentially expressed genes applying this threshold is given in Figure 6.5 on page 146. The first three time points show a rising trend for up regulated genes, whereas the number of down regulated genes exceeds except for the two hours measurement the number of up regulated genes. All time points show a sufficient number of differentially expressed genes. Hence, the selected preset activation profile prototypes in Table 5.10 are attainable in the simulation study based on this data set.

The Figures B.17 to B.32 in the appendix show four accuracy measures for every threshold algorithm in combination with the proposed smoothing algorithms for the corresponding smoothing parameter pairs (see Table B.6). Generally, the effect of all smoothing methods on the specificity (1-FPR) is small as demonstrated by

5.4 Simulation Study to Evaluate the Smoothing Algorithms

the flat surface of the corresponding plots. The same holds for the ACC, since the number of TNs is by far larger than the number of TPs. The largest range of smoothing effect on sensitivity can be observed for the weighted inverse normal score mean smoothing (IN) in the 2T-GSA and the 2S-GSA profile algorithms, although the maximal sensitivity values are reached by the weighted arithmetic mean smoothing (AM). The effect on the PPV_{p90} measure is small but similar for all methods: a stronger smoothing in the *fill* direction (i.e. turning non-significant positions to significant) decreases the PPV_{p90}, whereas an increasing smoothing in the *wipe* direction (i.e. turn a significant position between two non-significant positions to not significant) has only a minor effect on the PPV_{p90}. This is exemplary shown in Figure 5.5 for the IN smoothing following the 2T-GSA profile algorithm. There is no smoothing algorithm or even smoothing parameter combination, which improves the PPV_{p90} value. This does not mean that correctly identified profiles are biased by the smoothing and a wrong (e.g. not preset) profile results, since at the same smoothing a gain in sensitivity can be observed. The loss in PPV_{p90} can be explained by a smoothing effect, which annotates preset profiles to not preset gene sets (with less than 90 % gene coverage with preset sets) in a larger number than smoothing preset gene sets to their correct preset profile.

The parameter pair that maximizes the accuracy (ACC) is given in Table 5.11 for each profile algorithm and smoothing method in the simulation based on the aldosterone heart data set. In general, the smoothing increases sensitivity, specificity and accuracy in relation to the profile algorithms without smoothing, whereas the PPV_{p90} is decreased by all smoothing methods. The extent of the improvements varies between the different profile and smoothing algorithms. The weighted arithmetic mean smoothing (AM) performs with the highest sensitivity in all three applied activation profile algorithms. It increases the TPR by 11.06 % for the 2T-GSA profile algorithm, by 3.83 % for the 1S-GSA algorithm and by 27.04 % for the 2S-GSA algorithm (see Table 5.11). The distance to an hypothetical oracle algorithms knowing always whether to smooth and in which direction (wipe out or fill in significant positions) is comparatively small for 2T-GSA (2.91 %) and 1S-GSA (1.12 %), whereas for the 2S-GSA occurs a larger deviation of 7.32 % (compare Table 5.11 and best TPR value in Figure 5.5).

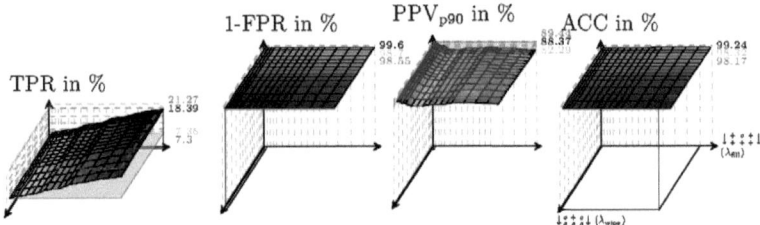

Figure 5.5: Accuracy Plots for the AH data with profile algorithm 2T-GSA and smoothing algorithm IN. The x- and z-axis illustrate increasing smoothing extent, i.e. the objective GSP grows in positive direction from 2.5 % to 90 % achieved by the correspondingly chosen smoothing parameters λ_{fill} and λ_{wipe}. The vertical y-axis indicates the sensitivity (TPR), specificity (1-FPR), PPV_{p90} or accuracy (ACC) in % as result of 2500 simulation runs with the parameter combination of the smoothing method IN, given in Table B.6. Maximum (dark), minimum (light), best (only TPR) values are given in addition to the values without smoothing (framed gray area) right to the surfaces. The sensitivity increases with increasing smoothing in the *fill* direction, while the PPV_{p90} decreases in this direction. The specificity (1-FPR) and ACC increase to a small extent in the *wipe* direction of smoothing.

The improvement in specificity is not maximal for the AM smoothing, but with 99.50 % for the 2T-GSA (99.64 % for 1S-GSA and 98.98 % for 2S-GSA) the values are very close to the best performing methods (SD: 99.61 %, IX: 99.77 % for 1S-GSA and IX: 99.21 % for 2S-GSA), whereas these smoothing methods perform poorly in the sensitivity. The PPV_{p90} measure is decreased by all smoothing methods to a moderate extent (to a minimum of 80.31 % for GM smoothing and 1S-GSA profiles). The observation that the number of not preset gene sets with a preset profile increases more than the number of correctly identified preset gene sets is due to the fact that the maximum possible number of TPs is small (i.e. limited by the number of preset sets) in comparison with the potential number of FPs (all not preset gene sets). The SD smoothing in combination with the 2T-GSA profile algorithm and the IX smoothing for the remaining two algorithms yield the smallest loss in PPV_{p90}, but the corresponding TPR values are very low. Hence, the AM algorithm that reduces the PPV_{p90} by 4.84 % for the 2T-GSA profiles, 3.65 % for the

5.4 Simulation Study to Evaluate the Smoothing Algorithms

Table 5.11: Best parameter combination of the applied smoothing methods in the simulation to validate the smoothing on the aldosterone heart data set. The left column shows the type of algorithm. The next columns show the smoothing method and the parameter combination also including the corresponding general smoothing proportion in %. The resulting values of the sensitivity or true positive rate (TPR), the specificity or 1 - false positive rate (FPR), the PPV_{p90} and the accuracy (ACC) are shown in the most right columns. The shown parameter pair within one smoothing algorithm was selected by maximal ACC. Maximum values are faded.

	data: AH	λ_{fill}	in%	λ_{wipe}	in%	TPR	1-FPR	PPV_{p90}	ACC
	none					7.30	98.55	89.44	98.17
	GE					13.62	99.22	87.76	98.87
	AM	0.0021	40.0	80.0904	90.0	18.36	99.50	84.60	99.16
	GM	0.3295	90.0	0.2770	90.0	16.99	99.59	85.50	99.25
2T-GSA	IN	0.3166	80.0	0.4647	90.0	16.89	99.59	85.59	99.24
	IX	0.3183	90.0	0.2625	90.0	9.09	99.60	83.89	99.23
	FD	0.7936	90.0	27.9746	80.0	17.02	99.18	83.73	98.84
	FS	0.0412	20.0	0.4502	90.0	16.67	99.59	86.76	99.24
	SU	0.9203	7.5	0.9265	7.5	17.94	99.55	84.40	99.21
	SD	1.0059	20.0	0.9802	20.0	7.32	99.61	88.02	99.22
	none					3.71	98.93	87.14	98.54
	AM	0.0181	25.0	80.0904	90.0	7.54	99.64	83.49	99.26
1S-GSA	GM	0.5778	80.0	0.2770	90.0	4.85	99.76	80.31	99.36
	IN	0.4608	70.0	0.4647	90.0	4.91	99.75	80.83	99.36
	IX	23.0178	2.5	0.2625	90.0	3.83	99.77	83.89	99.37
	none					29.72	97.61	85.75	97.33
	AM	0.0046	35.0	80.0904	90.0	56.76	98.98	80.40	98.80
2S-GSA	GM	0.3295	90.0	0.2770	90.0	54.98	99.17	80.99	98.99
	IN	0.3166	80.0	0.4647	90.0	54.91	99.16	81.03	98.98
	IX	0.3183	90.0	0.2625	90.0	37.17	99.21	81.12	98.95

1S-GSA profiles and 5.35 % for the 2S-GSA profiles seems to be acceptable regarding the advantage in sensitivity, although the selected AM smoothing results in the smallest increase in ACC except of the GE smoothing.

The weighted arithmetic mean smoothing (AM) algorithm is used in the following for the analysis of the aldosterone heart data set with parameters, which correspond to a smoothing of 40 % in the fill direction and a smoothing of 90 % in the wipe direction of smoothing in combination with the 2T-GSA profile algorithm (for 1S-GSA: 25 % and for 2S-GSA: 35 % in the fill direction) as listed in Table 5.15 on page 125.

Ovary Development Data Set

This data set consists of six time points in contrast to seven time points in the other studies. The number of differentially expressed genes is much higher (2431 to 3971 while controlling a TS-ABH FDR of 5 %) than for the aldosterone heart data set and the two skin healing data sets (see Figure 6.6 on page 150). This allows for more preset gene sets as stated in Table 5.10. Furthermore, the high number of differentially expressed genes is recycled in the simulation study and hence a false positive enrichment for not preset gene sets may be more likely.

The Figures B.33 to B.48 in the appendix show four accuracy measures for every GSA-type profile algorithm in combination with the proposed smoothing algorithms for the corresponding smoothing parameter pairs (see Table B.6). Analogously to the aldosterone heart data set, the effect of all smoothing methods on the specificity and accuracy is small (maximum range is 4.93 % for specificity and 5.09 % for ACC in the combination of 2S-GSA with IN smoothing) as demonstrated by the flat surface of the corresponding plots (e.g. Figure 5.6). For every combination of smoothing and profile algorithm the sensitivity, specificity and accuracy lie above the value without smoothing, while the PPV_{p90} is decreased below the level without smoothing, in general. The largest range of smoothing effect on sensitivity is again observed for the weighted inverse normal score mean smoothing (IN) and all three profile algorithms, but the maximum TPR is observed for the weighted arithmetic mean smoothing (AM) for all profile types. The decreasing of the PPV_{p90} occurs mainly in the *fill* direction of smoothing as demonstrated for the weighted inverse normal score mean smoothing (IN) algorithm in Figure 5.6 in an exemplary fashion. There are some other patterns for other smoothing methods, for instance an increasing PPV_{p90} with growing smoothing extent in the *fill* direction for the FD smoothing in combination with the 2T-GSA profile algorithm (maximum still below the value without smoothing, see Figure B.37 on page 236).

5.4 Simulation Study to Evaluate the Smoothing Algorithms

Table 5.12: Best parameter combination of the applied smoothing methods in the simulation to validate the smoothing on the ovary development data set. The left column shows the type of algorithm. The next columns show the smoothing method and the parameter combination also including the corresponding general smoothing proportion in %. The resulting values of the sensitivity or true positive rate (TPR), the specificity or 1 - false positive rate (FPR), the PPV_{p90} and the accuracy (ACC) are shown in the most right columns. The shown parameter pair within one smoothing algorithm was selected by maximal ACC. Maximum values are faded.

	data: OD	λ_{fill}	in%	λ_{wipe}	in%	TPR	1-FPR	PPV_{p90}	ACC
	none					9.73	96.20	81.25	95.12
	GE					15.57	97.95	83.42	96.92
	AM	0.0047	35.0	86.3824	90.0	27.53	98.44	79.72	97.56
2T-GSA	GM	0.2618	90.0	0.2635	90.0	25.06	99.23	79.96	98.30
	IN	0.1302	90.0	0.6924	90.0	26.22	99.15	79.60	98.24
	IX	0.2874	90.0	0.2289	90.0	13.61	99.26	79.85	98.20
	FD	0.7758	80.0	16.9952	80.0	26.24	99.09	78.96	98.18
	FS	0.3471	45.0	0.3467	90.0	26.22	99.05	79.21	98.14
	SU	0.1882	80.0	0.0965	90.0	26.10	99.05	79.42	98.14
	SD	0.0332	90.0	0.1059	80.0	26.79	98.52	78.13	97.62
	none					19.33	93.78	85.17	92.85
1S-GSA	AM	0.0047	35.0	86.3824	90.0	35.48	97.58	75.88	96.80
	GM	0.2618	90.0	0.2635	90.0	32.74	98.76	76.87	97.94
	IN	0.1302	90.0	0.6924	90.0	34.02	98.65	76.28	97.84
	IX	0.2874	90.0	0.2289	90.0	23.38	98.82	82.12	97.88
	none					17.67	92.67	81.44	91.73
2S-GSA	AM	0.0078	30.0	86.3824	90.0	41.56	97.11	75.95	96.42
	GM	0.2618	90.0	0.2635	90.0	39.11	98.57	76.57	97.83
	IN	0.2149	80.0	0.6924	90.0	39.61	98.43	76.82	97.69
	IX	0.2874	90.0	0.2289	90.0	24.59	98.64	79.86	97.72

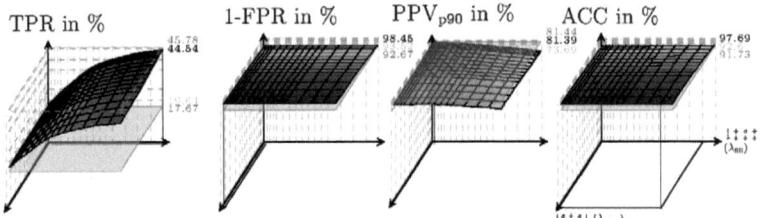

Figure 5.6: Accuracy Plots for the OD data with profile algorithm 2S-GSA and smoothing algorithm IN. The x– and z-axis illustrate increasing smoothing extent, i.e. the objective GSP grows in positive direction from 2.5 % to 90 % achieved by the correspondingly chosen smoothing parameters λ_{fill} and λ_{wipe}. The vertical y-axis indicates the sensitivity (TPR), specificity (1-FPR), PPV$_{\text{p90}}$ or accuracy (ACC) in % as result of 2500 simulation runs with the parameter combination of the smoothing method IN, given in Table B.6. Maximum (dark), minimum (light), best (only TPR) values are given in addition to the values without smoothing (framed gray area) right to the surfaces. The sensitivity increases with increasing smoothing in the *fill* direction, while the PPV$_{\text{p90}}$ decreases with increasing smoothing in the *fill* direction. The specificity (1-FPR) and accuracy (ACC) increase to a small extent in the *wipe* direction of smoothing.

The parameter pair that maximizes the accuracy (ACC) is given in Table 5.12 for each profile algorithm and smoothing algorithm in the simulation based on the ovary development data set. Analogously to the other data sets, the smoothing increases sensitivity, specificity and accuracy in relation to the profile algorithms without smoothing. Whereas the PPV$_{\text{p90}}$ is decreased by the smoothing methods except for GE smoothing, which shows a 2.17 % higher PPV$_{\text{p90}}$. The extent of the improvements in TPR, 1-FPR and ACC varies between the different profile and smoothing algorithms. The weighted arithmetic mean smoothing (AM) performs with the highest sensitivity in all three applied activation profile algorithms. The TPR is increased by 17.8 % for the 2T-GSA profile algorithm, by 16.15 % for the 1S-GSA algorithm and 23.89 % for the 2S-GSA algorithm in relation to the corresponding algorithm without smoothing. The distance to a hypothetical oracle algorithm knowing always whether to smooth and in which direction (wipe out or fill in significant positions) is for all three profile algorithms on a similar moder-

5.4 Simulation Study to Evaluate the Smoothing Algorithms

ate level (2T-GSA: 4.21%, 1S-GSA: 4.19%, 2S-GSA: 4.22%). The improvement in specificity by the AM smoothing is clearer (2T-GSA: 2.24%, 1S-GSA: 3.82%, 2S-GSA: 4.44%) than in the aldosterone heart data set. The weighted inverse χ^2 score mean smoothing (IX) procedure yields maximal specificity improvement with 3.06% for the 2T-GSA (5.04% for 1S-GSA and 5.97% for 2S-GSA), but the IX algorithm performs poorly in the sensitivity.

The loss in PPV_{p90} for the segmentation profile algorithms (1S-GSA, 2S-GSA) is higher than for the 2T-GSA algorithm. The GE smoothing algorithm results even in an increase of the PPV_{p90} for the 2T-GSA profile algorithm. Applying the 2T-GSA profile algorithm in combination with the selected AM smoothing reduces the PPV_{p90} from 81.25% without smoothing to 79.72%. The AM smoothing in combination with the 1S-GSA procedure yields a PPV_{p90} decrease of 9.29% or in combination with the 2S-GSA profile algorithm a PPV_{p90} decrease of 5.49%. The PPV_{p90} level is generally smaller than for the first simulation data set. The in relation higher number of FPs can be explained by the higher number of differentially expressed genes, which are all recycled in each simulation step.

The sensitivity performance is the reason for applying only the weighted arithmetic mean smoothing (AM) algorithm in the following. The parameters differ for the three activation profile algorithms and correspond to a smoothing of 35% GSP for 2T-GSA and 1S-GSA in the *fill* direction of smoothing, whereas the 2S-GSA algorithm should have an objective GSP of 30% in this direction. In the *wipe* smoothing direction the smoothing parameter λ_{wipe} is supposed to yield a GSP of 90% for all three profile algorithms. The parameter values are shown in Table 5.15.

Skin Healing Data Set

The low number of differentially expressed genes as shown in Figure 6.7 is the particular characteristic of this data set. The number of differentially up expressed genes (controlling a TS-ABH FDR of 5%) does not exceed 132 and the number of down expressed genes is maximally 30 and minimally 2 at the seven time points in the study. This hampers to select continuous preset profiles, which may be restored in the simulation setting and identified by the profile algorithms. The used preset profiles are focused on prototypes with up regulation as presented in Table 5.10.

Furthermore, the low number of up expressed genes at the early time points may be almost completely assigned to the preset gene sets, what results in a generally higher sensitivity. Whereas, the later time points are not considered in preset profiles and may lead together with the down regulated genes to a high number of false findings, since one single differentially expressed gene probably leads to a significant enrichment of the corresponding gene set.

Figures B.49 to B.64 in the appendix show four accuracy measures for every profile algorithm in combination with the proposed smoothing procedures for the corresponding smoothing parameter pairs (see Table B.6). Analogously to the other data sets, the effect of all smoothing methods on the specificity and accuracy is small as demonstrated by the flat surface of the corresponding plots. The sensitivity of the 1S-GSA profile algorithm is remarkably low ($<0.14\,\%$) for every smoothing method. This reveals the danger of ignoring significance on the gene level by this profile algorithm. The two-threshold methods perform with a much higher sensitivity on this data set. The increasing TPR in dependence of an increasing smoothing in the *fill* direction can be observed analogously to the other data sets, in particular for the AM, SU, SD and FD smoothing procedures. The PPV_{p90} value is increased by all smoothing methods above the level without smoothing, although the PPV_{p90} level is clearly lower than for the other examined data sets. In contrast to the results for the other data sets, the increase in PPV_{p90} is affected mainly with smoothing in the *wipe* direction for the skin healing simulations at least in the parameter regions with a low GSP (see Figure 5.7). The low PPV_{p90} values can be explained with the relatively low number of differentially expressed genes. This in turn means that already one or two of these genes may cause a significant enrichment, since all gene values are recycled in each simulation step. A second reason is the simple preset profile +oooooo, which is very likely to result by chance for not preset gene sets. The smoothing in the *wipe* direction may erase some of the incorrectly identified (false positive) gene sets with preset profiles. Hence, the smoothing algorithms are able to improve the PPV_{p90}.

The parameter pair that maximizes the ACC is given in Table 5.13 for each profile algorithm and smoothing algorithm in the simulation based on the skin

5.4 Simulation Study to Evaluate the Smoothing Algorithms

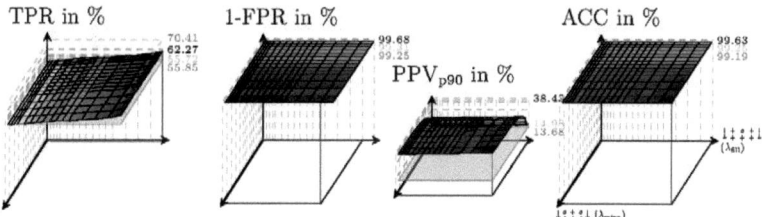

Figure 5.7: Accuracy Plots for the SH data with profile algorithm 2T-GSA and smoothing algorithm GM. The x- and z-axis illustrate increasing smoothing extent, i.e. the objective GSP grows in positive direction from 2.5 % to 90 % achieved by the correspondingly chosen smoothing parameters λ_{fill} and λ_{wipe}. The vertical y-axis indicates the sensitivity (TPR), specificity (1-FPR), PPV_{p90} or accuracy (ACC) in % as result of 2500 simulation runs with the parameter combination of the smoothing method IN, given in Table B.6. Maximum (dark), minimum (light), best (only TPR) values are given in addition to the values without smoothing (framed gray area) right to the surfaces. The sensitivity increases with increasing smoothing in the *fill* direction, while the PPV_{p90} increases mainly in the *wipe* directions of smoothing. The specificity (1-FPR) and accuracy (ACC) increase to a small extent in the *wipe* direction of smoothing.

healing data set. The 1S-GSA profile algorithm has in contrast to the results of the simulation based on the other data sets very poor results, which is due to the fact that this method ignores the significance on the gene level and identifies enrichment on the basis of quantiles of extreme gene expression differences. In this data set the number of significantly differentially expressed genes is very small in comparison with the number of genes related to the used quantiles. Hence, too many genes were considered as differential in the enrichment test and the true gene sets (spiked only with the smallest significant number of significantly differentially expressed genes) cannot be identified. This is a general weakness of the 1S-GSA profile algorithm, which affects especially studies with low numbers of differentially expressed genes. For both other profile algorithms, the sensitivity is increased by all smoothing procedures. In contrast to the simulation studies based on the first two data sets the 2T-GSA algorithm leads to higher sensitivity values than the 2S-GSA algorithm. This occurs probably due to the optimistic assumption, that the 2T-GSA

Table 5.13: Best parameter combination of the applied smoothing methods in the simulation to validate the smoothing on the skin healing data set. The left column shows the type of algorithm. The next columns show the smoothing method and the parameter combination also including the corresponding general smoothing proportion in %. The resulting values of the sensitivity or true positive rate (TPR), the specificity or 1 - false positive rate (FPR), the PPV_{p90} and the accuracy (ACC) are shown in the most right columns. The shown parameter pair within one smoothing algorithm was selected by maximal ACC. Maximum values are faded.

	data: SH	λ_{fill}	in%	λ_{wipe}	in%	TPR	1-FPR	PPV_{p90}	ACC
	none					55.85	99.25	13.68	99.19
	GE					60.17	99.31	14.28	99.26
	AM	0.0002	90.0	8.8592	90.0	65.93	99.67	37.81	99.63
	GM	0.8438	90.0	0.1428	90.0	61.93	99.68	38.43	99.63
2T-GSA	IN	0.5093	90.0	0.2135	90.0	63.54	99.68	38.52	99.63
	IX	0.8103	90.0	0.1504	90.0	56.45	99.68	37.26	99.62
	FD	0.4954	90.0	4.5754	80.0	65.58	99.68	38.71	99.63
	FS	0.0872	60.0	0.4312	80.0	65.34	99.68	38.99	99.63
	SU	0.1129	90.0	0.1147	90.0	65.94	99.65	34.79	99.60
	SD	0.0827	90.0	0.0826	90.0	65.21	99.64	34.61	99.60
	none					0.10	99.96	2.41	99.82
	AM	0.0008	70.0	8.8592	90.0	0.05	100.00	18.31	99.86
1S-GSA	GM	35.1406	2.5	0.2950	70.0	0.02	100.00	10.64	99.86
	IN	19.6587	2.5	0.3712	80.0	0.02	100.00	10.64	99.86
	IX	35.8762	2.5	3.8835	7.5	0.02	100.00	10.64	99.86
	none					41.00	99.25	9.94	99.17
	AM	0.0004	80.0	8.8592	90.0	49.30	99.71	32.48	99.64
2S-GSA	GM	0.8438	90.0	0.1428	90.0	45.52	99.71	31.99	99.64
	IN	0.5093	90.0	0.2135	90.0	46.51	99.71	32.22	99.64
	IX	0.8103	90.0	0.1504	90.0	41.60	99.71	30.46	99.63

knows the real significance border and the 2S-GSA algorithm has to adjust for the different segments, which indicate significant differential expression. The increase of sensitivity for the weighted arithmetic mean smoothing (AM) algorithm is ranked on second place concerning the 2T-GSA algorithm, but its improvement in relation without smoothing is with 10.08 % only 0.01 % worse than for the SU method, which

5.4 Simulation Study to Evaluate the Smoothing Algorithms

performs best. In the 2S-GSA simulation runs, the AM increases the TPR by 8.3 %, which is clearly better than the three competitors. The distance of the AM TPR to a hypothetical oracle algorithm knowing always whether to smooth and in which direction (wipe out or fill in significant positions) is relatively small with around 90 % of the possible optimum. The specificity is on a very high level for all selected combinations of smoothing method and parameters and hence the corresponding improvement is not as clear as in the other data sets. The maximum observed specificity is yielded by the GM, IN, IX, FD and FS for the 2T-GSA algorithm (99.68 %) or by AM, GM, IN and IX for the 2S-GSA algorithm (99.71 %). The maximal ACC value is analogously achieved by a couple of methods (99.63 % for AM, GM, IN, FD, FS with 2T-GSA profiles and 99.64 % for AM, GM, IN with 2S-GSA profiles). The FS smoothing performs best for PPV_{p90} measure in combination with the 2T-GSA algorithm, but Figure B.54 in the appendix reveals that the smoothing parameter seems to hardly affect the accuracy performance of the four accuracy measures. The AM smoothing is the best PPV_{p90} performer for the unreliable 1S-GSA profile algorithm and the 2S-GSA algorithm. The AM procedure improves the PPV_{p90} measure significantly by 24.13 % for the 2T-GSA profiles, by 15.9 % in combination with the 1S-GSA algorithm and 22.54 % for the 2S-GSA profiles.

The 1S-GSA profile algorithm seems not to work properly in this data set as demonstrated by the poor sensitivity performance. Therefore, in the following on the skin healing data smoothing is considered only for the 2T-GSA and 2S-GSA profile algorithms. The weighted arithmetic mean smoothing (AM) algorithm seems to be a proper choice in both cases. The parameters applied for the two activation profile methods correspond to a GSP of 90 % in the *fill* direction and the 2T-GSA algorithm and 80 % for the 2S-GSA respectively, whereas the smoothing extent in the *wipe* direction is supposed to result in a GSP of 90 %.

Tongue Healing Data Set

Although the skin healing data and the tongue healing data were collected from the same study (see section 6.5), their numbers of differentially expressed genes vary greatly. Even at the last time point, with lowest number of differentially expressed genes in the tongue healing data set, there are more genes identified as differentially expressed as in the maximal time point in the skin healing data (compare Figures 6.7 and 6.8). This allows for a higher number and more variants of preset profiles as stated in Table 5.10.

Figures B.65 to B.80 in the appendix show four accuracy measures for every threshold algorithm in combination with the proposed smoothing algorithms for the corresponding smoothing parameter pairs (see Table B.6). Analogously to the other data sets, the effect of all smoothing methods on the specificity and accuracy is small, as demonstrated by the flat surface of the corresponding plots. Similar to the previous considerations does the *fill* direction of smoothing affect the sensitivity, while the *wipe* direction has a minor influence on specificity and accuracy see Figure 5.8 (with exception of IX and FS algorithms where both parameters affect TPR and ACC to a very small extent).

The improvement in sensitivity in dependence on increasing smoothing in the *fill* direction can be observed analogously to the other data sets, in particular for the AM, IN, SU, SD and FD smoothing procedures. The PPV_{p90} values are decreased below the level without smoothing by all methods except for some smoothing parameters in the combination of the 1S-GSA profile algorithm with AM smoothing as is shown in Figure 5.8. In general, the shape of the PPV_{p90} surface is inconsistent for the smoothing algorithms, e.g. the PPV_{p90} is decreasing with increasing smoothing in both directions of smoothing for the SD procedure (see Figure B.72 on page 249) while the PPV_{p90} of FD smoothing seems to be influenced only by the *fill* direction of smoothing (see Figure B.69 on page 248).

The parameter pair that maximizes the accuracy (ACC) is given in Table 5.14 for each profile algorithm and smoothing algorithm in the simulation based on the tongue healing data set. The sensitivity is maximal with the AM smoothing in all profile algorithm simulations, i.e. in relation to the case without smoothing the true positive rate (TPR) is 22.29 % higher for the 2T-GSA algorithm, 13.14 % higher for the 1S-GSA algorithm and 22.83 % higher for the 2S-GSA algorithm. The distance to a hypothetical oracle algorithm knowing always whether to smooth and in which direction (wipe out or fill in significant positions) is relatively large with 5.3 % (2T-GSA), 5.54 % (1S-GSA) and 8.08 % (2S-GSA). The maximal improvement

5.4 Simulation Study to Evaluate the Smoothing Algorithms

Table 5.14: Best parameter combination of the applied smoothing methods in the simulation to validate the smoothing on the tongue healing data set. The left column shows the type of algorithm. The next columns show the smoothing method and the parameter combination also including the corresponding general smoothing proportion in %. The resulting values of the sensitivity or true positive rate (TPR), the specificity or 1 - false positive rate (FPR), the PPV_{p90} and the accuracy (ACC) are shown in the most right columns. The shown parameter pair within one smoothing algorithm was selected by maximal ACC. Maximum values are faded.

	data: TH	λ_{fill}	in%	λ_{wipe}	in%	TPR	1-FPR	PPV_{p90}	ACC
	none					19.01	95.42	89.94	96.31
	GE					29.74	96.89	88.90	96.44
	AM	0.0098	40.0	25.3714	90.0	41.30	98.45	86.14	98.07
	GM	0.5755	90.0	0.2147	90.0	34.18	98.66	86.09	98.23
2T-GSA	IN	0.2984	90.0	0.3634	90.0	37.87	98.65	85.90	98.25
	IX	0.6211	90.0	0.1865	90.0	23.52	98.69	86.95	98.19
	FD	0.2908	50.0	30.2826	80.0	39.17	98.64	85.25	98.24
	FS	0.0457	25.0	0.5012	90.0	36.85	98.67	88.41	98.26
	SU	0.4502	50.0	0.0950	90.0	37.96	98.47	86.56	98.07
	SD	0.0827	90.0	0.0826	90.0	37.02	98.26	85.14	97.86
	none					2.94	95.66	76.73	95.73
	AM	0.0428	25.0	25.3714	90.0	16.08	99.05	78.20	98.50
1S-GSA	GM	3.2836	25.0	0.2147	90.0	9.59	99.37	71.70	98.77
	IN	1.7584	30.0	0.3634	90.0	10.39	99.35	72.49	98.76
	IX	37.5894	2.5	0.1865	90.0	8.59	99.39	70.94	98.79
	none					18.29	93.43	84.32	94.31
	AM	0.0169	35.0	25.3714	90.0	41.12	97.61	78.88	97.23
2S-GSA	GM	0.5755	90.0	0.2147	90.0	37.89	97.92	79.21	97.52
	IN	0.4691	80.0	0.3634	90.0	38.29	97.91	79.26	97.52
	IX	0.6211	90.0	0.1865	90.0	24.66	97.98	79.59	97.49

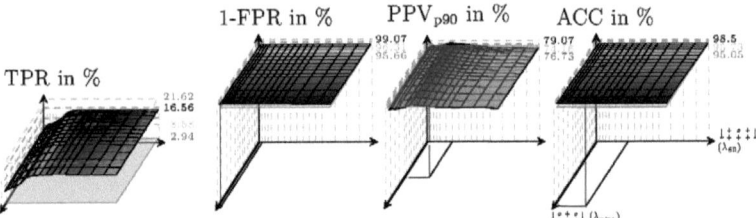

Figure 5.8: Accuracy Plots for the TH data with profile algorithm 1S-GSA and smoothing algorithm AM. The x- and z-axis illustrate increasing smoothing extent, i.e. the objective GSP grows in positive direction from 2.5 % to 90 % achieved by the correspondingly chosen smoothing parameters λ_{fill} and λ_{wipe}. The vertical y-axis indicates the sensitivity (TPR), specificity (1-FPR), PPV$_{p90}$ or accuracy (ACC) in % as result of 2500 simulation runs with the parameter combination of the smoothing method AM, given in Table B.6. Maximum (dark), minimum (light), best (only TPR) values are given in addition to the values without smoothing (framed gray area) right to the surfaces. The sensitivity increases with greater smoothing in the *fill* direction, while the ACC and specificity increase to a small extent with λ_{wipe}. The PPV$_{p90}$ varies in the *fill* directions of smoothing and is above the value without smoothing only for a ralatively small parameter range.

in specificity is yielded by the weighted inverse χ^2 score mean smoothing (IX) procedure, but its sensitivity performance is worst among all ACC selected smoothing methods. The AM smoothing increases the specificity by 3.03 % in the 2T-GSA profile algorithm simulation, by 3.39 % in the simulation linked to the 1S-GSA algorithm and 4.18 % in the 2S-GSA simulation. The ACC is not maximal for the selected AM smoothing, but it yields a clear improvement (very close to the best methods) of 1.76 % (2T-GSA), 2.77 % (1S-GSA) and 2.92 % (2S-GSA). The PPV$_{p90}$ value can only be improved for the combination of AM smoothing with the 1S-GSA profiles (+1.47 %), but in general the PPV$_{p90}$ values are smaller for the 1S-GSA profile algorithm than for the procedure 2S-GSA, and the latter are smaller than the PPV$_{p90}$ values of the 2T-GSA method. The 2T-GSA and the 2S-GSA profile algorithms work under the optimistic case that the threshold for the true significantly differentially expressed genes is known in contrast to the quantile definition of 1S-GSA. Despite the fact that the segmentation test approaches

5.4 Simulation Study to Evaluate the Smoothing Algorithms

(1S-GSA and 2S-GSA) adjust for their different definitions of differential expression they identify more false positive values than the 2T-GSA method as can be concluded from the PPV_{p90} values for the AM smoothing and the 2T-GSA and the 2S-GSA profiles. This can be explained by the construction of the simulation study, which uses exactly the enrichment threshold of the 2T-GSA procedure to spike in enriched gene sets. Nevertheless, the applied smoothing algorithms improve the case without smoothing for all three algorithms as should be shown by this simulation.

The AM smoothing algorithm shows by far the best sensitivity performance for all three profile algorithms. Therefore, the weighted arithmetic mean smoothing (AM) procedure is exclusively used in combination with the tongue healing data in the following. The smoothing parameters correspond to a GSP of 90 % for the *wipe* direction of smoothing and for the other direction the objective GSP depends on the profile algorithm, i.e. 40 % for the 2T-GSA procedure, 25 % for the 1S-GSA method and 35 % for the 2S-GSA algorithm. The smoothing parameters satisfying these requests are listed in Table 5.15.

Overall Conclusions per Smoothing Method

The simulation to compare the smoothing methods yields results, which are very complex due to the large number of algorithms, smoothing parameter values and the four different evaluation data sets. Figure 5.9 shows a performance heat map of the best ACC performers for each pair of smoothing algorithm and data set (see Tables 5.11, 5.12, 5.13, and 5.14). Especially the results obtained from the SH based simulations stand out. The AM smoothing is in most cases performing with the best sensitivity (TPR) and IX smoothing is superior in specificity (1-FPR). Regarding the accuracy, the smoothing algorithms with moderate sensitivity like GM and IN perform better, whereas the PPV_{p90} value is with exception of the SH results maximal without any smoothing. The simulation study results are analyzed separately for each method in more detail below.

Table 5.15 shows the selected smoothing methods in combination with the corresponding smoothing parameters selected from the results of the simulation to evaluate the smoothing methods. In general, a strong smoothing in the *wipe* direction seems to be advantageous in all determined simulation settings in order to obtain a high accuracy. A moderate extent of smoothing seems to be more reliable in the

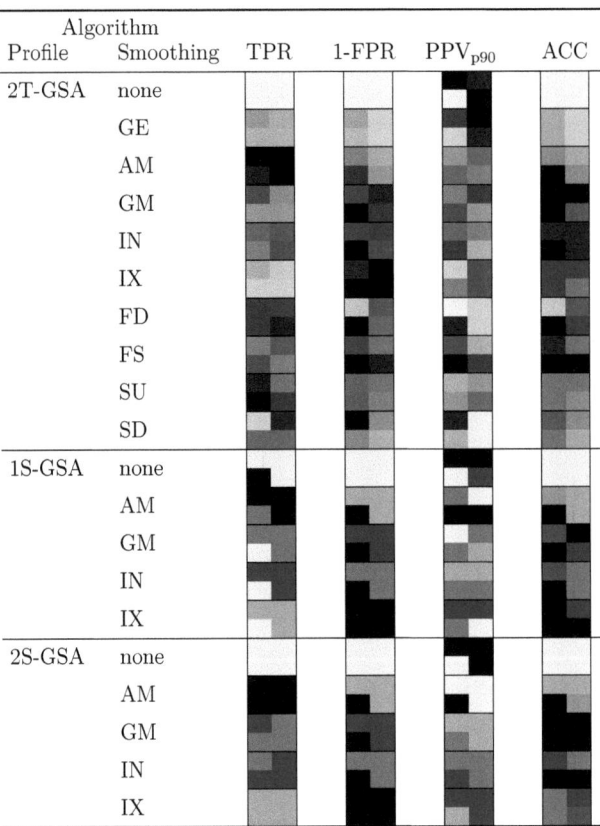

Figure 5.9: Heat Map of the smoothing algorithm performance in terms of sensitivity (TPR), specificity(1-FPR), PPV_{p90} and accuracy (ACC) across all four data sets. The performance of the smoothing method applying the accuracy maximizing parameter combinations (compare Tables 5.11, 5.12, 5.13, and 5.14) is ranked for each accuracy measure and profile algorithm. The colored rectangle for each combination of smoothing algorithm and accuracy measure is subdivided into four shaded fields wherein the upper left corner marks the performance in the simulation based on the AH data set. The other three fields represent clockwise the simulation results for the OD, TH and SH experiments. High values are shown in dark and the smallest value is illustrated in light gray.

5.4 Simulation Study to Evaluate the Smoothing Algorithms

Table 5.15: Smoothing methods and smoothing parameters selected on basis of the simulation study to evaluate the smoothing methods. Parameter values have to be adopted to the used enrichment significance level (α_{sets}) and the expected number of differential genes. See Table 5.16 for the smoothing parameters applied in application chapter 6.

Data	Algorithms Profile	Smoothing	Parameters (GSP in %) λ_{fill}	λ_{wipe}
AH	2T-GSA	AM	0.0021 (40)	80.0904 (90)
	1S-GSA	AM	0.0181 (25)	80.0904 (90)
	2S-GSA	AM	0.0046 (35)	80.0904 (90)
OD	2T-GSA	AM	0.0047 (35)	86.3824 (90)
	1S-GSA	AM	0.0047 (35)	86.3824 (90)
	2S-GSA	AM	0.0078 (30)	86.3824 (90)
SH	2T-GSA	AM	0.0002 (90)	8.8592 (90)
	1S-GSA			
	2S-GSA	AM	0.0004 (80)	8.8592 (90)
TH	2T-GSA	AM	0.0098 (40)	25.3714 (90)
	1S-GSA	AM	0.0428 (25)	25.3714 (90)
	2S-GSA	AM	0.0169 (35)	25.3714 (90)

other direction at least for the weighted arithmetic mean smoothing (AM). The specific values for the smoothing parameters applied in chapter 6 are determined by the analogous greedy search procedure as the simulation study parameters (see p. 105), but with the significance limit set to $\alpha_{\text{sets}} = 0.01$ in accordance with the selected FDR limit in the application part. The eight hexbin plots in Figure 5.10 show the results from the greedy search algorithm for the AM smoothing algorithm. Table 5.16 on the following page lists the used parameter values for analyzing the four data sets. The parameters are determined suitable for the enrichment test significance border of $\alpha_{\text{sets}} = 0.01$ for the TS-ABH FDR q-values. A generalization of finding suitable parameters for the weighted mean approaches for analyzing other data sets is given at the end of section 5.5.

The weighted arithmetic mean smoothing (AM) algorithm is used for all data sets because of its outstanding sensitivity performance in the simulations based

Table 5.16: Smoothing parameters applied in the data analysis in chapter 6 and derived from the simulation results (see Table 5.15) in consideration of the q-value enrichment test limit of $\alpha_{\text{sets}} = 0.01$ in the used data sets.

Data	Algorithms Profile	Algorithms Smoothing	Parameters (GSP in %) λ_{fill}	Parameters (GSP in %) λ_{wipe}
AH	2S-GSA	AM	0.1252 (35)	3.4166 (90)
OD	2S-GSA	AM	0.0462 (30)	11.2813 (90)
SH	2T-GSA	AM	0.0738 (90)	0.2290 (90)
TH	2S-GSA	AM	0.1388 (35)	3.5679 (90)

on the aldosterone heart, ovary development and tongue healing data and still relative good sensitivity in the difficult SH data set. The AM smoothing did not yield best specificity and accuracy in the simulation study (except for the skin healing simulation), but leads always to a clear improvement. Therefore, this type of smoothing can be recommended in general, in particular if the number of differentially expressed genes is not too small.

The smoothing by forcing continous differential expression status for all genes in the gene set (GE) is the only smoothing algorithm without any parameter to determine the smoothing extent. It was able to improve the accuracy values in all four simulations. The GE smoothing was always among the best performing methods regarding the PPV_{p90} values with exception of the difficult SH simulation. The GE smoothing should be considered in order to avoid the effort of determining good smoothing parameters for another method, because of its simplicity and the fact that any smoothing turns out to be better than no smoothing.

The weighted geometric mean smoothing (GM) shows an outstanding performance for the ACC values at least for the 2S-GSA profile algorithm. The same is true in combination with 2T-GSA profiles for the simulation with the aldosterone heart and the ovarian development data. Unfortunately, the sensitivity is clearly below the values for the AM and IN smoothing methods. Hence, the GM smoothing can be recommended if the focus lies on a maximal ACC and the number of differentially expressed genes is large.

5.4 Simulation Study to Evaluate the Smoothing Algorithms 127

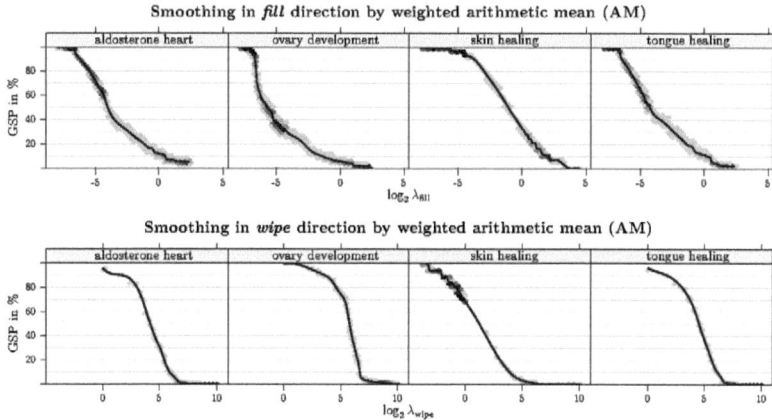

Figure 5.10: Hexbin Plots of the general smoothing proportion (GSP) in dependence of the logarithmized smoothing parameters λ_{fill} and λ_{wipe} for the weighted arithmetic mean smoothing and the four used data sets (with $\alpha_{\text{sets}} = 0.01$). Hexbin Plots use shaded hexagons to represent the number of observations on the covered area (darker areas symbolize more points). This form of visualization is chosen in order to emphasize the functional relationship and mitigate the impression of variation in a normal scatter plot. The smoothing parameter values, which correspond with the objective GSPs in the solid spline curve are used in the data analysis in chapter 6 and are listed in Table 5.15.

The weighted inverse normal score mean smoothing (IN) shows in general a similar response to the variation of the two smoothing parameters as the AM algorithm, but does never achieve maximal or almost maximal sensitivity values in any simulation, but IN outperforms the AM procedure in terms of specificity and ACC in most cases. It can be concluded to use IN as an alternative for the AM smoothing.

The weighted inverse χ^2 score mean smoothing (IX) performs with the worst sensitivity among all applied smoothing methods, but still slightly increases the TPR in relation to the profile analyses without smoothing. In contrast to the sensitivity the specificity is in all cases maximized by this procedure. However, the bad sensitivity and the non-response on parameter changes make this method unsuitable for a reasonable smoothing.

The smoothing on basis of the (relative) distance to significance in the enrichment test (FD) is except for the aldosterone heart data set among the three most sensitive smoothing methods in Tables 5.11 to 5.14. The restricted smoothing parameter range (see Figure B.15) prevents a general recommendation in favor for this method in combination with the 2T-GSA activation profile algorithm.

The smoothing using a shift of (not) differentially expressed genes in the Fisher enrichment test (FS) shows the best performance in terms of specificity, accuracy and PPV_{p90} for the difficult simulation setting based on the skin healing data set (see Table 5.13) and is not far behind the best performer of the other studies. The generally restricted range of the smoothing parameter is accompanied by a low effect of the smoothing parameters on the analyzed accuracy measures. On the one hand it seems that the FS smoothing performs well in the simulation independent from the actually chosen parameter and on the other hand the differential expression status of a single gene seems to be in most cases the key factor for a smoothing event. All in all the FS smoothing is a simple and effective smoothing algorithm, which can be used without much care for selecting the *optimal* smoothing parameters. The method's strength lies in analyzing data sets with few differentially expressed genes, but the disadvantage is its limitation to the 2T-GSA algorithm, since this principle cannot be easily transferred to the segmentation procedures.

The smoothing using sequential tests for enrichment with undirected differential expression (SU) has an inconsistent performance for the different simulations. While showing the best sensitivity in the skin healing simulation, it has only middle ranks for the other data sets and accuracy measures. In contrast to the other simulations, the simulation parameters seem to have no effect for the aldosterone heart simulation. This indicates that an application of this smoothing algorithm on a study without replicates needs more effort to identify smoothing parameters which influence the smoothing extent. This restricts the general usability of this smoothing method.

The smoothing using sequential tests for enrichment with directed differential expression (SD) performs with the exception of the ovary development simulation worse than the SU algorithm, although it uses more detailed information in a similar way. It suffers from the same disadvantages as the SU smoothing algorithm and is due to its worse performance less recommended.

5.5 Joint Conclusions from Both Simulation Studies

This section provides a brief summary of the two comprehensive simulation studies structured according to type of algorithm and smoothing for the three GSA-type profile algorithms.

The Modified STEM Method
The STEM procedure is applied in a slightly modified manner in order to capture both similar gene expression trajectories (identified by correlation) and similar expression levels (identified by mean distance). STEM does only slightly benefit from a raise in signal strength, because its construction as a clustering method, the only such method in the comparison. Only larger gene sets spiked in (virtually) completely with genes following the model pattern are reliably detected as the comparison simulation study revealed and even in those cases the FDR was extraordinary high. However, cluster based methods and STEM in particular might be better suited to detect transcription factors or common gene regulatory principles than gene set activation.

The maSigFun Procedure
The regression based procedure in combination with the chosen R^2 level of 0.5 is very reliable in identifying small gene sets with a common (to all included genes) and clear gene expression pattern virtually without FPs. These restrictions are not desirable for an exploratory activation profile algorithm, which searches for activation within a large gene set universe, but may be useful in studies conducted to identify those small parallel expressed gene sets (e.g. regularized by the same transcription factor).

The 1S-GSA Algorithm

The segmentation test idea applied for the estimation of gene set activation profiles results in an outstanding sensitivity for detecting spiked in expression trajectories from significant gene sets identified by the GSA-type algorithms, by the maSigFun procedure and to a remarkable extent by STEM (larger than STEM itself). The extraordinary high sensitivity comes at the expense of a very high FDR. The sensitivity grows in general with increasing number of informative (i.e. differentially expressed) genes and slightly proportional to the number of genes in the set. The drawback of the method is the missing consideration of significance in the tested segments as the simulation studies based on the SH data set reveal. Due to its tendency to identify many FPs the 1S-GSA profile algorithm must be applied with care, in particular for gene expression studies with only few differentially expressed genes (< 200 per time point).

The 2T-GSA Method

The simple idea of transferring the widespread enrichment test idea in form of subsequent enrichment tests on the activation profile analysis of time series data has shown a generally good performance in both simulation studies. The FDR lies clearly below the values of the GSA-type competitors, but the same holds for the sensitivity, which is detracting its usefulness. The consideration of the significance of differential expression makes this procedure more reliable in particular for studies with a low number of differentially expressed genes. Although not conducted to compare the GSA-type methods, the resulting accuracy measures of the second simulation study provide a hint that the 2T-GSA profile algorithm is superior to the segmentation procedures in the wound healing data sets. This should be generalizable to situations with a low to moderate number of differentially expressed genes at the study time points.

The 2S-GSA Procedure

The promising approach of combining the sensitive segmentation test procedure with a significance threshold to avoid a large proportion of false positives falls a bit below expectations. The FDR performance is clearly better than in the 1S-GSA algorithm,

5.5 Joint Conclusions from Both Simulation Studies

but worse than for 2T-GSA. In particular, the ability to detect STEM source signals is much weaker than for the competing segmentation method. Nevertheless, the 2S-GSA activation profile algorithm does outperform the simple 1S-GSA procedure in combination with a reasonable smoothing for the smoothing simulations based on the AH and OD data sets. This is supposed to be generally valid for gene expression time series studies with a large number of differentially expressed genes.

Smoothing by Enforcing Gene Wise Continuity

This naive type of smoothing provides a reasonable improvement in accuracy, whereas the distance to the best performing smoothers is quite large. The restriction that the GE smoothing is only combinable with 2T-GSA profiles is advantageous in a way, since both procedures are simple (no smoothing parameters) and straight forward (easy to explain to a user).

Smoothing by Sequential Testing

Both smoothing algorithms that use undirected (e.g. no distinction of up and down regulation) (SU) or directed (SD) tests of differential expression between the smoothing position and neighboring time points need the most effort to be calculated and to determine adequate smoothing parameters. On the one hand is one of these sophisticated procedures the best performer in sensitivity regarding the SH based simulation, but on the other hand the PPV_{p90} value lies clearly below the level of competitors with a comparable sensitivity. The difficult determination of reliable smoothing parameter (e.g. within AH simulation) and the difficulty to extend the smoothing algorithm to other profile algorithms than 2T-GSA with manageable effort limit the general applicability of these smoothing methods.

Smoothing According to Distance to Significance or by Shifting Genes

The common idea of FD and FS smoothing is to soften the strict significance border of the enrichment tests if a gene set is considered for smoothing at a certain time point. This hard limit is only existent in the 2T-GSA algorithm and therefore these smoothers are limited to 2T-GSA profiles. Regarding the PPV_{p90} values the FS smoothing outperforms the more sophisticated FD smoothing for the ACC-optimized parameter selection, but this cannot be stated for the other considered measures. All in all, both methods show medium performance and the simulation study revealed no hint why to prefer these smoothing algorithms.

Smoothing by Applying Weighted (Score) Means

Applicable on level of p- or q-values the idea of weighted means (directly or on scores) is the most flexible of all proposed, which can in addition become more flexible through a wide range of comparatively easy to compute smoothing parameters. Despite this flexibility is in common to the four smoothing algorithms, the performances are quite different. While the (weighted) mean of the χ^2-scores seems to be the natural choice, because of the close relation of the hyper geometrical and the χ_1^2-distribution, the sensitivity is the worst among all smoothing algorithms in the simulation study. In comparison, the weighted inverse normal score mean smoothing (IN) performed clearly better than IX or GM (with exception of the SH simulation), but still with obvious distance to the procedure averaging directly on the significance measures. The reason for this performance lack should be searched in the different level of decision (p- or q-value) and averaging (Normal-scores), which allows less power in smoothing extent than the direct proceeding on the significance measures. The AM smoothing outperforms the other weighted mean approaches in terms of TPR in all simulation settings. It is preferred in the following, because of its high sensitivity performance and fast parameter determination.

Generalized Function for the Smoothing Parameter Determination of AM, GM and IN Smoothing

The comprehensive procedure to determine the smoothing parameters as described in section 5.4 (pp. 104) increases the computation effort and time. An easy way to obtain reasonable parameters would be desirable. The smoothing extent is in addition to the smoothing parameter dependent on the number of genes in the gene universe, the gene set sizes and the number of differentially expressed genes. Assuming that the determined OD gene set universe is representative for many types of studies, Table 5.17 gives the parameters for functions to determine smoothing parameters in dependence of the general smoothing proportion (GSP), the total number of genes G, the proportion of differentially expressed genes $\tilde{p}_{ag}$ in % and

5.5 Joint Conclusions from Both Simulation Studies

the significance limit to determine a gene set as significantly enriched α_{sets}. The functions are estimated from a linear model fit to the results from the extensive application of the greedy search algorithm from the first part of section 5.4, e.g. for the *fill* direction of AM smoothing:

$$\lambda_{\text{fill}}^{\text{AM}}(\text{GSP}, G, \tilde{p}_{\text{ag}}, \alpha_{\text{sets}}) = \exp\left(\mu_{\text{fill}}^{\text{AM}} + \beta_{\text{fill}}^{\text{AM,GSP}} \cdot \log \text{GSP} \right.$$
$$+ \beta_{\text{fill}}^{\text{AM},G} \cdot \log G$$
$$+ \beta_{\text{fill}}^{\text{AM},\tilde{p}_{\text{ag}}} \cdot \log \tilde{p}_{\text{ag}}$$
$$\left. + \beta_{\text{fill}}^{\text{AM},\alpha_{\text{sets}}} \cdot \log \alpha_{\text{sets}}\right).$$

Table 5.17: Coefficients of the smoothing parameter generating functions based on a greedy search algorithm and the OD gene set universe (see section 6.1). R^2 denotes the measure of determination in the regression model.

smoothing	direction	μ	$\beta^{\cdot,\text{GSP}}$	$\beta^{\cdot,G}$	$\beta^{\cdot,\tilde{p}_{\text{ag}}}$	$\beta^{\cdot,\alpha_{\text{sets}}}$	R^2
AM	fill	3.499	−0.991	−0.020	−0.183	0.212	0.93
AM	wipe	−0.195	−0.878	0.002	0.370	−0.676	0.88
GM	fill	0.645	−0.826	0.00015	−0.277	0.064	0.56
GM	wipe	−0.106	−0.623	−0.011	0.231	−0.162	0.66
IN	fill	−6.527	−2.843	−0.026	−0.919	0.524	0.79
IN	wipe	−0.012	−0.799	−0.017	0.337	−0.143	0.93

The smoothing parameters resulting from the functions determined by the regression coefficients in Table 5.17 should be considered as approximation regarding the domain of

$$\text{GSP} \in (0,1), G \in [10^3, 5 \cdot 10^4], \tilde{p}_{\text{ag}} \in [0.01, 0.25] \text{ and } \alpha_{\text{sets}} \in [10^{-4}, 0.25],$$

which was the domain for the greedy search algorithm. The wide domain results in quite large differences of those values derived by the approximation with the function in relation to the determined smoothing parameters stated in Table 5.16 and applied in application. Smoothing oriented at these parameters should be considered at least for the *wipe* direction in order to discard the often high number of false discoveries with a single significant position. Additionally, it is important for the application of these parameter generating functions that the gene set universe is not too different in terms of gene set size distribution from the used gene set universe in the OD data set. This condition should be fulfilled for all typical GO gene set definitions, but on gene set universes with a greater proportion of small sets (e.g. only KEGG sets) more caution must be exercised in applying the parameter generating functions.

CHAPTER 6

Application of the Gene Set Activation Profiles Estimation on Gene Expression Time Series Experiments

The laboratory mouse (*mus musculus*) is one of the most popular model organisms in biology. The proposed methods for the activation profile estimation of gene sets are applied on three freely available mouse experiments. The raw data sets are freely available in the GEO data base. The three experiments hybridized Affymetrix mouse 430 2.0 microarrays in different time series designs. The chip contains expression data for 45,101 probe sets. Not all probe sets can be used in the proposed algorithms due to the gene set focus and the non-uniqueness of some gene to probe set annotations. The details of the filtering are presented in section 6.1 and an overview about the total numbers of genes and sets per data set is given in Table 6.1.

A brief introduction for the data sets and the results of the application of the selected algorithm in combination with the selected smoothing according to the results of the simulation studies (see Table 5.16) for the estimation of temporal activation of gene sets is presented in sections 6.3, 6.4 and 6.5.

6.1 Filtering the Gene Universe

From the three microarray experiments described in the last three sections of chapter 6, four different gene expression data sets are created to demonstrate and validate the existing and newly proposed methods for estimating gene set activation profiles. The first data set examines the effect of Aldosterone on the gene expression in heart cells. The second study was conducted to study the embryonic development of the mouse ovary. The last experiment focuses on the molecular biological differences in the healing of skin and mucosa (on the tongue) cells after wounding. In this thesis data for both injuries is analyzed separately, i.e. there is one data set corresponding to the skin healing and one related to the tongue healing.

Affymetrix mouse 430 2.0 microarrays were hybridized and scanned to obtain the data for the four data sets. Using data of the same chip simplifies the annotation of the gene identifiers to the Affymetrix probe sets and consequently the annotation of the probe sets to the different kinds of gene set definitions. Hence, the source of differences between the four data sets is not the chip type, but the kind of experiment.

Table 6.1 gives a summary of the four data sets concerning the experimental design, the number of used genes, and the number of gene sets used in simulation studies (see chapter 5) and for estimation of gene set activation profiles. Although all four data sets are generated from the same chip type, the number of used probe sets

Table 6.1: Summary of the four data sets used for the estimation of temporal activation profiles of gene sets.

	Mouse ovarian	Aldosterone effect	Skin healing	Tongue healing
time points T	6	7	7	7
replicates M	3	1	3	3
reference	single experiment	single time series	replicated control	replicated control
probe sets J	12,976	12,974	12,979	12,976
GO BP sets	6378	6378	6378	6378
KEGG sets	220	220	220	220
BioCarta sets	214	212	212	212
Reactome sets	256	257	257	256
BioCyc sets	150	145	147	151
total sets	7218	7212	7214	7217

6.1 Filtering the Gene Universe   137

(here equivalently denoted with genes) and gene sets varies between the experiments. These differences occur due to the probe set filtering which is explained in the following. A scheme of filtering the Affymetrix probe sets and creating the analyzed gene sets is given in Figure 6.1 on the next page.

Entrez Gene Uniqueness Filtering

The Affymetrix mouse 430 2.0 microarray includes a total of 45,101 probe sets. Not all of these features represent a real gene product, e.g. some are used for quality assessment of the chip data and some are included because of similar gene products in other species. Some genes occur in variants, for instant single nucleotide polymorphisms (SNPs), in which only a single nucleic base distinguishes between the different DNA-sequences keeping the other bases constant. This may influence the function of the gene, but must not. However, these circumstances caused the manufacturer Affymetrix to include a multiple of probe sets on the chip for interrogating the gene expression of a single gene in some cases. An unconsidered reproduction of the same information unit would heavily bias the result in an analysis that focuses on group enrichment strategies. One way to reduce the probe set universe to unique gene product features is the mapping of the Affymetrix probe set identifiers to the Entrez Gene identifiers and the subsequent filtering to a one to one mapping.

As shown in Figure 6.1, for 38,535 chip probe sets there is a mapping to exactly one of 20,877 Entrez Gene IDs available. This mapping is unique for 11,133 features. The remaining 9744 Entrez Gene IDs have at least two chip features annotated (27,402 probe sets). The first filtering step accounts for the probe set quality, which is characterized by the probe set identifier suffix (Yu, F. Wang, et al. 2007). The highest reliability have probe sets with the _at suffix that represent unique transcript variants. The suffixes _a_at and _s_at characterize probe sets that recognize multiple alternative transcripts from a single gene. The suffix _x_at is not specific for a single gene and hence the least reliable. The filtering process keeps only probe sets with the highest available suffix class (i.e. _at > _a_at > _s_at > _x_at). For 2255 Entrez Gene IDs only a single feature results from this procedure. The remaining 18,848 probe sets annotate to a total of 7489 Entrez

6 Application of Gene Set Activation Profile Estimation

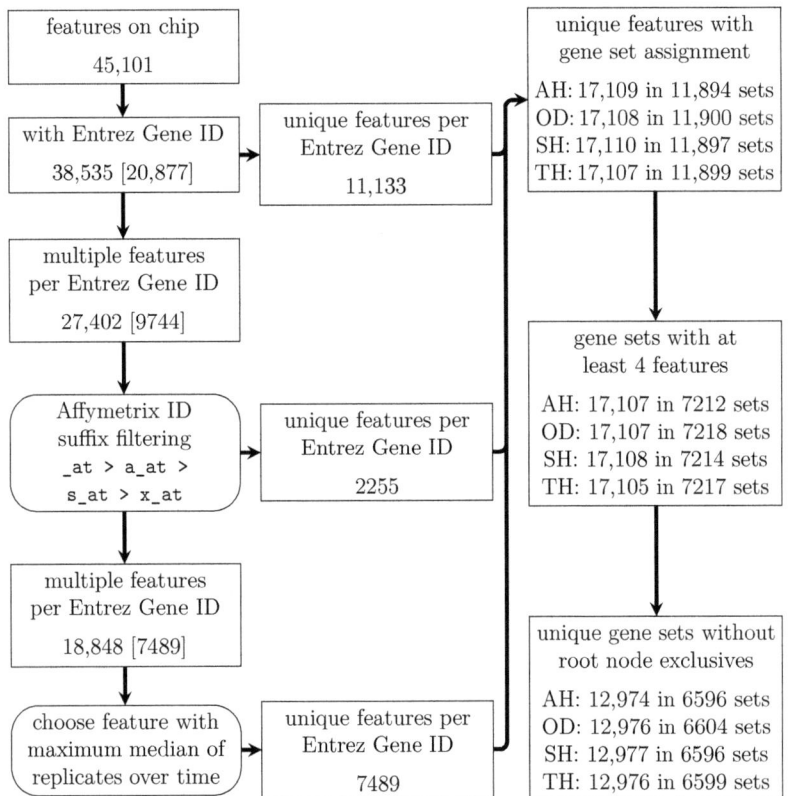

Figure 6.1: Scheme of filtering microarray features (Affymetrix probe sets) in order to obtain unique Entrez Gene representants with an gene set annotation according to the definition of GO biological process, KEGG, BioCyc, Reactome and Biocarta. The numbers vary slightly between the data sets aldosterone heart (AH), ovary development (OD), skin healing (SH) and tongue healing (TH). The number of unique Entrez Gene IDs is printed in brackets.

6.1 Filtering the Gene Universe

Gene IDs. These undergo the second filtering step. Since here a one-to-one mapping strategy is applied in contrast to an averaging strategy a unique Affymetrix probe set must represent a single Entrez Gene ID. The maximum of the time point wise medians (of the replicates per time point) is chosen as criterion to decide which feature is used as representative for the Entrez Gene ID. This is done in order to keep the probe set with the highest signal of transcript presence in the experiment. This leads in total to 20,877 probe set features with unique Entrez Gene ID.

In this thesis the gene set definitions from the GO biological processes ontology, KEGG pathways, BioCarta pathways, Reactome pathways and the MetaCyc data base for mice are used (see section 2.2). In dependence on the data set a total of 17,107 (TH) to 17,110 (SH) probe sets have an annotation to 11,894 (AH) to 11,900 (OD) gene sets (compare Figure 6.1). The differences between the data sets occur due to the preceding filtering steps and the fact that the gene set annotations is not based on the Entrez Gene IDs, but on Affymetrix IDs. Therefore, some genes are selected by the filtering procedures, which do not have a gene set annotation or even another annotation than another probe set with the same Entrez Gene ID. The idea of analyzing gene sets is only meaningful if the set size is not too small in order to reasonably summarize the information of the included genes. Here, a gene set must include at least four genes to be considered in the analysis. This results in 17,105 (TH) to 17,108 (SH) probe sets linked to 7212 (AH) to 7218 (OD) gene sets. The last step skips those genes, which are only annotated to the biological processes root node (GO:0008150). This is the most general gene set and would include virtually all probe sets, which prevents this group from becoming significantly enriched in an enrichment test. At the end of the filtering procedure there are 12,974 (AH) to 12,977 (SH) probe sets available grouped into 6596 (AH and SH) to 6604 (OD) unique gene sets, i.e. this number of gene sets omits those including exactly the same genes. Although the numbers do not vary much, the number of probe sets used in all four data sets is only 11,154 due to the second filtering step. The Venn diagram in Figure 6.2 presents the overlap of probe sets with respect to the four used data sets. The number of gene sets depending on the definition (i.e. GO, KEGG, BioCarta, Reactome or BioCyc) is given in Table 6.1 on page 136.

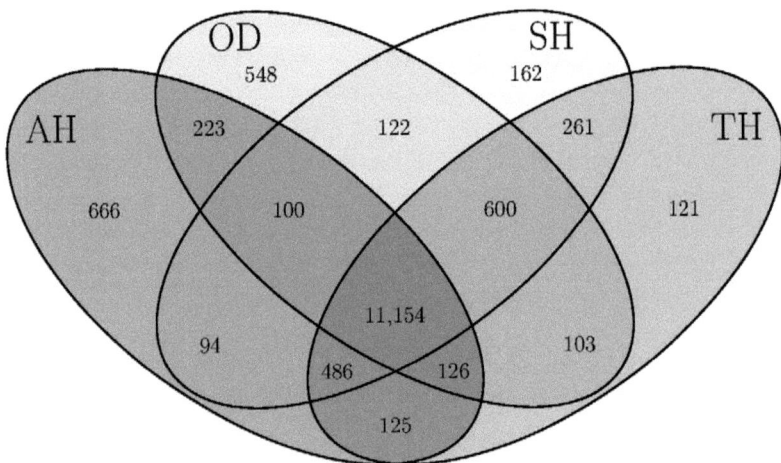

Figure 6.2: Venn diagram, presenting the overlap in used Affymetrix mouse 430 2.0 probe sets after filtering regarding the four exemplary data sets: Aldosterone heart (AH), ovary development (OD), skin healing (SH) and tongue healing (TH). The majority of here distinguished probe sets are annotated to the same gene (with respect to the Entrez Gene ID).

6.2 Summary Over All Profile Algorithms and Data Sets

Five ways for the estimation of gene set activation profiles from gene expression time series are proposed in this thesis. As the smoothing does not result in new active gene sets, Table 6.2 lists the results from applying the five competing algorithms described in chapter 4 without smoothing for the GSA-type algorithms. The diagonal states the number of exclusively identified sets per method in the corresponding row, which is maximal in all four application cases for the 1S-GSA algorithm. The identification regards to any non-zero activation profile, significant enrichment from model cluster (STEM) or significant regression model (maSigFun). In particular the extraordinary high number of identified sets in relation to the competing methods for the SH data set reveals the risk to report to many sets as significant when ignoring the

6.2 Summary Over All Profile Algorithms and Data Sets

Table 6.2: Table of intersection proportions for the five profile algorithms applied on the four examined data sets, e.g. for the SH data the identified 1S-GSA gene sets include 75 % of the four gene sets found by the maSigFun algorithm, wheras a single gene set is exclusively identified by the maSigFun approach. Entries in shaded cells are absolute numbers. Omitted values are 0.

data	profile algorithm	included proportion of gene sets reported by					Total
		2T-GSA	1S-GSA	2S-GSA	STEM	maSigFun	
AH	2T-GSA	0	43.00	70.16			91
	1S-GSA	97.80	98	84.68			207
	2S-GSA	95.60	50.72	17			124
	STEM				1		1
	maSigFun					0	0
OD	2T-GSA	8	76.72	79.62	34.33		1177
	1S-GSA	97.71	71	97.32	35.07	66.67	1499
	2S-GSA	98.56	94.60	21	35.82	66.67	1457
	STEM	3.91	3.14	3.29	83		134
	maSigFun		0.13	0.14		1	3
SH	2T-GSA	1	4.19	22.09			58
	1S-GSA	98.28	1056	98.80	56.91	75.00	1361
	2S-GSA	94.83	18.07	3	13.01		249
	STEM		5.14	6.43	53		123
	maSigFun		0.22			1	4
TH	2T-GSA	4	53.85	76.85	7.69	66.67	838
	1S-GSA	97.73	508	93.26	61.54	66.67	1521
	2S-GSA	96.66	64.63	56	15.38	66.67	1054
	STEM	0.12	0.53	0.19	5		13
	maSigFun	0.24	0.13	0.19		1	3

significance of differential expression as done by 1S-GSA. This can be observed to a smaller extent for the other data sets, too. In general, gene sets stated by 1S-GSA include a large part of the gene sets with significant profiles reported by 2T-GSA and 2S-GSA. The maSigFun procedure yields in all studies the smallest number of significant gene sets. Furthermore, in concordance with the simulation study, at least for the OD and TH data set two thirds of the reported sets are also identified by the segmentation test algorithms. The total number of STEM determined gene sets is not proportional to the GSA-type algorithms, since the totals for the wound healing data sets are small for STEM and TH data and high for the SH data set and for 2T-GSA or 2S-GSA the other way around. The proportion of STEM stated sets, which are also identified by the competing profile algorithms is only for the wound healing data examples and the 1S-GSA algorithm above 50 % whereas in the OD experiment all GSA type algorithms cover at least one third of the STEM sets.

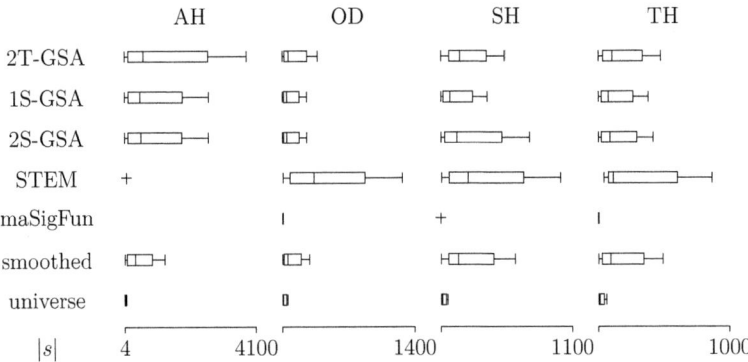

Figure 6.3: Boxplots of gene set sizes for all significant profiles and all profile algorithms (first five without smoothing and the applied smoothing according to Table 5.16). The boxplots at bottom show the gene set sizes in the gene set universe $\check{S}$. Outliers are omitted.

Figure 6.3 shows boxplots of the gene set sizes of the reported sets with a significant activation profile categorized according to profile algorithm and data set. In distinction to gene sets reported by the maSigFun procedure, for the STEM algorithm small gene sets are detected less frequently. The chosen minimum set size of four genes seems to be not sufficient for the STEM algorithm, which is another disadvantage in addition to its inferiority in terms of reproducibility in the first simulation study. The 2T-GSA algorithm shows apart from the SH data set the broadest inter quartile range of set sizes. This can be observed due to the smaller total number of reported significant profiles, whereas the additional discoveries of the segmentation algorithms are often among the more prevalent small gene sizes as

6.2 Summary Over All Profile Algorithms and Data Sets

Table 6.3: Gene sets with significant activation profile after smoothing according to procedures stated in Table 5.16 on page 126 and subdivided by gene set definition for all four application data sets. The value p_{def} describes the proportion of significant profiles from all gene sets of the same gene set definition type (e.g. the two significant KEGG sets are 0.91 % of all KEGG sets in the gene set universe). The proportion of the significant sets of one definition of the whole gene set universe $\check{S}$ is denoted with $p_{\check{S}}$ (e.g. the seven significant BioCyc gene sets of the OD profile analysis account for 0.1 % of the gene set universe).

Data		GO BP	KEGG	Reactome	BioCarta	BioCyc	Total
AH	significant	99	3	2	1	0	105
	p_{def} in %	1.55	1.36	0.78	0.47	0	
	$p_{\check{S}}$ in %	1.37	0.04	0.03	0.01	0	
OD	significant	1128	81	72	12	7	1300
	p_{def} in %	17.69	36.82	28.12	5.61	4.67	
	$p_{\check{S}}$ in %	15.63	1.12	1	0.17	0.1	
SH	significant	44	2	4	0	0	50
	p_{def} in %	0.69	0.91	1.56	0	0	
	$p_{\check{S}}$ in %	0.61	0.03	0.06	0	0	
TH	significant	772	38	19	13	0	842
	p_{def} in %	12.1	17.27	7.42	6.13	0	
	$p_{\check{S}}$ in %	10.7	0.53	0.26	0.18	0	
Total (definition)		6378	220	212-214	256-257	145-151	

the boxplots of the gene set sizes of the whole gene set universes suggest. Whereas the boxplots for the replicated data sets become (slightly) wider due to applying the selected smoothing, the opposite occurs for the AH data set without replicates. In the latter case it turns out that the smoothing procedure discards mainly large sets with only one significant position and keeps the majority of sets with less genes (compare Figure 6.4 on the next page).

Table 6.3 reports the resulting significant gene set profiles after smoothing (with parameters from Table 5.16) categorized according to the five different types of gene set definition. The Gene Ontology biological process gene sets have the highest proportion regarding the whole gene set universe ($p_{\check{S}}$) in all four application examples.

The same is true only for AH data in relation to the available number of gene sets of the same gene set definition. This proportion is maximal for the KEGG gene set definition in case of the OD and TH data and for Reactome in case of the SH data. Gene sets resulting from the BioCyc definition are only reported with significant activation profiles in the OD experiment. This can be explained with the distribution of gene set sizes. The BioCyc gene set definition leads to gene sets with a maximum size of 20 genes, which is on the one hand small in comparison to the alternative gene set definitions and on the other hand only a very clear enrichment with differentially expressed genes at the examined time points will be identified by the profile algorithms, which include a TS-ABH FDR adjustment for the high total number of applied tests.

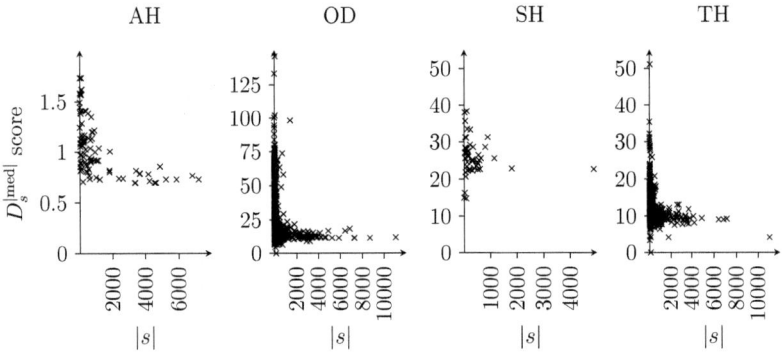

Figure 6.4: Scatter plot of gene set sizes and corresponding $D_s^{|\text{med}|}$ score of gene sets with significant reported gene set activation profile. The scale is different for the AH plot, because the fold change is used as measure for differential expression instead of the D_g^{schrink} statistic in the case of a non-replicated time series experiment.

In the detailed analysis of the gene set activation profiles in the following sections, a ranking based on the $D_s^{|\text{med}|}$ and D_s^{med} scores is used to sort the identified gene set activation profiles according to their extent of differential expression. Although both scores use only the statistics of significantly differentially expressed genes at identified enriched positions, the high score values can be observed only for gene

6.2 Summary Over All Profile Algorithms and Data Sets

sets with small numbers of genes annotated. Figure 6.4 shows four scatter plots for the link between the $D_s^{|\text{med}|}$ score and the gene set size. The highest scattering with the higher score values can be observed for gene set sizes below 1000 among all four experiments. The absolute highest scores result for the profiles of the OD data sets, which can be taken as indication for the intensity of the molecular biologic differences during the time series experiment. The relatively small difference between the two wound healing data sets in central location of the score values cannot be explained by the different applied activation profile algorithms. The smoothed 2T-GSA procedure was applied on the SH data, whereas the more sophisticated smoothed 2S-GSA method was used on the TH data set, but the score values result from the statistics of differential gene expression in both cases. Hence, the few identified significant SH activation profiles include genes with stronger differential expression in relation to the reference than the majority of the reported significant profiles from the TH study. There is a single zero score value in each of the scatter plots of the OD and SH data sets. This is an artifact of the 2S-GSA algorithm in combination with smoothing. The statistic value of differentially expressed genes are only stored at non-zero profile positions to minimize the computational effort. In these two cases an activation profile like oo+o+o was smoothed to ooo+oo and hence at the former non-significant enriched fourth position the statistics to calculate the scores had not been saved. Therefore, the score is calculated to zero. This is an infrequent phenomenon due to the implementation of the segmentation algorithms and can be ignored since the scores are used only for a ranking of the identified significant activation profiles.

The following three sections discuss the resulting activation profiles separately for each experiment in more detail in order to take into account the individual characteristics of the studies and results.

6.3 Aldosterone Effect on Mouse Heart Gene Expression

The first gene expression study analyses the genetic mechanisms in the mouse heart following an administration of the steroid hormone aldosterone under a 3% NaCl diet. The experiment was constructed as two-group-comparison. The subjects of the experimental group were injected with a physiologic dose of aldosterone (10 µg/kg). The subjects of the control group got an injection with the vehicle only. Five male mice of each experimental group were scarified at 0.5, 1, 2, 3, 4, 5 and 12 hours after injection. The RNA of all five mice hearts was extracted and hybridized on a single Affymetrix mouse 430 2.0 array. Hence, a single measurement for each time point and group is available in the data set. The raw data was published in November of 2005 as GEO series GSE3440. Originally, the study was analyzed with the CAGED algorithm (Turchin, Guo, et al. 2006). Turchin, Guo, et al. (2006) report 12 genes with a similar gene expression trajectory over time (down regulated between 1 h and 3 h). The differential expression of representative genes was successfully validated by quantitative real time RT-PCR.

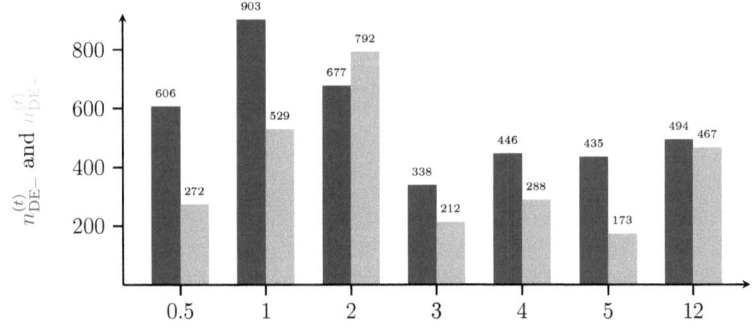

Figure 6.5: Bar chart of the total numbers of down ($n_{\mathrm{DE}-}^{(t)}$) and up ($n_{\mathrm{DE}+}^{(t)}$) expressed genes with respect to the untreated reference exceeding a fold change of 1.5 for the aldosterone effect on mouse heart data set.

6.3 Aldosterone Effect on Mouse Heart Gene Expression

Figure 6.5 shows the total number of differentially expressed genes categorized in up and down regulation if a fold change (FC) threshold of 1.5 is applied. The FC limit is lower than the observed gene expression ratios for the 12 reported genes in the original contribution, but those genes are among the identified differentially expressed genes. The modus in down regulation is achieved one hour after injection while the number of up regulated genes increases until two hours and clearly falls after this time point. The eventually falsely reported differentially expressed genes will only have a minor effect on the reported gene sets, since the activation profile analysis is focused on the gene set level and enrichment tests are used.

Table 6.4 lists the resulting gene set activation profiles from the 2S-GSA algorithm (TS-ABH FDR limit: $\alpha_{\text{sets}} = 0.01$) smoothed with AM method and smoothing parameters $\lambda_{\text{fill}} = 0.1252$ and $\lambda_{\text{wipe}} = 3.4166$. Despite the quite large numbers of differentially expressed genes in Figure 6.5, there is no hint for an enrichment with differential expression at the last time point (12 h) for any gene set. The total number of non-zero profiles is with 105 quite small, but consistent with the weak signals in terms of the extent of differential expression (FC values). 30 gene sets show an enrichment with down regulated gene expression at the second time point (1 h). Although the maximum of up regulated genes occurs at the third time point (2 h) as shown in Figure 6.5, the second most frequent profile is ooo+ooo with 25 gene sets and there are only 15 gene sets with the profile oo+oooo, i.e. an enrichment with up regulation at the third time point. The analysis does not yield any continuous differentially expressed profiles in contrast to the activation profile analyses on the

Table 6.4: The 13 2S-GSA activation profiles after smoothing with AM ($\lambda_{\text{fill}} = 0.1252$, $\lambda_{\text{wipe}} = 3.4166$) resulting from the AH data set.

Profiles	#	Profiles	#	Profiles	#
ooooooo	7107	oo+oooo	11	-o-oooo	1
o-ooooo	30	o-o+ooo	8	o-+oooo	1
ooo+ooo	25	-oooooo	7	o-o-ooo	1
ooo-ooo	15	ooooo+o	4	ooo+o-o	1
				oooo+oo	1

Table 6.5: Top 12 of D_s^{med}-ranked up regulated and top 8 down regulated gene set activation profiles resulting from a 2S-GSA profile algorithm combined with AM smoothing on the AH data. The single set stated significant with the STEM procedure is reported at bottom.

ID	D_s^{med} Rank	Profile	Description	$\|s\|$	2T-GSA	1S-GSA	STEM	maSigFun
GO:0035456	1	ooo+ooo	response to interferon-beta	15	✓	✓	✗	✗
GO:0035458	2	ooo+ooo	cellular response to interferon-beta	14	✓	✓	✗	✗
GO:0051856	3	ooo+ooo	adhesion to symbiont	8	✓	✓	✗	✗
GO:0044403	4	ooo+ooo	symbiosis, encompassing mutualism through parasitism	87	✗	✓	✗	✗
GO:0051825	5	ooo+ooo	adhesion to other organism involved in symbiotic interaction	8	✓	✓	✗	✗
GO:0001562	8	ooo+ooo	response to protozoan	17	✓	✓	✗	✗
GO:0042832	9	ooo+ooo	defense response to protozoan	15	✓	✓	✗	✗
GO:0044419	10	ooo+ooo	interspecies interaction between organisms	104	✓	✓	✗	✗
GO:0051702	11	ooo+ooo	interaction with symbiont	37	✗	✓	✗	✗
REACT:GRB2	12	ooooo+o	genes involved in GRB2	10	✓	✓	✗	✗
GO:0009605	31	o-ooooo	response to external stimulus	796	✗	✓	✗	✗
GO:0048870	29	o-ooooo	cell motility	750	✓	✓	✗	✗
GO:0051674	30	o-ooooo	localization of cell	750	✓	✓	✗	✗
GO:0040011	28	o-ooooo	locomotion	898	✓	✓	✗	✗
GO:0002523	27	o-ooooo	leukocyte migration involved in inflammatory response	7	✗	✗	✗	✗
GO:0050900	26	o-ooooo	leukocyte migration	144	✗	✓	✗	✗
GO:0006935	15	o-ooooo	chemotaxis	324	✗	✓	✗	✗
GO:0042330	16	o-ooooo	taxis	325	✗	✓	✗	✗
GO:0030595	6	o-ooooo	leukocyte chemotaxis	94	✗	✓	✗	✗
GO:0060326	7	o-ooooo	cell chemotaxis	120	✗	✓	✗	✗
KEGG:00982		ooooooo	drug metabolism - cytochrome P450	67	✗	✗	✓	✗

6.3 Aldosterone Effect on Mouse Heart Gene Expression

three replicated data sets. The AM smoothing algorithm was applied only to two profiles in the *fill* direction, but did not result in a continuous profile, whereas the smoothing in *wipe* direction withdraws 19 gene sets from the list of non-zero profiles.

Table 6.5 lists 21 gene sets with a non-zero activation profiles with together with the gene set size, a brief description of the gene set definition and the information which profile algorithms reported the gene set with a significant profile. The table is sorted according to the extreme ranges of the D_s^{med} score introduced in section 4.4. A complete list of results can be found in Table B.7 in the Appendix. The single gene set identified exclusively by the STEM algorithm is shown at bottom in Table 6.5. This KEGG gene set is enriched in a cluster, which starts with down expression with respect to reference and increases thereafter (STEM model profile: $-2/3 - 1/3\,0\,0\, -1/3\,0\,0$). The majority of high ranked gene sets enriched with up regulated genes in Table 6.5 are related to an immune response similar to those occurring as reaction on an infection or contact with a symbiont. All these gene sets share the common profile ooo+ooo and are recognized at least by the two segmentation type algorithms. The eight gene sets with a down regulation in their activation profile are all reported with the same profile: o-ooooo. With exception of the GO:0009605 set with description "response to external stimulus" all gene sets are involved in biological processes related to cell movement. In relation to the listed genes with up regulation the gene set sizes are clearly larger. Only the interesting 7 genes including set GO:0002523 ("leukocyte migration involved in inflammatory response") is an exception of this general rule. As summary over all reported gene set activation profiles (see Table B.7) it turns out that one hour after aldosterone injection (in relation to injection without the steroid hormone) the response to an external stimulus (e.g. "leukocyte migration", "immune response") is down regulated, but three hours after the injection the immune response occurs due to an increased expression of genes in related gene sets ("response to interferon-beta", "immune response"). This observation is supported by the gene sets with o-o+ooo activation profile, which are for instance GO:0006950 ("response to stress"), GO:0050896 ("response to stimulus"), GO:0006952 ("defense response") or GO:0006954 ("inflammatory response"). The only significant gene set with a connection to the steroid hormone injection is GO:0032371 ("regulation of sterol transport") with the reported activation profile oo+oooo.

The analysis of the AH data set has proven to be challenging due to the weak signals (differential expression) and the lack of replicated measurements in the study. Turchin, Guo, et al. (2006) discuss that the blood aldosterone concentration also increases in the control group due to the stress of the injection, which probably mitigates the differences on the gene expression level. Nevertheless, the resulting gene set activation profiles show a reasonable linkage to the experiment and the inflammatory processes known in connection with the salt diet, although the observation period was limited to twelve hours, which may lead to missing long-term effects.

6.4 Embryonic Mouse Ovary Development Experiment

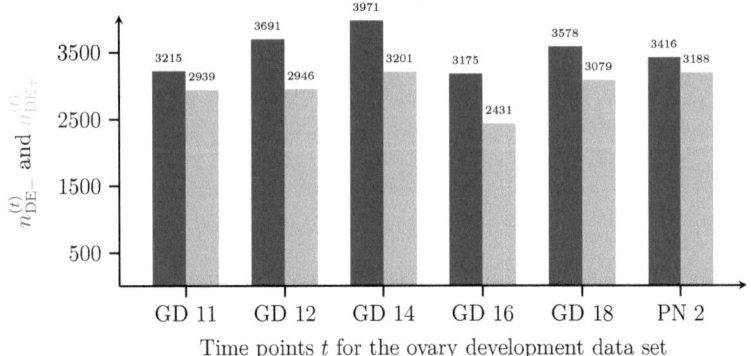

Figure 6.6: Bar chart of the total numbers of significantly down ($n_{\mathrm{DE}-}^{(t)}$) and up ($n_{\mathrm{DE}+}^{(t)}$) regulated genes with respect to the mixed reference measurement at postnatal day PN 0.5 controlling a TS-ABH FDR of 0.05 for the embryonic ovary development data set.

The GEO series GSE5334 includes 19 raw data .cel-files of hybridized Affymetrix mouse 430 2.0 arrays. The study was conducted to learn more about the molecular

6.4 Embryonic Mouse Ovary Development Experiment

biological processes in the ovary development. The RNA from the pair of ovaries from three different female mice was hybridized on the microarrays at each of the six time points: gestational days GDs 11, 12, 14, 16, 18 and postnatal day PN 2. A single chip was hybridized as reference with a mix of the whole body RNA of four female and four male mice at approximately 12 hours after birth. Unfortunately, there is no original contribution, which reports results originating from this data set. The analyzed tissue in this experiment is narrowly defined, which allows to focus on the gene set activation profiles closely related to the ovary and its biological processes.

Figure 6.6 shows the absolute numbers of differentially expressed genes per time point and direction of regulation. Differential expression is determined with the D_g^{shrink} statistic (see section 4.2) while controlling the TS-ABH FDR across all tests at a level of $\alpha_{\text{genes}} = 0.05$. The number of significantly up and down regulated genes is clearly above 2000 across all time points as expected in a development experiment in contrast with a single static measurement (reference after birth). The numbers of differentially expressed genes are high in comparison with the other experiments analyzed in this contribution. The large number of differentially expressed genes must not inevitably lead to an extraordinary large number of a priori defined gene sets with a significant activation profile, although this is observed in the following.

Table 6.6 reveals the most frequent activation profiles resulting from a 2S-GSA profile analysis combined with an AM smoothing with parameters $\lambda_{\text{fill}} = 0.0462$ and $\lambda_{\text{wipe}} = 11.2813$. The TS-ABH FDR is controlled at level 0.01 across all enrichment

Table 6.6: The 14 most frequent 2S-GSA activation profiles after smoothing with AM ($\lambda_{\text{fill}} = 0.0462$, $\lambda_{\text{wipe}} = 11.2813$) resulting from the OD data set.

Profiles	#	Profiles	#	Profiles	#
oooooo	5918	-ooooo	33	+ooooo	21
------	478	---ooo	30	ooo---	21
++++++	256	+++ooo	27	--oooo	18
++oooo	55	ooooo-	25	+++oo+	16
oo----	44	oo++++	22	others	254

tests. The maximum frequency among the 1300 non-zero activation profiles is achieved by the continuous down regulated profile (------) followed by the profile where all time points show a significant enrichment with significantly up expressed genes (++++++). Generally, profiles with a continuous segment are in the majority in contrast to the results from the AH profile analysis. Those continuous profiles are very likely not generated by chance and hence provide a more reliable insight in the gene set activity during the examined time series experiment than single position or irregularly alternating profiles do.

The large number of non-zero activation profiles makes it difficult to provide a deep insight in the analysis results. Table 6.7 lists 26 genes sets – ten to each extreme of the ranking according to the D_s^{med} scores from section 4.4 and another six hand-picked from the STEM results. The top 100 according to the $D_s^{|\mathrm{med}|}$ score are recorded in Table B.8 in the Appendix (the complete result table is available in the electronic version of the document). There are six GO gene sets whose term description reveals a direct link to the female reproductive system among the ten top ranked gene sets with up regulated positions in their profiles. Since also piRNA is closely related to the germ line cell (Lau, Seto, et al. 2006) this number can be enlarged to seven. The top ranked gene set with up regulation GO:0007066 includes only four genes, has the term description: "female meiosis sister chromatid cohesion" and its enrichment with significantly up regulated genes turns out to occur already at gestational day 14, whereas the other reproduction related gene sets become active after birth or in case of GO:0009566 ("fertilization") at GD 18. The appearance of the well-fitting gene sets on a high position in the ranking emphasizes the adequacy of both the activation profile algorithm and the ranking method. This observation gives confidence that new findings are based on the underlying molecular genetic processes and not by chance, although there is no biological view in form of an original contribution based on this data set available for comparison purposes. The top ranked gene sets are relatively small (< 100) with some exceptions, which was a general finding in all data sets due to the score calculation (see section 6.2). The majority of significant sets is characterized with an activation profile with one or more positions enriched with down regulated genes and the ten top ranked profiles

6.4 Embryonic Mouse Ovary Development Experiment

of this class in Table 6.7 obtain in summary a higher rank due to the $D_s^{|\text{med}|}$ score than the profiles with up regulated positions. Interestingly, among the top ten sets with a down regulated position there are only three GO sets, but three Reactome sets and four sets from the BioCarta gene set definition. The extraordinary high total number of continuously down and up regulated profiles across all examined time points can be explained with the type of reference. The mixed-tissue reference sample was taken after the embryonic development process finished and hence the molecular genetic differences must be larger than if the reference had been another embryonic tissue (e.g. brain or muscle). One of the 20 top ranked gene sets was also identified by the maSigFun procedure (GO:0034587 "piRNA metabolic process"), whereas the STEM algorithm reports rather larger gene sets as referred in section 6.2. Six of these gene sets, which are also identified by all GSA-type algorithms, are added to the top ranked profiles in Table 6.7. The GO term descriptions are mainly related to general developmental issues like proliferation, migration or motility. The largest gene set – GO:0048734 – listed with 2646 genes encompasses the quite general description "system development".

The analysis of the ovary development experiment shows that the enrichment method is able to identify appropriate gene sets with a reasonable profile and the applied ranking supports the researcher by the identification of meaningful profiles even when the list of significant findings is large.

Table 6.7: Top 10 of D_s^{med}-ranked up regulated and top 10 down regulated gene set activation profiles resulting from a 2S-GSA profile algorithm combined with AM smoothing on the OD data. The six hand-picked additional sets at the bottom show interesting gene sets with lower ranking.

ID	D_s^{med} Rank	Profile	Description	$\|s\|$	2T-GSA	1S-GSA	STEM	maSigFun
GO:0007066	10	oo+ooo	female meiosis sister chromatid cohesion	4	✗	✓	✗	✗
GO:0015671	60	o+oooo	oxygen transport	6	✗	✓	✗	✗
GO:0048608	71	ooooo+	reproductive structure development	202	✗	✓	✗	✗
GO:0009566	74	oooo++	fertilization	82	✗	✓	✗	✗
GO:0022602	75	ooooo+	ovulation cycle process	74	✗	✓	✗	✗
GO:0042698	76	ooooo+	ovulation cycle	78	✗	✓	✗	✗
GO:0048511	79	ooooo+	rhythmic process	160	✗	✓	✗	✗
GO:2000194	83	ooooo+	regulation of female gonad development	6	✗	✓	✗	✗
GO:0007130	104	oo++++	synaptonemal complex assembly	13	✓	✓	✗	✗
GO:0034587	108	oo++++	piRNA metabolic process	9	✗	✓	✗	✓
REACT: COMMON	11	------	genes involved in common pathway	13	✓	✓	✗	✗
BioCarta: INTRINSIC	9	------	intrinsic prothrombin activation pathway	16	✓	✓	✗	✗
BioCarta: EXTRINSIC	8	------	extrinsic prothrombin activation pathway	13	✗	✓	✗	✗

continued on next page ...

... continued from previous page

ID	Rank	Profile	Description	$|s|$	2T-GSA	1S-GSA	STEM	maSigFun
BioCarta: AMI	7	------	acute myocardial infarction	15	✓	✓	✗	✗
GO:0050790	6	o-oooo	regulation of catalytic activity	1427	✗	✓	✗	✗
BioCarta: FIBRINOLYSIS	5	------	fibrinolysis pathway	10	✗	✓	✗	✗
GO:0042542	4	--oooo	response to hydrogen peroxide	52	✗	✓	✗	✗
REACT:GRB2	2	--oooo	genes involved in GRB2	11	✗	✓	✗	✗
REACT: P130CAS	3	--oooo	genes involved in p130Cas linkage to MAPK signaling for integrins	11	✗	✓	✗	✗
GO:0070301	1	-ooooo	cellular response to hydrogen peroxide	30	✗	✓	✗	✗
GO:0003002	1216	o+oooo	regionalization	285	✓	✓	✓	✗
GO:0008283	867	oo----	cell proliferation	1207	✓	✓	✓	✗
GO:0016477	952	oo----	cell migration	695	✓	✓	✓	✗
GO:0030154	907	oo----	cell differentiation	2202	✓	✓	✓	✗
GO:0048870	888	oo-oo-	cell motility	750	✓	✓	✓	✗
GO:0048731	719	o-----	system development	2646	✓*	✓	✓	✗

6.5 Gene Expression During Wound Healing in Skin and Tongue

The GEO series GSE23006 includes the data of a wound healing experiment in mice. The study was conducted to analyze the gene expression during wound healing in two different tissues, skin and oral mucosa on the mouse tongue (L. Chen, Arbieva, et al. 2010). The differences in the healing of both tissues is well known in mice and human, for instance the oral mucosa heals more rapidly and with less or without any scar formation. In contrast to the original research question focused on the comparison between the two tissues, in this thesis the gene expression time series is analyzed separately for both tissues. For each tissue 24 Affymetrix mouse 430 2.0 arrays were hybridized. Three biological replicates are available before wounding and at time points 6 h, 12 h, 24 h, 3 d, 5 d, 7 d and 10 d after wounding. Analogously to the original contribution, the measurements before wounding are used as reference in the following analyses. The new focus are the changes in gene expression of a priori defined gene sets during wound healing either in mouse skin or mouse oral mucosa (tongue).

In the original analysis by L. Chen, Arbieva, et al. (2010) a one-way ANOVA test was used to identify differential expression with respect to the unwounded reference and accounting for a FDR limit of 5 %. In contrast to the high number of differentially expressed genes in the original contribution, the set of differentially expressed genes identified by the D_g^{shrink} statistic in this thesis is small. This observation illustrates the effect of the instable variance estimation based on only a few replicates. L. Chen, Arbieva, et al. (2010) use a further filtering step on the gene list, which aligns the total numbers of differentially expressed genes in both approaches. In contrast to the original paper, the number of differential expressed probe sets was higher applying the D_g^{shrink} statistic for the analysis of the data set investigating the gene expression during tongue healing in mice (TH) than in the data set investigating the gene expression during skin healing in mice (SH), although the pattern across the examined time points are very similar (compare Figures 6.7 and 6.8).

The original analysis uses the web based application DAVID (Sherman, D. W. Huang, et al. 2007) to assess gene set enrichment and functional annotation analysis (i.e. the grouping of main gene functions among a long gene list) for five k-means clusters in both experimental subgroups. Although the type of gene set analysis differs greatly from the proposed gene set activation profile estimations in this thesis, accordances and nonconformities in the identified gene sets are reported in the following two separate subsections for each type of wounding.

Skin Healing

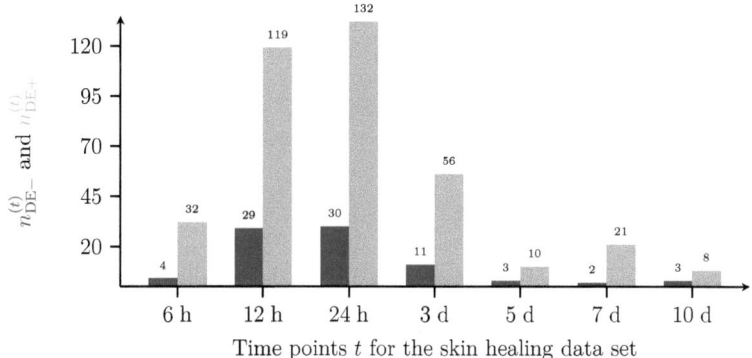

Figure 6.7: Bar chart of the total numbers of significantly down ($n_{\mathrm{DE}-}^{(t)}$) and up ($n_{\mathrm{DE}+}^{(t)}$) regulated genes with respect to the uninjured skin reference measurement at time point 0 controlling a TS-ABH FDR of 0.05 for the mouse skin healing data set.

The total numbers of significantly down and up regulated genes per time point for the SH data is shown in Figure 6.7. In comparison with the findings of the other three data sets and particular with the associated TH data, the total numbers per time point are extraordinarily low. Starting with a low number at 6 h after wounding (down: 4; up: 32), a steep raise is observed for time points 12 h and 24 h, whereas the maximum is higher for up regulation (132) than for down expression (30). The second half of the time series shows only a few differentially expressed genes. Hence, this data set is an example for a profile analysis based on only few differentially expressed genes, which must not be a sign of low quality, since the underlying molecular genetic processes can be more refined for the examined tissue. Nevertheless, the activation profile analysis should report reasonable gene sets, even if the results are less robust due to the low number of genes which may cause significant enrichment.

The resulting profiles from applying the 2T-GSA profile algorithm (FDR limit $\alpha_{\mathrm{sets}} = 0.01$) with a subsequent AM smoothing ($\lambda_{\mathrm{fill}} = 0.0738$ and $\lambda_{\mathrm{wipe}} = 0.229$)

Table 6.8: The 10 2T-GSA activation profiles after smoothing with AM ($\lambda_{\text{fill}} = 0.0738$, $\lambda_{\text{wipe}} = 0.2290$) resulting from the SH data set.

Profiles	#	Profiles	#	Profiles	#
ooooooo	7167	+++oooo	9	oo++ooo	3
o++oooo	11	++++ooo	6	o−ooooo	2
oo+oooo	10	o+ooooo	6	ooooo+o	2
				ooo+ooo	1

are summarized in Table 6.8. In accordance with the observed distribution of differentially expressed genes (Figure 6.6) the profiles stating an enrichment with up regulated genes in the first half of the time series represent the majority of identified profiles. An enrichment with down regulated genes is reported only for two gene set activation profiles featuring a single significant position at 12 h. There are only two profiles, which refer to an enrichment with up regulated genes with respect to the unwounded reference tissue at a late time point, i.e. seven days after wounding. All other 46 reported profiles show at least at one of the three time points with more than 50 significantly up regulated genes (12 h to 3 d) an enrichment position. In contrast to the profiles from the AH data set, here the continuous profiles account for the major proportion, whereas around one half of all profiles with an enrichment at 6 h are a product of smoothing.

Table 6.9 lists 25 gene sets with their reported activation profile, the term description from the gene set definition source, the gene set size, the rank according to the $D_s^{|\text{med}|}$ score and the information which profile algorithm identifies this gene set in addition to the chosen 2T-GSA method. The top 20 gene sets are chosen according to their extreme position in the D_s^{med} score ranking and the hand picked five sets at the bottom of the table show interesting or typical findings of the competing STEM and maSigFun algorithms. The key word description of the two top ranked gene sets are mentioned with a similar expression pattern in the original contribution of L. Chen, Arbieva, et al. (2010). The gene sets KEGG:04620 ("Toll-like receptor signaling pathway"), GO:0006954 ("inflammatory response"), GO:0009611 ("response to wounding") and GO:0006935 ("chemotaxis") are also reported by L.

6.5 Gene Expression During Wound Healing in Skin and Tongue

Chen, Arbieva, et al. (2010) in their cluster based enrichment analysis with an early up regulation (compare also the full list of activation profiles in Table B.9 in the Appendix). The same holds for the gene sets related to *chemokine* and *cytokine*. The gene set activation profile analysis reports in addition gene sets related to cell motion (e.g. GO:004011 "locomotion" or GO:0060326 "cell chemotaxis") or the late up regulated "keratinization" process (GO:0031424). The gene set sizes of the top ranked sets are slightly larger than for the other data sets, although a single differentially expressed gene in the set would lead to a significant enrichment and the proportion of small sets is by far larger than the proportion of medium and large sets (see Figure 6.3). The set sizes of the significant STEM sets are quite large as typical for the findings of STEM. The identified sets are usually also detected by one or both segmentation algorithms and are mainly connected to *cell division* or related processes (four sets are given in Table 6.9). The maximum gene set size of the significant maSigFun profiles is 7, for the set REACT:DEATH, see Table 6.9. This small size for reported sets is typical for the maSigFun procedure across all applied analyses.

The activation profile analysis with the 2T-GSA algorithm has on the one hand proved to be a little conservative in comparison to the other GSA-type algorithm or the STEM procedure in terms of number of significant findings. On the other hand, the feared effect of reporting thousands of gene sets with only a single differentially expressed gene included (due to the small overall number of differentially expressed genes) did not occur. The TS-ABH FDR adjustment and the smoothing protect from an overwhelming number of (potentially false) discoveries even in the case of a very low number of differentially expressed genes.

Table 6.9: Top 18 of D_s^{med}-ranked up regulated and the only two down regulated gene set activation profiles resulting from a 2T-GSA profile algorithm combined with AM smoothing on the SH data. The five hand-picked additional sets at the bottom show interesting gene sets identified by STEM or maSigFun.

ID	D_s^{med} Rank	Profile	Description	$\|s\|$	1S-GSA	2S-GSA	STEM	maSigFun
KEGG:04620	1	oo+oooo	Toll-like receptor signaling pathway	93	✓	✓	✗	✗
GO:0031424	2	ooooo+o	keratinization	25	✓	✓	✗	✗
REACT: CHEMOKINE	3	+++oooo	genes involved in chemokine receptors bind chemokines	36	✓	✓	✗	✗
REACT: PEPTIDE	4	+++oooo	genes involved in peptide ligand-binding receptors	115	✓	✓	✗	✗
REACT: GPCR	5	o++oooo	genes involved in GPCR ligand binding	250	✓	✓	✗	✗
GO:0050715	7	o++oooo	positive regulation of cytokine secretion	46	✓	✓	✗	✗
GO:0050714	8	o++oooo	positive regulation of protein secretion	71	✓	✓	✗	✗
GO:0040011	6	o++oooo	locomotion	898	✓	✓	✗	✗
KEGG:04060	9	+++oooo	cytokine-cytokine receptor interaction	219	✓	✓	✗	✗
GO:0006954	10	o++oooo	inflammatory response	341	✓	✓	✗	✗
GO:0030593	12	++++ooo	neutrophil chemotaxis	44	✓	✓	✗	✗
GO:0050663	13	o++oooo	cytokine secretion	72	✓	✓	✗	✗

continued on next page ...

6.5 Gene Expression During Wound Healing in Skin and Tongue

... continued from previous page

ID	Rank	Profile	Description	$|s|$	1S-GSA	2S-GSA	STEM	maSigFun
GO:0050707	14	o++oooo	regulation of cytokine secretion	61	✓	✓	✗	✗
GO:0050708	15	o++oooo	regulation of protein secretion	103	✓	✓	✗	✗
GO:0051222	16	o++oooo	positive regulation of protein transport	138	✓	✓	✗	✗
GO:0030595	17	++++ooo	leukocyte chemotaxis	94	✓	✓	✗	✗
GO:0060326	18	++++ooo	cell chemotaxis	120	✓	✓	✗	✗
REACT:STRIATED	19	ooooo+o	genes involved in striated muscle contraction	20	✓	✓	✗	✗
GO:0000910	50	o-oooo	cytokinesis	77	✓	✓	✗	✗
GO:0033205	48	o-oooo	cell cycle cytokinesis	27	✓	✓	✗	✗
GO:0010558		ooooooo	negative regulation of macromolecule biosynthetic process	845	✓	✗	✓	✗
GO:0032502		ooooooo	developmental process	3505	✓	✗	✓	✗
GO:0051301		ooooooo	cell division	405	✓	✓	✓	✗
GO:0007067		ooooooo	mitosis	278	✓	✓	✓	✗
REACT:DEATH		ooooooo	death receptor signalling	7	✓	✗	✗	✓

Tongue Healing

The total numbers of differentially expressed genes in both directions – up and down regulation – for the skin healing data set is shown separately for each examined time point in Figure 6.8. The D_g^{shrink} test statistic controlled at a TS-ABH FDR limit

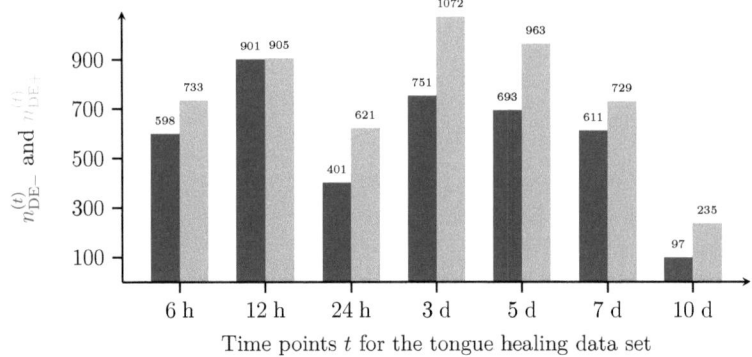

Figure 6.8: Bar chart of the total numbers of significantly down ($n_{\text{DE}-}^{(t)}$) and up ($n_{\text{DE}+}^{(t)}$) regulated genes with respect to the uninjured skin reference measurement at time point 0 controlling a TS-ABH FDR of 0.05 for the mouse tongue healing data set.

of 5 % yields a relatively uniform distribution with exception of the results at 10 d where the total number of differentially expressed genes is clearly smaller. At 12 h after the tongue injury the detection of up and down regulated genes is nearly equal, but for all other time points the up direction clearly dominates. In comparison with the above analyzed data sets, the total numbers are in a medium range, hence the determined activation profiles should not be affected by a lack or an abundance of differentially expressed features.

The 2S-GSA profile algorithm uses a segmentation test on the as significant stated genes from Figure 6.8 separately per time point, whereas the TS-ABH FDR is limited to 1 % across all enrichment tests and the AM smoothing method is used to smooth the profiles to more reliable and continuous profiles ($\lambda_{\text{fill}} = 0.1388$ and $\lambda_{\text{wipe}} = 3.5679$). Gene set activation profiles enriched up regulated genes at one or more positions are predominant among both the most frequent profiles listed in Table 6.10 and the whole set of profiles. Continuous profiles with up regulated positions at the early time points represent the nine most frequently occurring

6.5 Gene Expression During Wound Healing in Skin and Tongue

categories. The only profile with down regulated positions among the fourteen most abundant profiles in Table 6.10 is o-o----, although the smoothing evaluates a change of this profile in both directions. This finding is on the one hand a strong hint that this is a true biological pattern or at least the hybridization at the third time point (24 h) was weaker for some reason so that some true down regulated genes are missed. On the other hand it is shown that the smoothing algorithm is well balanced and does not correct every considered position.

The total number of 842 gene sets with a significant activation profile is too large to discuss all profiles in detail. Analogously to the previous analyzed data sets, the ranking according to the both extremes of the D_s^{med} score is used to identify gene set profiles with a clear difference to the reference across the reported enriched positions. A list of the top ten gene sets for each extreme is given in Table 6.11 and the top 100 activation profiles are recorded in the Appendix in Table B.10 with a link to the complete list in the electronic version of this document. The top ranked gene sets enriched with up regulated genes in their activation profiles include again only a small number of genes (clearly below 100), whereas there are three gene sets with more than 100 genes in the top ten of profiles with down regulation (see Table 6.11). The gene set MetaCyc:PWY-5687 ("pyrimidine ribonucleotides interconversion") in Table 6.11 containing four genes is exclusively detected by the maSigFun procedure, but does not show an easily accessible connection to the known wound healing processes. Nor the other two maSigFun exclusive sets do. The same holds for the three other gene sets, which were identified (also) by the STEM algorithm, but

Table 6.10: The 14 most frequent 2S-GSA activation profiles after smoothing with AM ($\lambda_{fill} = 0.1388$, $\lambda_{wipe} = 3.5679$) resulting from the TH data set.

Profiles	#	Profiles	#	Profiles	#
oooooooo	6375	+++++oo	53	o-o----	25
++++ooo	105	o++oooo	49	++ooooo	18
+++oooo	77	+oooooo	37	ooo+ooo	18
++++++o	59	o+++ooo	31	ooooo+o	17
+++++++	54	oo++ooo	27	others	272

show a very inconspicuous 2S-GSA profile. In analogy to the results from the other three data sets, STEM has the tendency to report larger gene sets than the other competing algorithms.

There are noticeable many interrupted activation profiles among the top ten ranked profiles with down regulated positions. A direct and specific link to the wound healing process cannot be stated without deeper biologic knowledge. The impression that the smoothing was not successful at the most suitable profile positions is biased due to the fact that the top ranked profiles exhibit a very strong differential expression and enrichment. Hence, the profiles with successful smoothing appear further down the ranked list.

Table 6.11: Top 10 of D_s^{med}-ranked up regulated and top 10 down regulated gene set activation profiles resulting from a 2S-GSA profile algorithm combined with AM smoothing on the TH data. The four hand-picked additional sets at the bottom show interesting gene sets identified by STEM or maSigFun.

ID	D_s^{med} Rank	Profile	Description	$\|s\|$	2T-GSA	1S-GSA	STEM	maSigFun
KEGG:04621	1	o++oooo	NOD-like receptor signaling pathway	51	✗	✓	✗	✗
GO:0035457	2	+++oooo	cellular response to interferon-alpha	4	✗	✓	✗	✗
GO:0010573	3	++++ooo	vascular endothelial growth factor production	16	✓	✓	✗	✗
GO:0010574	4	++++ooo	regulation of vascular endothelial growth factor production	16	✓	✓	✗	✗
GO:0001660	5	o+++ooo	fever generation	12	✓	✓	✗	✗
GO:0031649	6	o+++ooo	heat generation	17	✓	✓	✗	✗

continued on next page . . .

... continued from previous page

ID	Rank	Profile	Description	$|s|$	2T-GSA	1S-GSA	STEM	maSigFun
GO:0071354	7	++oooooo	cellular response to interleukin-6	10	✗	✓	✗	✗
GO:0031622	8	o+++ooo	positive regulation of fever generation	8	✗	✓	✗	✗
GO:0031652	9	o+++ooo	positive regulation of heat generation	10	✓	✓	✗	✗
REACT: STRIATED	10	ooo++++	genes involved in striated muscle contraction	20	✓	✓	✗	✗
GO:0030071	253	ooooo-o	regulation of mitotic metaphase/anaphase transition	22	✗	✓	✗	✗
GO:0007346	248	o-oo--o	regulation of mitotic cell cycle	218	✓	✓	✗	✗
GO:0031577	241	o-oo---	spindle checkpoint	22	✓	✓	✗	✗
GO:0090068	214	o-oo---	positive regulation of cell cycle process	90	✓	✓	✗	✗
KEGG:04614	209	o-ooooo	renin-angiotensin system	17	✗	✗	✗	✗
GO:0051640	208	o-o----	organelle localization	118	✗	✓	✗	✗
GO:0045786	186	ooooo-o	negative regulation of cell cycle	277	✓	✓	✗	✗
GO:0051303	162	oooo--o	establishment of chromosome localization	19	✓	✓	✗	✗
GO:0050000	163	oooo--o	chromosome localization	19	✓	✓	✗	✗
GO:0051310	139	oooo--o	metaphase plate congression	16	✗	✗	✗	✗

continued on next page ...

ID	Rank	Profile	Description	$\|s\|$	2T-GSA	1S-GSA	STEM	maSigFun
... continued from previous page								
MetaCyc: PWY-5687		ooooooo	pyrimidine ribonucleotides interconversion	4	✗	✗	✗	✓
GO:0006818		ooooooo	hydrogen transport	89	✗	✗	✓	✗
KEGG:00280	797	o-ooooo	Valine, leucine and isoleucine degradation	47	✓	✓	✓	✗
REACT: INTEGRATION		ooooooo	genes involved in integration of energy metabolism	137	✗	✓	✓	✗

Among the top ten ranked profiles with up regulated positions only the gene sets GO:0035457 and GO:71354 are related to the inflammatory response identified in *early up-regulated* class of the original paper by L. Chen, Arbieva, et al. (2010), but all identified gene sets of the original contribution can be confirmed in the full list of significant gene set activation profiles. These are for instance from the *early up-regulated* class: GO:0006954 ("inflammatory response"), GO:0009611 ("response to wounding"), KEGG:04620 ("Toll-like receptor signaling pathway"), GO:0001816 ("cytokine production"), KEGG:04060 ("cytokine-cytokine receptor interaction") or GO:0006935 ("chemotaxis"), which are all reported with the common (smoothed) activation profile ++++++o. There are also main agreements among the reported enrichments in the *late up-regulated* class. The term key words of the gene set activation profiles GO:0030198 (ooo++++, "extracellular matrix organization"), GO:0032964 (oooo+++, "collagen biosynthetic process"), KEGG:04512 (o-o++++, "ECM-receptor interaction") and GO:0007154 (±++++++ "cell communication") had been reported by L. Chen, Arbieva, et al. (2010) in their original k-means cluster enrichment analysis. The last profile contains a ± position, which occurs only in less than 5 % of the profiles and usually in large gene sets (e.g. 3547 genes in

6.5 Gene Expression During Wound Healing in Skin and Tongue

GO:0007154) with a more comprehensive GO term definition. The perfectly reasonable gene set terms with a significant enrichment with clearly differentially expressed genes: GO:0010573 (++++ooo, "vascular endothelial growth factor production") and GO:0031649 (o+++ooo, "heat generation") are among the new findings from the 2S-GSA profile algorithm not mentioned by the original paper (L. Chen, Arbieva, et al. 2010), but fit well in the wider context of the wound healing process (Gurtner, Werner, et al. 2008). This may be understood as evidence that the exploratory gene set activation profile algorithm is able to contribute to the identification of new molecular genetic processes gene expression time series studies in general and in the by far not completely understood process of wound healing in particular.

CHAPTER 7

Discussion and Summary

This contribution emphasizes the increasing importance of gene expression time series and predefined gene sets with known functionality in the analysis of molecular genetic processes. The need for an exploratory and explicit gene set analysis approach on this type of data is not yet recognized in the relevant literature. The available methods focus either on a single gene analysis or on the identification of co-regulated clusters. Gene set analysis is popular in two-group comparison experiments, but rather neglected in gene expression time series. Here, three gene set activation profile algorithms are proposed against the background of well established methods from the two groups gene set analysis. The algorithms are compared with two state-of-the-art techniques for the analysis of gene expression time series, although the intention is different for all approaches.

Three algorithms are proposed for the analysis of predefined Gene Ontology (GO), KEGG, BioCarta, Reactome, BioCyc or alternative definitions of gene sets in gene expression time series, which include commonly only a small number of time points (4 to 12) and replicates (1 to 5). The three introduced algorithms identify in the first step all differentially expressed genes with respect to a reference (e.g. by fold change or shrinkage t-test) per time point. The second step applies two enrichment tests for each time point, which identify either a significant enrichment with significant up regulated genes (+), a significant enrichment with down expressed genes (-), both (±) or no enrichment (o). The concatenation of these symbols across all time points results in the gene set activation profile. The three proposed profile algorithms differ only in the type of enrichment test (Fisher test: 2T-GSA, segmentation test: 1S-GSA and segmentation test satisfying a significance limit: 2S-GSA). The FDR is controlled at every step of the algorithm for the large number of tests. The STEM procedure introduced by Ernst, Nau, and Bar-Joseph (2005) clusters all gene expression vectors according to data independent chosen model profiles and uses a Fisher test to identify those gene sets, which are significantly enriched with genes from significant clusters. M. Nueda, Sebastián, et al. (2009) describe the maSigFun algorithm, which uses a two step regression procedure to estimate a linear model in dependence of the time for the gene set on basis on the standardized gene expression vectors annotated to the gene set.

The first of two extensive simulation studies compares two widespread algorithms – STEM and maSigFun – with the new profile algorithms. A main result is that maSigFun favors small sets with genes virtually perfectly fitting to a linear model, whereas STEM identifies due to its clustering mechanism mainly larger gene sets with a high proportion of genes close to the common model profile. The three introduced activation profile algorithms show a reasonable accuracy. The accuracy improves with increasing extent of truly differentially expressed gene trajectories. The limitation should be noted that the threshold free variant suffers from a high number of false positives if the number of true differential expression signals is low. It is important to note that the intention of the five methods differs. The maSigFun procedure is successful in the identification of strictly co-regulated genes (and a common transcription factor), but with the limitation that the co-regulated gene sets must be known in advance. The method's statistical power is not very high due to its strict control of type-I-error, which is mitigated in this thesis by the control of the FDR. The STEM algorithm, and even more all cluster based methods, also intends to identify functionally related or co-regulated genes, but without the primary objective to use the available knowledge about such structures. It is intrinsic to those methods that the identification of smaller clusters as significant is more difficult than the detection of large gene sets following the same expression trajectory in time. The subsequent analysis of a priori defined gene sets on the identified cluster suffers from the same impairment, since usually only the significant clusters are used in this second step.

The second simulation study reveals the success of nine different smoothing methods in combination with the proposed gene set activation profile algorithms. The smoothing does not identify additional gene sets, but a strong smoothing in the *wipe* direction (i.e. reject up or down positions) improves sensitivity by reducing the number of false positives. The additional smoothing in the *fill* direction provides more reasonable activation profiles if the assumption of continuity in biological processes is fulfilled for the current experiment. This assumption may not hold for rare time series studies with cell types known to be "out of control" or mis-regulated like in cancer cells. Other examples compromising the continuity assumption are designs of experiments in which the time interval between the selected time points are too large to cover the main molecular genetic processes. Nevertheless, the smoothing parameters are optimized by the simulation results and smoothing leads to reasonable results in applications, where the selected smoothing method does not *fill* every suitable non continuous activation profile. An adaptive choice of the smoothing parameter in dependence on the number of significant genes per time point and the current gene set size could lead to further improvements. This is a promising starting point for future research. The proposed ranking of the reported gene sets turned out to yield a useful sorting of the significant results. Among the high ranked positions are known processes from the original contributions or gene sets with functions that fit very well to the time series experiment.

The application of the activation profile algorithm accounting for significance in differential expression proves to report gene sets known from the initial original contribution. Furthermore, more reasonable gene sets with an enrichment with up or down regulated genes across the observed time period are identified. The different characteristics (e.g. type of reference) of the used data sets can be seen as evidence for the wide application range for the gene set activation profile algorithms. In particular, the 2S-GSA procedure in combination with a suitable smoother equips the researcher with a useful exploratory tool. The algorithm identifies the molecular genetic processes on a gene set level and allows to benefit from the a priori available knowledge. This works effectively and reliably in a situation with much less observations than needed for a reasonable modeling approach. The profile algorithms can be directly applied on gene expression multi-dose studies. The procedures for examination of differential expression and gene set analysis may be easily exchanged with other methods more suitable in the user's opinion.

The quality of each method, which involves a priori knowledge depends on the quality of the used annotations. In this thesis only data from hybridized Affymetrix arrays is used, but the proposed algorithms are directly transferable on time series data generated by other high throughput techniques. Even the annotation of the probes and probe sets on the microarray can be improved or at least quality controlled as proposed by Yu, F. Wang, et al. (2007). Moreover, the gene set annotations are in a permanent change due to the steady inclusion of current research results. On the one hand, this has a negative effect on the comparison to earlier analyses. On the other hand, the used gene set information is as up to date as possible, which improves the reliability and usefulness of the results. The here applied Bioconductor annotation packages are only updated biannually, but Kumar, Holm, and Toronen (2013) recently published the software GOParGenPy, which provides an easy way to use the most recent GO annotation data. Quality problems of gene set annotations and other pitfalls in the microarray analysis were first discussed by Khatri and Drăghici (2005) and Allison, Cui, et al. (2006).

The desirable property to account for all dependencies referring to genes, time points and gene sets is difficult or impossible due to the very limited numbers of observations in typical gene expression time series experiments. However, some newer methods provide models for the gene set enrichment using covariance matrices (Y.-T. Huang and X. Lin 2013, e.g.). This topic is another promising research field together with the decorrelating characteristic of the rotation test approach, which was only briefly discussed in this thesis.

Unfortunately, the true functional molecular genetic processes are not known for gene expression time series and therefore a second confirmation of analysis results is common and usually done by RT-PCR, protein level measurements or a second microarray experiment with new cases in the same design of experiment. Since these procedures are time and cost intensive, a good exploratory analysis of the data allows to focus on the either new, relevant or unexpected functions. The proposed gene set profile algorithm simplify the complex gene expression time series and gene (set) annotation information to a symbolic profile, indicating enrichment with up or down regulated genes per time point position. The loss in information due to averaging and dimension reduction reduces noise and allows for more interpretable, meaningful and more robust results. Figure 7.1 illustrates the way from the expression values to the gene set activation profile for one set of genes stated significant in the tongue healing data set.

It has to be mentioned that the independence assumption in the enrichment set, which uses the genes instead of the observation units as sampling model is rightly criticized as unrealistic (e.g. by Goeman and Bühlmann 2007). The alternative in form of resampling-based methods (with observations as sampling model) would lead to a discrete p-value distribution with a few values, due to the common lack of replicates. Hence, a strong evidence for enrichment at least after adjustment for FDR would be impossible to obtain. The formulation of a concrete null hypothesis for the profile algorithms was intentionally omitted. The competitive hypothesis ("gene in the set are at most as often differentially expressed as the genes outside the set") of the enrichment test is used subsequently for each comparison with the reference. Two alternatives are for instance the self-contained test ("No gene in the set is differentially expressed") or the most restrictive modeling approach that assumes all genes in the set following the same pattern. Both have proven to result in too many or respectively virtually no findings, which both are not desirable in an exploratory analysis.

Figure 7.1 shows the outline of the proposed and applied gene set activation profile algorithms for the concrete example of the gene set GO:0031649 ("heat generation"). The set was stated with profile o+++ooo in the application section 6.5. The plots highlight the fact that not all genes in the set contribute to the activation profile pattern. Furthermore, the activation of the function would have been missed if the assumption that all genes in the set follow the same pattern had been made. This is the main difference compared to most competing algorithms for detecting gene expression changes on the level of gene sets in the literature. The example shows

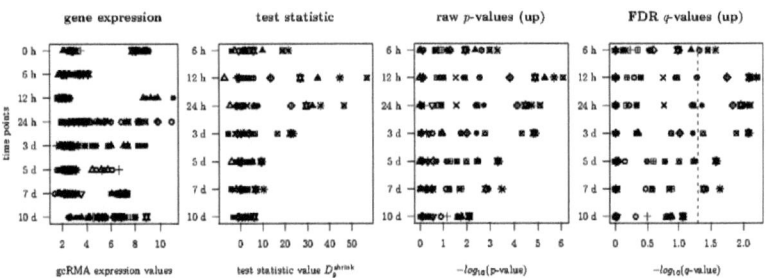

Figure 7.1: Plots as outline for the GSA-type gene set activation profile analysis at the example of gene set GO:0031649 ("heat generation", 17 genes). The gcRMA corrected expression values are used for a time point wise t-test comparison to the common reference (0 h). The raw p-values are transformed to TS-ABH FDR q-values adjusting for the high number of tests. The dashed line displays the significance threshold for up regulation ($\alpha_{genes} = 0.05$). The five to six genes exceeding this limit lead to significant enrichments (compare activation profile in Table 6.11) in the segmentation test at time points 12 h to 3 d.

also that a small number of significantly differentially expressed genes may lead to significant positions in the gene set activation profile. The enrichment test depends in addition on the overall number of differentially expressed genes in the data set at the current time point. The smoothing algorithm did not lead to a change of the activation profile, which seems to be a plausible decision regarding the p-values plot. All in all, Figure 7.1 provides an insight on the noisiness and the difficulty of making the *activity*-decision for a smaller gene set. Hence, the proposed gene set activity profiles algorithms are a usable exploratory tool for the gene set analysis on gene expression time series experiments.

Bibliography

Ackermann, M. and K. Strimmer (2009).
A general modular framework for gene set enrichment analysis.
BMC Bioinformatics 10.

Allison, D. B., X. Cui, G. P. Page, and M. Sabripour (2006).
Microarray data analysis: from disarray to consolidation and consensus.
Nature Reviews Genetics 7.1, pp. 55–65.

Angelini, C., L. Cutillo, D. De Canditiis, M. Mutarelli, and M. Pensky (2008).
BATS: a Bayesian user-friendly software for Analyzing Time Series microarray experiments.
BMC Bioinformatics 9.1, p. 415.

Angelini, C., D. De Canditiis, M. Mutarelli, and M. Pensky (2007).
A Bayesian Approach to Estimation and Testing in Time-course Microarray Experiments.
Statistical Applications in Genetics and Molecular Biology 6.1, pp. 1–31.

Ashburner, M., C. A. Ball, J. A. Blake, D. Botstein, H. Butler, J. M. Cherry, A. P. Davis, K. Dolinski, S. S. Dwight, J. T. Eppig, M. A. Harris, D. P. Hill, L. Issel-Tarver, A. Kasarskis, S. Lewis, J. C. Matese, J. E. Richardson, M. Ringwald, G. M. Rubin, and G. Sherlock (2000).
Gene Ontology: tool for the unification of biology.
Nature Genetics 25.1, pp. 25–29.

Bansal, M., G. D. Gatta, and D. Di Bernardo (2006).
Inference of Gene Regulatory Networks and Compound Mode of Action from Time Course Gene Expression Profiles.
Bioinformatics 22.7, pp. 815–822.

Bar-Joseph, Z., G. Gerber, D. Gifford, T. Jaakkola, and I. Simon (2002).
"A new approach to analyzing gene expression time series data".
In: *Proceedings of the Annual International Conference on Computational Molecular Biology*,
Pp. 39–48.

Bar-Joseph, Z. (2004).
Analyzing Time Series Gene Expression Data.
Bioinformatics 20.16, pp. 2493–2503.

Bar-Joseph, Z., G. Gerber, I. Simon, D. K. Gifford, and T. S. Jaakkola (2003).
Comparing the continuous representation of time-series expression profiles to identify differentially expressed genes.
Proceedings of the National Academy of Sciences of the United States of America 100.18, pp. 10146–10151.

Ben-Dor, A., R. Shamir, and Z. Yakhini (1999).
Clustering Gene Expression Patterns.
Journal of Computational Biology 6.3-4, pp. 281–297.

Benjamini, Y. and Y. Hochberg (1995).
Controlling the False Discovery Rate: A Practical and Powerful Approach to Multiple Testing.
Journal of the Royal Statistical Society. Series B (Methodological) 57.1, pp. 289–300.

Benjamini, Y., A. M. Krieger, and D. Yekutieli (2006).
Adaptive Linear Step-up Procedures That Control the False Discovery Rate.
Biometrika 93.3, pp. 491–507.

Bonferroni, C. E. (1936).
Teoria statistica delle classi e calcolo delle probabilità.
Pubblicazioni del R Istituto Superiore di Scienze Economiche e Commerciali di Firenze 8, pp. 3–62.

Brown, M. P. S., W. N. Grundy, D. Lin, N. Cristianini, C. W. Sugnet, T. S. Furey, M. Ares, and D. Haussler (2000).
Knowledge-Based Analysis of Microarray Gene Expression Data by Using Support Vector Machines.
Proceedings of the National Academy of Sciences 97.1, pp. 262–267.

Chen, J. and K. Chang (2008).
Discovering Statistically Significant Periodic Gene Expression.
International Statistical Review 76.2, pp. 228–246.

Chen, L., Z. H. Arbieva, S. Guo, P. T. Marucha, T. A. Mustoe, and L. A. DiPietro (2010).
Positional differences in the wound transcriptome of skin and oral mucosa.
BMC Genomics 11, p. 471.

Cheng, C., X. Ma, X. Yan, F. Sun, and L. M. Li (2006).
MARD: a new method to detect differential gene expression in treatment-control time courses.
Bioinformatics 22.21, pp. 2650–2657.

Cho, R. J., M. J. Campbell, E. A. Winzeler, L. Steinmetz, A. Conway, L. Wodicka, T. G. Wolfsberg, A. E. Gabrielian, D. Landsman, D. J. Lockhart, and R. W. Davis (1998).
A Genome-Wide Transcriptional Analysis of the Mitotic Cell Cycle.
Molecular Cell 2.1, pp. 65–73.

Chou, J. W., T. Zhou, W. K. Kaufmann, R. S. Paules, and P. R. Bushel (2007).
Extracting gene expression patterns and identifying co-expressed genes from microarray data reveals biologically responsive processes.
BMC Bioinformatics 8.1, p. 427.

Chuan Tai, Y. and T. P. Speed (2009).
On Gene Ranking Using Replicated Microarray Time Course Data.
Biometrics 65.1, pp. 40–51.

Chudova, D., C. Hart, E. Mjolsness, and P. Smyth (2003).
Gene expression clustering with functional mixture models.
Advances in Neural Information Processing 16.

Coffey, N. and J. Hinde (2011).
Analyzing Time-Course Microarray Data Using Functional Data Analysis - A Review.
Statistical Applications in Genetics and Molecular Biology 10.1, pp. 1–32.

Conesa, A., M. J. Nueda, A. Ferrer, and M. Talon (2006).
maSigPro: a method to identify significantly differential expression profiles in time-course microarray experiments.
Bioinformatics 22.9, pp. 1096–1102.

Costa, I. G., A. Schönhuth, C. Hafemeister, and A. Schliep (2009).
Constrained Mixture Estimation for Analysis and Robust Classification of Clinical Time Series.
Bioinformatics 25.12, pp. i6–i14.

Costa, I. G., A. Schönhuth, and A. Schliep (2005).
The Graphical Query Language: A Tool for Analysis of Gene Expression Time-Courses.
Bioinformatics 21.10, pp. 2544–2545.

Cui, X. and G. A. Churchill (2003).
Statistical tests for differential expression in cDNA microarray experiments.
Genome Biology 4.4, p. 210.

Dalma-Weiszhausz, D., J. Warrington, E. Tanimoto, and C. Miyada (2006).
The Affymetrix GeneChip® Platform: An Overview.
Vol. 410.

Das, D., Z. Nahlé, and M. Q. Zhang (2006).
Adaptively inferring human transcriptional subnetworks.
Molecular Systems Biology 2.1.

Di Camillo, B., G. Toffolo, S. K. Nair, L. J. Greenlund, and C. Cobelli (2007).
Significance analysis of microarray transcript levels in time series experiments.
BMC Bioinformatics 8.Suppl 1, S10.

Dørum, G., L. Snipen, M. Solheim, and S. Sæbø (2009).
Rotation testing in gene set enrichment analysis for small direct comparison experiments.
Statistical Applications in Genetics and Molecular Biology 8.1.

Edgar, R., M. Domrachev, and A. Lash (2002).
Gene Expression Omnibus: NCBI gene expression and hybridization array data repository.
Nucleic Acids Research 30.1, pp. 207–210.

Efron, B. and R. Tibshirani (2007).
On Testing the Significance of Sets of Genes.
The Annals of Applied Statistics 1.1, pp. 107–129.

Eisen, M. B., P. T. Spellman, P. O. Brown, and D. Botstein (1998).
Cluster analysis and display of genome-wide expression patterns.
Proceedings of the National Academy of Sciences 95.25, pp. 14863–14868.

ElBakry, O., M. O. Ahmad, and M. Swamy (2012).
Identification of Differentially Expressed Genes for Time-Course Microarray Data Based on Modified RM ANOVA.
Computational Biology and Bioinformatics 9.2, pp. 451–466.

Ernst, J. and Z. Bar-Joseph (2006).
STEM: A tool for the analysis of short time series gene expression data.
BMC Bioinformatics 7.

Ernst, J., G. Nau, and Z. Bar-Joseph (2005).
Clustering short time series gene expression data.
Bioinformatics 21.Suppl 1.

Goeman, J. J. and P. Bühlmann (2007).
Analyzing gene expression data in terms of gene sets: methodological issues.
Bioinformatics 23.8, pp. 980–987.

Gurtner, G. C., S. Werner, Y. Barrandon, and M. T. Longaker (2008).
Wound repair and regeneration.
Nature 453.7193, pp. 314–321.

Hafemeister, C., I. G. Costa, A. Schönhuth, and A. Schliep (2011).
Classifying Short Gene Expression Time-Courses with Bayesian Estimation of Piecewise Constant Functions.
Bioinformatics 27.7, pp. 946–952.

Hastie, T., R. Tibshirani, M. Eisen, A. Alizadeh, R. Levy, L. Staudt, W. Chan, D. Botstein, and P. Brown (2000).
'Gene shaving' as a method for identifying distinct sets of genes with similar expression patterns.
Genome Biology 1.2, research0003.1–research0003.21.

Heard, N. A., C. C. Holmes, and D. A. Stephens (2006).
A quantitative study of gene regulation involved in the immune response of anopheline mosquitoes: An application of Bayesian hierarchical clustering of curves.
Journal of the American Statistical Association 101.473, pp. 18–29.

Hirose, O., R. Yoshida, S. Imoto, R. Yamaguchi, T. Higuchi, D. S. Charnock-Jones, C. Print, and S. Miyano (2008).
Statistical inference of transcriptional module-based gene networks from time course gene expression profiles by using state space models.
Bioinformatics 24.7, pp. 932–942.

Holter, N. S., A. Maritan, M. Cieplak, N. V. Fedoroff, and J. R. Banavar (2001).
Dynamic Modeling of Gene Expression Data.
Proceedings of the National Academy of Sciences 98.4, pp. 1693–1698.

Hong, F. and H. Li (2006).
Functional Hierarchical Models for Identifying Genes with Different Time-Course Expression Profiles.
Biometrics 62.2, pp. 534–544.

Hosack, D. A., G. Dennis, B. T. Sherman, H. C. Lane, and R. A. Lempicki (2003).
Identifying biological themes within lists of genes with EASE.
Genome Biology 4.10.

Huang, Y.-T. and X. Lin (2013).
Gene set analysis using variance component tests.
BMC Bioinformatics 14.1, p. 210.

Hummel, M., R. Meister, and U. Mansmann (2008).
GlobalANCOVA: exploration and assessment of gene group effects.
Bioinformatics 24.1, pp. 78–85.

Hvidsten, T., A. Lægreid, and J. Komorowski (2003).
Learning rule-based models of biological process from gene expression time profiles using gene ontology.
Bioinformatics 19.9, pp. 1116–1123.

Irizarry, R., B. Bolstad, F. Collin, L. Cope, B. Hobbs, and T. Speed (2003).
Summaries of affymetrix GeneChip probe level data RID B-8085-2009.
Nucleic Acids Research 31.4.

Kalaitzis, A. A. and N. D. Lawrence (2011).
A Simple Approach to Ranking Differentially Expressed Gene Expression Time Courses through Gaussian Process Regression.
BMC Bioinformatics 12.1, p. 180.

Kaminski, N. and Z. Bar-Joseph (2007).
A Patient-Gene Model for Temporal Expression Profiles in Clinical Studies.
Journal of Computational Biology 14.3, pp. 324–338.

Kanehisa, M. and S. Goto (2000).
KEGG: Kyoto Encyclopedia of Genes and Genomes.
Nucleic Acids Research 28.1, pp. 27–30.

Kanehisa, M., S. Goto, Y. Sato, M. Furumichi, and M. Tanabe (2012).
KEGG for integration and interpretation of large-scale molecular data sets.
Nucleic Acids Research 40.Database issue, pp. D109–114.

Karp, P. D., C. A. Ouzounis, C. Moore-Kochlacs, L. Goldovsky, P. Kaipa, D. Ahrén, S. Tsoka, N. Darzentas, V. Kunin, and N. López-Bigas (2005).
Expansion of the BioCyc collection of pathway/genome databases to 160 genomes.
Nucleic Acids Research 33.19, pp. 6083–6089.

Khatri, P. and S. Drăghici (2005).
Ontological Analysis of Gene Expression Data: Current Tools, Limitations, and Open Problems.
Bioinformatics 21.18, pp. 3587–3595.

Kim, J. and J. H. Kim (2007).
Difference-based clustering of short time-course microarray data with replicates.
BMC Bioinformatics 8.1, p. 253.

Kim, K. I. and M. A. van de Wiel (2008).
Effects of dependence in high-dimensional multiple testing problems.
BMC Bioinformatics 9.1, p. 114.

Kim, S.-Y. and D. J. Volsky (2005).
PAGE: Parametric Analysis of Gene Set Enrichment.
BMC Bioinformatics 6.1, p. 144.

Kumar, A. A., L. Holm, and P. Toronen (2013).
GOParGenPy: a high throughput method to generate Gene Ontology data matrices.
BMC Bioinformatics 14.1, p. 242.

Langsrud, Ø. (2005).
Rotation tests.
Statistics and Computing 15.1, pp. 53–60.

Lau, N. C., A. G. Seto, J. Kim, S. Kuramochi-Miyagawa, T. Nakano, D. P. Bartel, and R. E. Kingston (2006).
Characterization of the piRNA complex from rat testes.
Science Signaling 313.5785, p. 363.

Liu, T., N. Lin, N. Shi, and B. Zhang (2009).
Information criterion-based clustering with order-restricted candidate profiles in short time-course microarray experiments.
BMC Bioinformatics 10.1, p. 146.

Liu, X. and M. C. K. Yang (2009).
Identifying temporally differentially expressed genes through functional principal components analysis.
Biostatistics 10.4, pp. 667–679.

Luan, Y. and H. Li (2003).
Clustering of Time-Course Gene Expression Data Using a Mixed-Effects Model with B-Splines.
Bioinformatics 19.4, pp. 474–482.

Ma, P., W. Zhong, and J. Liu (2009).
Identifying Differentially Expressed Genes in Time Course Microarray Data.
Statistics in Biosciences 1.2, pp. 144–159.

Magni, P., F. Ferrazzi, L. Sacchi, and R. Bellazzi (2008).
TimeClust: A Clustering Tool for Gene Expression Time Series.
Bioinformatics 24.3, pp. 430–432.

Matthews, L., G. Gopinath, M. Gillespie, M. Caudy, D. Croft, B. de Bono, P. Garapati, J. Hemish, H. Hermjakob, B. Jassal, A. Kanapin, S. Lewis, S. Mahajan, B. May, E. Schmidt, I. Vastrik, G. Wu, E. Birney, L. Stein, and P. D'Eustachio (2009).
Reactome knowledgebase of human biological pathways and processes.
Nucleic Acids Research 37.Database, pp. D619–D622.

Miller, M. B. and Y.-W. Tang (2009).
Basic Concepts of Microarrays and Potential Applications in Clinical Microbiology.
Clinical Microbiology Reviews 22.4, pp. 611–633.

Minguez, P., F. Al-Shahrour, and J. Dopazo (2006).
A function-centric approach to the biological interpretation of microarray time-series.
Genome Informatics 17.2, pp. 57–66.

Nueda, M. J., A. Conesa, J. A. Westerhuis, H. C. J. Hoefsloot, A. K. Smilde, M. Talón, and A. Ferrer (2007).
Discovering Gene Expression Patterns in Time Course Microarray Experiments by ANOVA–SCA.
Bioinformatics 23.14, pp. 1792–1800.

Nueda, M., P. Sebastián, S. Tarazona, F. García-García, J. Dopazo, A. Ferrer, and A. Conesa (2009).
Functional assessment of time course microarray data.
BMC Bioinformatics 10.Suppl 6, S9.

Opgen-Rhein, R. and K. Strimmer (2007).
Accurate Ranking of Differentially Expressed Genes by a Distribution-Free Shrinkage Approach.
Statistical Applications in Genetics and Molecular Biology 6.1, pp. 1–18.

Park, T., S.-G. Yi, S. Lee, S. Lee, D.-H. Yoo, J.-I. Ahn, and Y.-S. Lee (2003).
Statistical tests for identifying differentially expressed genes in time-course microarray experiments.
Bioinformatics 19.6, pp. 694–703.

Parkinson, H., M. Kapushesky, N. Kolesnikov, G. Rustici, M. Shojatalab, N. Abeygunawardena, H. Berube, M. Dylag, I. Emam, A. Farne, et al. (2009).
ArrayExpress update–from an archive of functional genomics experiments to the atlas of gene expression.
Nucleic Acids Research 37.Database, pp. D868–D872.

Peddada, S. D., E. K. Lobenhofer, L. Li, C. A. Afshari, C. R. Weinberg, and D. M. Umbach (2003).
Gene Selection and Clustering for Time-Course and Dose–response Microarray Experiments Using Order-Restricted Inference.
Bioinformatics 19.7, pp. 834–841.

R Development Core Team (2012).
R: A Language and Environment for Statistical Computing.
Vienna, Austria.

Ramakrishnan, N., S. Tadepalli, L. T. Watson, R. F. Helm, M. Antoniotti, and B. Mishra (2010).
Reverse Engineering Dynamic Temporal Models of Biological Processes and Their Relationships.
Proceedings of the National Academy of Sciences 107.28, pp. 12511–12516.

Ramoni, M. F., P. Sebastiani, and I. S. Kohane (2002).
Cluster Analysis of Gene Expression Dynamics.
Proceedings of the National Academy of Sciences 99.14, pp. 9121–9126.

Rivals, I., L. Personnaz, L. Taing, and M.-C. Potier (2007).
Enrichment or Depletion of a GO Category Within a Class of Genes: Which Test?
Bioinformatics 23.4, pp. 401–407.

Sacchi, L., R. Bellazzi, C. Larizza, P. Magni, T. Curk, U. Petrovic, and B. Zupan (2005).
TA-clustering: Cluster analysis of gene expression profiles through Temporal Abstractions.
International Journal of Medical Informatics 74.7–8, pp. 505–517.

Sahoo, D., D. L. Dill, R. Tibshirani, and S. K. Plevritis (2007).
Extracting Binary Signals from Microarray Time-Course Data.
Nucleic Acids Research 35.11, pp. 3705–3712.

Šášik, R., N. Iranfar, T. Hwa, and W. F. Loomis (2002).
Extracting Transcriptional Events from Temporal Gene Expression Patterns During Dictyostelium Development.
Bioinformatics 18.1, pp. 61–66.

Scharl, T., B. Grün, and F. Leisch (2010).
Mixtures of Regression Models for Time Course Gene Expression Data: Evaluation of Initialization and Random Effects.
Bioinformatics 26.3, pp. 370–377.

Schliep, A., I. G. Costa, C. Steinhoff, and A. Schonhuth (2005).
Analyzing gene expression time-courses.
IEEE/ACM Transactions on Computational Biology and Bioinformatics 2.3, pp. 179–193.

Schliep, A., A. Schonhuth, and C. Steinhoff (2003).
Using hidden Markov models to analyze gene expression time course data.
Bioinformatics 19.Suppl 1, pp. i255–i263.

Segal, E., M. Shapira, A. Regev, D. Pe'er, D. Botstein, D. Koller, and N. Friedman (2003).
Module networks: identifying regulatory modules and their condition-specific regulators from gene expression data.
Nature Genetics 34.2, pp. 166–176.

Al-Shahrour, F., J. Carbonell, P. Minguez, S. Goetz, A. Conesa, J. Tarrraga, I. Medina, E. Alloza, D. Montaner, and J. Dopazo (2008).
Babelomics: advanced functional profiling of transcriptomics, proteomics and genomics experiments.
Nucleic Acids Research 36, W341–W346.

Al-Shahrour, F., R. Díaz-Uriarte, and J. Dopazo (2005).
Discovering Molecular Functions Significantly Related to Phenotypes by Combining Gene Expression Data and Biological Information.
Bioinformatics 21.13, pp. 2988–2993.

Sharan, R., R. Shamir, et al. (2000).
"CLICK: a clustering algorithm with applications to gene expression analysis".
In: *Proceedings of the International Conference on Intelligent Systems for Molecular Biology*.
Vol. 8,
307–316.

Sherman, B. T., D. W. Huang, Q. Tan, Y. Guo, S. Bour, D. Liu, R. Stephens, M. W. Baseler, H. C. Lane, and R. A. Lempicki (2007).
DAVID Knowledgebase: a gene-centered database integrating heterogeneous gene annotation resources to facilitate high-throughput gene functional analysis.
BMC Bioinformatics 8, p. 426.

Sima, C., J. Hua, and S. Jung (2009).
Inference of Gene Regulatory Networks Using Time-Series Data: A Survey.
Current Genomics 10.6, pp. 416–429.

Smyth, G. (2005).
Limma: Linear Models for Microarray Data.
In: *Bioinformatics and Computational Biology Solutions Using R and Bioconductor*.
Statistics for Biology and Health.
Springer New York,
Pp. 397–420.

Smyth, G. K. (2004).
Linear Models and Empirical Bayes Methods for Assessing Differential Expression in Microarray Experiments.
Statistical Applications in Genetics and Molecular Biology 3.1, pp. 1–25.

Springer, T., K. Ickstadt, and J. Stoeckler (2011).
Frame potential minimization for clustering short time series.
Advances in Data Analysis and Classification 5.4, pp. 341–355.

Storey, J. D. and R. Tibshirani (2003).
Statistical significance for genomewide studies.
Proceedings of the National Academy of Sciences 100.16, pp. 9440–9445.

Storey, J. D., W. Xiao, J. T. Leek, R. G. Tompkins, and R. W. Davis (2005).
Significance analysis of time course microarray experiments.
Proceedings of the National Academy of Sciences of the United States of America
102.36, pp. 12837–12842.

Subramanian, A., P. Tamayo, V. K. Mootha, S. Mukherjee, B. L. Ebert, M. A. Gillette, A. Paulovich, S. L. Pomeroy, T. R. Golub, E. S. Lander, and J. P. Mesirov (2005).
Gene set enrichment analysis: A knowledge-based approach for interpreting genome-wide expression profiles.
Proceedings of the National Academy of Sciences of the United States of America
102.43, pp. 15545–15550.

Tamayo, P., D. Slonim, J. Mesirov, Q. Zhu, S. Kitareewan, E. Dmitrovsky, E. S. Lander, and T. R. Golub (1999).
Interpreting Patterns of Gene Expression with Self-Organizing Maps: Methods and Application to Hematopoietic Differentiation.
Proceedings of the National Academy of Sciences 96.6, pp. 2907–2912.

Tavazoie, S., J. D. Hughes, M. J. Campbell, R. J. Cho, and G. M. Church (1999).
Systematic determination of genetic network architecture.
Nature Genetics 22.3, pp. 281–285.

Tchagang, A. B., K. V. Bui, T. McGinnis, and P. V. Benos (2009).
Extracting biologically significant patterns from short time series gene expression data.
BMC Bioinformatics 10.1, p. 255.

Törönen, P., P. Pehkonen, and L. Holm (2009).
Generation of Gene Ontology benchmark datasets with various types of positive signal.
BMC Bioinformatics 10.1, p. 319.

Turchin, A., C. Z. Guo, G. K. Adler, V. Ricchiuti, I. S. Kohane, and G. H. Williams (2006).
Effect of acute aldosterone administration on gene expression profile in the heart.
Endocrinology 147.7, pp. 3183–3189.

Tusher, V. G., R. Tibshirani, and G. Chu (2001).
Significance analysis of microarrays applied to the ionizing radiation response.
Proceedings of the National Academy of Sciences 98.9, pp. 5116–5121.

Vastrik, I., P. D'Eustachio, E. Schmidt, G. Joshi-Tope, G. Gopinath, D. Croft, B. de Bono, M. Gillespie, B. Jassal, S. Lewis, L. Matthews, G. Wu, E. Birney, and L. Stein (2007).
Reactome: a knowledge base of biologic pathways and processes.
Genome Biology 8.3, R39.

Wang, L., X. Chen, R. Wolfinger, J. Franklin, R. Coffey, and B. Zhang (2009).
A Unified Mixed Effects Model for Gene Set Analysis of Time Course Microarray Experiments.
Statistical Applications in Genetics and Molecular Biology 8.1.

Wang, L., G. Chen, and H. Li (2007).
Group SCAD regression analysis for microarray time course gene expression data.
Bioinformatics 23.12, pp. 1486–1494.

Wang, L., M. F. Ramoni, and P. Sebastiani (2006).
Clustering Short Gene Expression Profiles.
In: *Research in Computational Molecular Biology*.
Vol. 3909.
Lecture Notes in Computer Science.
Springer Berlin / Heidelberg,
Pp. 60–68.

Wang, X., E. Dalkic, M. Wu, and C. Chan (2008).
Gene-module level analysis: Identification to Networks and Dynamics.
Current Opinion in Biotechnology 19.5, pp. 482–491.

Wang, X., M. Wu, Z. Li, and C. Chan (2008).
Short time-series microarray analysis: Methods and challenges.
BMC Systems Biology 2.1, p. 58.

Wei, Z. and H. Li (2008).
A Hidden Spatial-Temporal Markov Random Field Model for Network-Based Analysis of Time Course Gene Expression Data.
The Annals of Applied Statistics 2.1, pp. 408–429.

Wu, Z., R. Irizarry, R. Gentleman, F. Martinez-Murillo, and F. Spencer (2004).
A model-based background adjustment for oligonucleotide expression arrays.
Journal of the American Statistical Association 99.468, pp. 909–917.

Xu, X. L., J. M. Olson, and L. P. Zhao (2002).
A Regression-Based Method to Identify Differentially Expressed Genes in Microarray Time Course Studies and Its Application in an Inducible Huntington's Disease Transgenic Model.
Human Molecular Genetics 11.17, pp. 1977–1985.

Yao, F., H. Muller, and J. Wang (2005).
Functional data analysis for sparse longitudinal data.
Journal of the American Statistical Association 100.470, pp. 577–590.

Yeung, K. Y., C. Fraley, A. Murua, A. E. Raftery, and W. L. Ruzzo (2001).
Model-Based Clustering and Data Transformations for Gene Expression Data.
Bioinformatics 17.10, pp. 977–987.

Yoneya, T. and H. Mamitsuka (2007).
A Hidden Markov Model-Based Approach for Identifying Timing Differences in Gene Expression Under Different Experimental Factors.
Bioinformatics 23.7, pp. 842–849.

Yu, H., F. Wang, K. Tu, L. Xie, Y.-Y. Li, and Y.-X. Li (2007).
Transcript-level annotation of Affymetrix probesets improves the interpretation of gene expression data.
BMC Bioinformatics 8, p. 194.

Yuan, M. and C. Kendziorski (2006).
Hidden Markov Models for Microarray Time Course Data in Multiple Biological Conditions.
Journal of the American Statistical Association 101.476, pp. 1323–1332.

Zhang, K., H. Wang, A. C. Bathke, S. W. Harrar, H.-P. Piepho, and Y. Deng (2011).
Gene set analysis for longitudinal gene expression data.
BMC Bioinformatics 12.1, p. 273.

List of Figures

2.1	Sketch of the main areas of bioinformatics.	7
2.2	Construction and workflow of an Affymetrix gene expression microarray experiment.	10
2.3	Number of time points and replicates for time series experiments on Affymetrix arrays in the ArrayExpress data base.	13
2.4	Most common types of single gene time series data.	18
2.5	Possible types of references for the proposed methods.	18
4.1	Scheme of the three proposed profile algorithms.	33
5.1	Simulation algorithm flow chart of the simulation for the comparison of the competing profile algorithms.	68
5.2	Flow chart of the simulation for the validation of profile smoothing algorithms.	77
5.3	Heat map of activation algorithm comparison across all simulation studies.	100
5.4	Hexbin Plot of GSP for FS smoothing in dependence of λ_{fill}.	106
5.5	Main text accuracy plots for AH data, 2T-GSA profile algorithm and AM smoothing.	110
5.6	Main text accuracy plots for OD data, 2S-GSA profile algorithm and IN smoothing.	114
5.7	Main text accuracy plots for SH data, 2T-GSA profile algorithm and GM smoothing.	117
5.8	Main text accuracy plots for TH data, 1S-GSA profile algorithm and AM smoothing.	122
5.9	Heat map of smoothing algorithm performance in all simulation studies.	124
5.10	Hexbin Plots of GSP in dependence of λ_{fill} and λ_{wipe} for AM and $\alpha_{\text{sets}} = 0.01$.	127

6.1 Applied filtering scheme for microarray features on Affymetrix mouse 430 2.0 microarray. 138
6.2 Probe set venn diagram for the four used exemplary data sets after filtering procedure. 140
6.3 Gene set sizes boxplots of significant profiles. 142
6.4 Scatter plots of gene set size and $D_s^{|med|}$ score. 144
6.5 Total numbers of differentially expressed genes for the aldosterone heart data set. 146
6.6 Total numbers of differentially expressed genes for the ovary development data set. 150
6.7 Total numbers of differentially expressed genes for the skin healing data set. 157
6.8 Total numbers of differentially expressed genes for the tongue healing data set. 162

7.1 Example for the steps of the determination of a gene set activation profile (→:0031649). 174

B.1 Hexbin Plot of the general smoothing proportion (GSP) in dependence of the smoothing parameter λ_{fill} for the arithmetic mean smoothing and the four used data sets.. 215
B.2 Hexbin Plot of the general smoothing proportion (GSP) in dependence of the smoothing parameter λ_{fill} for the geometric mean smoothing and the four used data sets. 216
B.3 Hexbin Plot of the general smoothing proportion (GSP) in dependence of the smoothing parameter λ_{fill} for the inverse normal score smoothing and the four used data sets. 216
B.4 Hexbin Plot of the general smoothing proportion (GSP) in dependence of the smoothing parameter λ_{fill} for the inverse χ^2 score smoothing and the four used data sets. 216
B.5 Hexbin Plot of the general smoothing proportion (GSP) in dependence of the smoothing parameter λ_{fill} for the undirected sequential difference smoothing and the four used data sets. 217
B.6 Hexbin Plot of the general smoothing proportion (GSP) in dependence of the smoothing parameter λ_{fill} for the directed sequential difference smoothing and the four used data sets. 217
B.7 Hexbin Plot of the general smoothing proportion (GSP) in dependence of the smoothing parameter λ_{fill} for the distance to significance smoothing and the four used data sets. 217

B.8 Hexbin Plot of the general smoothing proportion (GSP) in dependence of the smoothing parameter λ_{fill} for the shift in contingency table smoothing and the four used data sets. 218

B.9 Hexbin Plot of the general smoothing proportion (GSP) in dependence of the smoothing parameter λ_{wipe} for the arithmetic mean smoothing and the four used data sets. 219

B.10 Hexbin Plot of the general smoothing proportion (GSP) in dependence of the smoothing parameter λ_{wipe} for the geometric mean smoothing and the four used data sets. 219

B.11 Hexbin Plot of the general smoothing proportion (GSP) in dependence of the smoothing parameter λ_{wipe} for the inverse normal score smoothing and the four used data sets. 219

B.12 Hexbin Plot of the general smoothing proportion (GSP) in dependence of the smoothing parameter λ_{wipe} for the inverse χ^2 score smoothing and the four used data sets. 220

B.13 Hexbin Plot of the general smoothing proportion (GSP) in dependence of the smoothing parameter λ_{wipe} for the undirected sequential difference smoothing and the four used data sets. 220

B.14 Hexbin Plot of the general smoothing proportion (GSP) in dependence of the smoothing parameter λ_{wipe} for the directed sequential difference smoothing and the four used data sets. 220

B.15 Hexbin Plot of the general smoothing proportion (GSP) in dependence of the smoothing parameter λ_{wipe} for the distance to significance smoothing and the four used data sets. 221

B.16 Hexbin Plot of the general smoothing proportion (GSP) in dependence of the smoothing parameter λ_{wipe} for the shift in contingency table smoothing and the four used data sets. 221

B.17 Accuracy Plots with data: AH, profile algorithm: 2T-GSA, smoothing algorithm: AM. Detailed figure description at section start on page 228. 229

B.18 Accuracy Plots with data: AH, profile algorithm: 2T-GSA, smoothing algorithm: GM. Detailed figure description at section start on page 228. 229

B.19 Accuracy Plots with data: AH, profile algorithm: 2T-GSA, smoothing algorithm: IN. Detailed figure description at section start on page 228. 230

B.20 Accuracy Plots with data: AH, profile algorithm: 2T-GSA, smoothing algorithm: IX. Detailed figure description at section start on page 228. 230

B.21 Accuracy Plots with data: AH, profile algorithm: 2T-GSA, smoothing algorithm: FD. Detailed figure description at section start on page 228. 230

B.22 Accuracy Plots with data: AH, profile algorithm: 2T-GSA, smoothing algorithm: FS. Detailed figure description at section start on page 228. 231

B.23 Accuracy Plots with data: AH, profile algorithm: 2T-GSA, smoothing algorithm: SU. Detailed figure description at section start on page 228. . . . 231
B.24 Accuracy Plots with data: AH, profile algorithm: 1S-GSA, smoothing algorithm: SD. Detailed figure description at section start on page 228. . . . 231
B.25 Accuracy Plots with data: AH, profile algorithm: 1S-GSA, smoothing algorithm: AM. Detailed figure description at section start on page 228. . . . 232
B.26 Accuracy Plots with data: AH, profile algorithm: 1S-GSA, smoothing algorithm: GM. Detailed figure description at section start on page 228. . . . 232
B.27 Accuracy Plots with data: AH, profile algorithm: 1S-GSA, smoothing algorithm: IN. Detailed figure description at section start on page 228. . . . 232
B.28 Accuracy Plots with data: AH, profile algorithm: 1S-GSA, smoothing algorithm: IX. Detailed figure description at section start on page 228. . . . 233
B.29 Accuracy Plots with data: AH, profile algorithm: 2S-GSA, smoothing algorithm: AM. Detailed figure description at section start on page 228. . . . 233
B.30 Accuracy Plots with data: AH, profile algorithm: 2S-GSA, smoothing algorithm: GM. Detailed figure description at section start on page 228. . . . 233
B.31 Accuracy Plots with data: AH, profile algorithm: 2S-GSA, smoothing algorithm: IN. Detailed figure description at section start on page 228. . . . 234
B.32 Accuracy Plots with data: AH, profile algorithm: 2S-GSA, smoothing algorithm: IX. Detailed figure description at section start on page 228. . . . 234
B.33 Accuracy Plots with data: OD, profile algorithm: 2T-GSA, smoothing algorithm: AM. Detailed figure description at section start on page 228. . . . 235
B.34 Accuracy Plots with data: OD, profile algorithm: 2T-GSA, smoothing algorithm: GM. Detailed figure description at section start on page 228. . . . 235
B.35 Accuracy Plots with data: OD, profile algorithm: 2T-GSA, smoothing algorithm: IN. Detailed figure description at section start on page 228. . . . 236
B.36 Accuracy Plots with data: OD, profile algorithm: 2T-GSA, smoothing algorithm: IX. Detailed figure description at section start on page 228. . . . 236
B.37 Accuracy Plots with data: OD, profile algorithm: 2T-GSA, smoothing algorithm: FD. Detailed figure description at section start on page 228. . . . 236
B.38 Accuracy Plots with data: OD, profile algorithm: 2T-GSA, smoothing algorithm: FS. Detailed figure description at section start on page 228. . . . 237
B.39 Accuracy Plots with data: OD, profile algorithm: 2T-GSA, smoothing algorithm: SU. Detailed figure description at section start on page 228. . . . 237
B.40 Accuracy Plots with data: OD, profile algorithm: 1S-GSA, smoothing algorithm: SD. Detailed figure description at section start on page 228. . . . 237
B.41 Accuracy Plots with data: OD, profile algorithm: 1S-GSA, smoothing algorithm: AM. Detailed figure description at section start on page 228. . . . 238

B.42 Accuracy Plots with data: OD, profile algorithm: 1S-GSA, smoothing algorithm: GM. Detailed figure description at section start on page 228.238
B.43 Accuracy Plots with data: OD, profile algorithm: 1S-GSA, smoothing algorithm: IN. Detailed figure description at section start on page 228.238
B.44 Accuracy Plots with data: OD, profile algorithm: 1S-GSA, smoothing algorithm: IX. Detailed figure description at section start on page 228.239
B.45 Accuracy Plots with data: OD, profile algorithm: 2S-GSA, smoothing algorithm: AM. Detailed figure description at section start on page 228.239
B.46 Accuracy Plots with data: OD, profile algorithm: 2S-GSA, smoothing algorithm: GM. Detailed figure description at section start on page 228.239
B.47 Accuracy Plots with data: OD, profile algorithm: 2S-GSA, smoothing algorithm: IN. Detailed figure description at section start on page 228.240
B.48 Accuracy Plots with data: OD, profile algorithm: 2S-GSA, smoothing algorithm: IX. Detailed figure description at section start on page 228.240
B.49 Accuracy Plots with data: SH, profile algorithm: 2T-GSA, smoothing algorithm: AM. Detailed figure description at section start on page 228.241
B.50 Accuracy Plots with data: SH, profile algorithm: 2T-GSA, smoothing algorithm: GM. Detailed figure description at section start on page 228.241
B.51 Accuracy Plots with data: SH, profile algorithm: 2T-GSA, smoothing algorithm: IN. Detailed figure description at section start on page 228.242
B.52 Accuracy Plots with data: SH, profile algorithm: 2T-GSA, smoothing algorithm: IX. Detailed figure description at section start on page 228.242
B.53 Accuracy Plots with data: SH, profile algorithm: 2T-GSA, smoothing algorithm: FD. Detailed figure description at section start on page 228.242
B.54 Accuracy Plots with data: SH, profile algorithm: 2T-GSA, smoothing algorithm: FS. Detailed figure description at section start on page 228.243
B.55 Accuracy Plots with data: SH, profile algorithm: 2T-GSA, smoothing algorithm: SU. Detailed figure description at section start on page 228.243
B.56 Accuracy Plots with data: SH, profile algorithm: 1S-GSA, smoothing algorithm: SD. Detailed figure description at section start on page 228.243
B.57 Accuracy Plots with data: SH, profile algorithm: 1S-GSA, smoothing algorithm: AM. Detailed figure description at section start on page 228.244
B.58 Accuracy Plots with data: SH, profile algorithm: 1S-GSA, smoothing algorithm: GM. Detailed figure description at section start on page 228.244
B.59 Accuracy Plots with data: SH, profile algorithm: 1S-GSA, smoothing algorithm: IN. Detailed figure description at section start on page 228.244
B.60 Accuracy Plots with data: SH, profile algorithm: 1S-GSA, smoothing algorithm: IX. Detailed figure description at section start on page 228.245

B.61 Accuracy Plots with data: SH, profile algorithm: 2S-GSA, smoothing algorithm: AM. Detailed figure description at section start on page 228.245
B.62 Accuracy Plots with data: SH, profile algorithm: 2S-GSA, smoothing algorithm: GM. Detailed figure description at section start on page 228.245
B.63 Accuracy Plots with data: SH, profile algorithm: 2S-GSA, smoothing algorithm: IN. Detailed figure description at section start on page 228.246
B.64 Accuracy Plots with data: SH, profile algorithm: 2S-GSA, smoothing algorithm: IX. Detailed figure description at section start on page 228.246
B.65 Accuracy Plots with data: TH, profile algorithm: 2T-GSA, smoothing algorithm: AM. Detailed figure description at section start on page 228.247
B.66 Accuracy Plots with data: TH, profile algorithm: 2T-GSA, smoothing algorithm: GM. Detailed figure description at section start on page 228.247
B.67 Accuracy Plots with data: TH, profile algorithm: 2T-GSA, smoothing algorithm: IN. Detailed figure description at section start on page 228.248
B.68 Accuracy Plots with data: TH, profile algorithm: 2T-GSA, smoothing algorithm: IX. Detailed figure description at section start on page 228.248
B.69 Accuracy Plots with data: TH, profile algorithm: 2T-GSA, smoothing algorithm: FD. Detailed figure description at section start on page 228.248
B.70 Accuracy Plots with data: TH, profile algorithm: 2T-GSA, smoothing algorithm: FS. Detailed figure description at section start on page 228.249
B.71 Accuracy Plots with data: TH, profile algorithm: 2T-GSA, smoothing algorithm: SU. Detailed figure description at section start on page 228.249
B.72 Accuracy Plots with data: TH, profile algorithm: 1S-GSA, smoothing algorithm: SD. Detailed figure description at section start on page 228.249
B.73 Accuracy Plots with data: TH, profile algorithm: 1S-GSA, smoothing algorithm: AM. Detailed figure description at section start on page 228.250
B.74 Accuracy Plots with data: TH, profile algorithm: 1S-GSA, smoothing algorithm: GM. Detailed figure description at section start on page 228.250
B.75 Accuracy Plots with data: TH, profile algorithm: 1S-GSA, smoothing algorithm: IN. Detailed figure description at section start on page 228.250
B.76 Accuracy Plots with data: TH, profile algorithm: 1S-GSA, smoothing algorithm: IX. Detailed figure description at section start on page 228.251
B.77 Accuracy Plots with data: TH, profile algorithm: 2S-GSA, smoothing algorithm: AM. Detailed figure description at section start on page 228.251
B.78 Accuracy Plots with data: TH, profile algorithm: 2S-GSA, smoothing algorithm: GM. Detailed figure description at section start on page 228.251
B.79 Accuracy Plots with data: TH, profile algorithm: 2S-GSA, smoothing algorithm: IN. Detailed figure description at section start on page 228.252

B.80 Accuracy Plots with data: TH, profile algorithm: 2S-GSA, smoothing algorithm: IX. Detailed figure description at section start on page 228.252

List of Tables

2.1	Gene Expression technologies and the frequency of their application for transcription profiling and time series expression experiments. .	8
4.1	Contingency table for Fisher's exact test in the gene set enrichment analysis.	37
4.2	Possible results of multiple testing for significance.	39
4.3	Subprofiles considered for smoothing in alternating activation profiles.	47
5.1	Totals of available prototype gene sets per data set for the simulation study, which compares the five profile algorithms.	69
5.2	TPR and FDR for identifying spiked-in activation profiles in the simulation to compare the five profile algorithms on basis of the AH data.	83
5.3	TPR and FDR for identifying spiked-in activation profiles in the simulation to compare the five profile algorithms on basis of the OD data.	85
5.4	TPR per algorithm and type of spiked-in profile in the simulation to compare the five profile algorithms on basis of the OD data.	86
5.5	TPR and FDR for identifying spiked-in activation profiles in the simulation to compare the five profile algorithms on basis of the SH data.	90
5.6	TPR per algorithm and type of spiked-in profile in the simulation to compare the five profile algorithms on basis of the SH data.	91
5.7	TPR and FDR for identifying spiked-in activation profiles in the simulation to compare the five profile algorithms on basis of the TH data.	95
5.8	TPR per algorithm and type of spiked-in profile in the simulation to compare the five profile algorithms on basis of the TH data.	96

5.9	TPR for exact gene set activation profile matching from the simulation study comparing the five profile algorithms.	102
5.10	Preset activation profile per data set used in the simulation study to validate the smoothing algorithms.	104
5.11	Best smoothing parameters in aldosterone heart simulation.	111
5.12	Best smoothing parameters in ovary development simulation.	113
5.13	Best smoothing parameters in skin healing simulation.	118
5.14	Best smoothing parameters in tongue healing simulation.	121
5.15	Smoothing methods and parameters chosen based on simulation results.	125
5.16	Smoothing parameters used in application.	126
5.17	Coefficients of the smoothing parameter generating functions. . . .	133
6.1	Summary of the four data sets used for the estimation of temporal activation profiles of gene sets.	136
6.2	Cross-table of all applied profile algorithms on all data sets.	141
6.3	Gene sets with significant activation profile according to application data set and gene set definition.	143
6.4	Activation profile frequency in AH data set (smoothed 2S-GSA). . .	147
6.5	List of the top ranked gene set activation profiles resulting from a 2S-GSA profile algorithm combined with AM smoothing on the AH data.	148
6.6	Activation profile frequency in OD data set (smoothed 2S-GSA). . .	151
6.7	List of the top ranked gene set activation profiles resulting from a 2S-GSA profile algorithm combined with AM smoothing on the OD data.	154
6.8	Activation profile frequency in SH data set (smoothed 2T-GSA). . .	158
6.9	List of the top ranked gene set activation profiles resulting from a 2T-GSA profile algorithm combined with AM smoothing on the SH data.	160
6.10	Activation profile frequency in TH data set (smoothed 2S-GSA). . .	163
6.11	List of the top ranked gene set activation profiles resulting from a 2S-GSA profile algorithm combined with AM smoothing on the TH data.	164
A.1	Example for the q-value calculation in the TS-ABH FDR procedure.	205
B.1	TPR per algorithm and type of spiked in profile in the simulation to compare the five profile algorithms on basis of the AH data.	210

B.2   ACC and NPV for identifying spiked in activation profiles in the simulation to compare the five profile algorithms on basis of the AH data. 211
B.3   ACC and NPV for identifying spiked in activation profiles in the simulation to compare the five profile algorithms on basis of the OD data. 212
B.4   ACC and NPV for identifying spiked in activation profiles in the simulation to compare the five profile algorithms on basis of the SH data. 213
B.5   ACC and NPV for identifying spiked in activation profiles in the simulation to compare the five profile algorithms on basis of the TH data. 214
B.6   Smoothing parameters used in simulation study to validate smoothing methods 222
B.7   All 105 significant gene set activation profiles resulting from a 2S-GSA profile algorithm combined with AM smoothing on the AH data. . . 253
B.8   Top 100 significant gene set activation profiles according to the $D_s^{|\text{med}|}$ score resulting from a 2S-GSA profile algorithm combined with AM smoothing on the OD data. The complete list is available (in electronic version only) 258
B.9   All 50 significant gene set activation profiles according to the $D_s^{|\text{med}|}$ score resulting from a 2S-GSA profile algorithm combined with AM smoothing on the OD data. 264
B.10 Top 100 significant gene set activation profiles according to the $D_s^{|\text{med}|}$ score resulting from a 2S-GSA profile algorithm combined with AM smoothing on the OD data. The complete list is available (in electronic version only) 267

APPENDIX A

FDR q-value in the Two-Stage Adaptive Benjamini Hochberg Linear Step-up Procedure

The calculation of the FDR q-value in the two-stage adaptive linear step-up procedure of Benjamini, Krieger, and Yekutieli (2006) (TS-ABH) is presented in the following. The value $q_{(i)}$ is defined as the smallest FDR bound, which is controlled by the two-stage-procedure while rejecting the i-th smallest p-value.

Table A.1 shows the calculation of the FDR q-values $q_{(i)}$ for an example with $n = 10$ tests. The number of rejected hypotheses in linear step-up procedure of the

Table A.1: Example for the q-value calculation in the TS-ABH FDR procedure.

(i)	$p_{(i)}$	1st stage			2nd stage		$q_{(i)}$
		$\tilde{q}^{I}_{(i)} = p_{(i)}n/i$	$\tilde{q}^{I}_{(i)}$	$k^{I}_{(i)}$	$\tilde{q}^{II}_{(i)}$	$k^{II}_{(i)}$	
1	0.001	0.010	0.0101	1	0.0111	1	0.0101
2	0.006	0.030	0.0309	3	0.0429	4	0.0215
3	0.007	0.023	0.0239	3	0.0333	3	0.0215
4	0.015	0.038	0.0390	4	0.0625	4	0.0270
5	0.200	0.400	0.6667	5	0.8000	7	0.3158
6	0.400	0.667	1.0000	6	1.0000	10	0.6667
7	0.500	0.714	1.0000	7	1.0000	10	0.6667
8	0.800	1.000	1.0000	10	1.0000	10	0.9048
9	0.900	1.000	1.0000	10	1.0000	10	0.9048
10	0.950	0.950	1.0000	10	1.0000	10	0.9048

first stage controlling a FDR of $\tilde{q}$ is given by

$$k = \max\{i\colon p_{(i)} \leq \tilde{q}/(1+\tilde{q}) \cdot i/n\}. \tag{a}$$

This is equivalent with

$$k = \max\{i\colon p_{(i)}n/i \leq \tilde{q}/1+\tilde{q}\}$$

and hence the values $\tilde{q}^{I}_{(i)} = p_{(i)}n/i$ show the lowest critical values, which lead to a rejection of $H_0^{(i)}$ in the first stage controlling a FDR limit of

$$\tilde{q}_{(i)} = \tilde{r}(\tilde{q}^{I}_{(i)}) = (1 - \tilde{q}^{I}_{(i)})^{-1} - 1 = (1 - p_{(i)}n/i)^{-1} - 1.$$

Since the first stage uses $\tilde{q}^{I}_{(i)} = \tilde{q}_{(i)}/(1+\tilde{q}_{(i)})$ as critical FDR value and not $\tilde{q}_{(i)}$ the recalculation by function $\tilde{r}$ is needed. Applying the FDR limit of $\tilde{q}_{(i)}$ in the TS-ABH procedure leads to $k^{I}_{(i)}$ rejections in the first stage in Table A.1.

The $k^{I}_{(i)}$ is used to calculate a new critical value $\tilde{q}^{II}_{(i)}$ in the second stage linear step-up procedure as product of the old critical value and the quotient $(n-k^{I}_{(i)})/k^{I}_{(i)}$:

$$\tilde{q}^{II}_{(i)} = \begin{cases} 1, & \text{if } k^{I}_{(i)} = n \\ \min\left\{1, \tilde{q}^{I}_{(i)} \frac{n-k^{I}_{(i)}}{k^{I}_{(i)}}\right\}, & \text{else.} \end{cases}$$

The number of rejected hypotheses in the second stage controlling a FDR of $\tilde{q}_{(i)}$ is given with

$$k^{II}_{(i)} = \max\left\{j\colon \tilde{q}^{I}_{(j)} \leq \tilde{q}^{II}_{(i)}\right\}$$

or in words as the last critical value of the first stage below the ith critical value of the second step.

In Table A.1 on the preceding page, the smallest p-value $p_{(1)}$ leads to the critical value $\tilde{q}^{I}_{(1)} = 0.010$, which is the smallest critical value in the first stage. Hence, $k^{I}_{(1)} = 1$ and the second step would allow a new critical value of $\tilde{q}^{II}_{(1)} = 0.0\overline{1}$. There are no other critical values of the first step, which are smaller than $\tilde{q}^{II}_{(1)}$. Therefore $k^{II}_{(1)} = 1$ and the corresponding FDR q-value of the TS-ABH procedure is $q_{(1)} = \tilde{q}_{(1)} = 0.0\overline{1}$.

The first stage critical value of the third hypothesis is smaller than the corresponding value for the second hypothesis. The step up procedure rejects the hypotheses with the increasing p-value, fulfilling condition in equation (a), hence both hypothesis can be rejected controlling a FDR of $\tilde{q}_{(3)} = 0.0239$. Actually, the minimal controlled

FDR bound, which allows to reject the third (and second) hypothesis, is even smaller and occurs when the first stage allows one ($k^I_{(1)} = 1$), but the second stage three rejections ($k^{II}_{(1)} = 3$). This value is calculated from $\tilde{q}^I_{(3)}$ and $k^I_{(1)}$ by

$$q_{(3)} = \tilde{r}\left(\frac{(n-k^I_{(1)})\tilde{q}^I_{(3)}}{n}\right) = \left(1 - \frac{(n-k^I_{(1)})\tilde{q}^I_{(3)}}{n}\right)^{-1} - 1 = 0.0215,$$

which leads to the smallest FDR bound controlled while rejecting the first hypothesis in the first stage and the next two hypotheses in the second stage. Analogously, all other q-values can be determined. Generally, the q-value can be computed by

$$q_{(i)} = \begin{cases} \min\{1, \min_j\{\tilde{r}\left(q^I_{(j)}\right) : i \leq k^{II}_{(j)}\}\}, & \text{if } i = 1 \\ \min\{1, \min_j\{\tilde{r}\left(q^I_{(j)}\right) : i \leq k^{II}_{(j)}\}, \tilde{r}\left(\frac{(n-k^I_{(i-1)})\tilde{q}^I_{(i)}}{n}\right)\}, & \text{if } i > 1. \end{cases}$$

APPENDIX B

Figures and Tables

Detailed tables for the simulation to compare the five profile algorithms

The following page shows the detailed true positive rates for identifying the five types of spiked in gene set with active profile with each of the competing profile algorithms based on the AH simulation.

Thereafter, four pages show the overall ACC and NPV values for the five algorithms in the simulation study to compare the profile algorithms.

Table B.1: TPR per algorithm and type of spiked in profile in the simulation to compare the five profile algorithms on basis of the AH data.

Aldosterone heart data $\|s\|_{SIM}$ p_{as}		algorithm	TPR with $p_{ag} = 0.2$			TPR with $p_{ag} = 0.6$			TPR with $p_{ag} = 1$		
			2T-GSA 1S-GSA 2S-GSA STEM maSigFun			2T-GSA 1S-GSA 2S-GSA STEM maSigFun			2T-GSA 1S-GSA 2S-GSA STEM maSigFun		
10	0.05	2T-GSA	8.9	8.6	3.7	25.2	21.1	3.9	40.0	35.7	4.0
		1S-GSA	24.2	26.2	17.1	40.2	38.7	29.5	59.4	60.2	51.5
		2S-GSA	12.5	13.3	6.7	33.3	28.5	8.7	52.4	47.9	13.4
		STEM	0.0	0.0	0.0	0.0	0.0	0.0	0.0	0.0	0.6
		maSigFun	0.0	0.0	0.0	0.0	0.0	0.0	0.0	0.0	0.0
	0.1	2T-GSA	9.0	8.2	3.9	21.8	20.2	3.9	36.8	36.2	4.6
		1S-GSA	34.0	33.8	26.9	55.0	55.0	36.3	76.9	75.7	57.2
		2S-GSA	13.3	12.8	6.5	30.1	28.4	8.5	48.7	47.3	11.5
		STEM	0.0	0.0	0.1	0.0	0.0	0.0	0.0	0.0	0.6
		maSigFun	0.0	0.0	0.0	0.0	0.0	0.0	0.0	0.0	0.0
	0.25	2T-GSA	5.4	4.3	2.0	12.0	12.2	1.6	23.2	22.8	2.4
		1S-GSA	33.3	33.3	23.2	52.9	50.1	31.6	68.0	69.5	50.8
		2S-GSA	8.4	7.2	2.6	20.7	18.4	3.6	33.8	33.3	6.2
		STEM	0.0	0.0	0.0	0.0	0.0	0.2	0.0	0.0	0.5
		maSigFun	0.0	0.0	0.0	0.0	0.0	0.0	0.0	0.0	0.0
	0.5	2T-GSA	0.3	0.4	0.2	2.4	2.2	0.0	7.7	5.6	0.5
		1S-GSA	18.8	18.3	15.9	28.9	31.8	18.6	46.4	43.8	29.5
		2S-GSA	0.6	0.5	0.3	3.1	2.7	0.0	9.8	8.0	1.0
		STEM	0.0	0.0	0.0	0.0	0.0	0.3	0.0	0.0	1.0
		maSigFun	0.0	0.0	0.0	0.0	0.0	0.0	0.0	0.0	0.0
20	0.05	2T-GSA	11.4	11.2	5.9	36.2	38.2	7.2	67.3	59.4	6.9
		1S-GSA	27.9	28.9	23.8	59.7	58.5	43.8	83.5	81.8	74.7
		2S-GSA	11.9	11.5	6.9	37.5	39.0	9.0	68.0	60.3	11.1
		STEM	0.0	0.0	0.0	0.0	0.0	0.0	0.0	0.0	1.3
		maSigFun	0.0	0.0	0.0	0.0	0.0	0.0	0.0	0.0	0.0
	0.1	2T-GSA	12.6	10.3	6.4	37.5	33.4	7.7	60.3	58.0	5.6
		1S-GSA	33.4	36.0	28.4	62.9	63.7	51.6	87.7	86.1	80.1
		2S-GSA	13.1	12.3	7.2	39.1	34.4	10.6	64.1	62.7	10.7
		STEM	0.0	0.0	0.0	0.0	0.0	0.2	0.0	0.0	1.5
		maSigFun	0.0	0.0	0.0	0.0	0.0	0.0	0.0	0.0	0.0
	0.25	2T-GSA	7.2	8.8	3.4	27.8	24.1	3.6	49.6	43.6	4.6
		1S-GSA	25.8	26.2	21.5	55.5	51.9	40.9	81.0	80.3	66.1
		2S-GSA	9.0	10.1	4.7	32.5	28.7	6.5	54.9	49.8	8.0
		STEM	0.0	0.2	0.1	0.0	0.0	0.8	0.0	0.0	2.2
		maSigFun	0.0	0.0	0.0	0.0	0.0	0.0	0.0	0.0	0.0
	0.5	2T-GSA	2.2	1.8	1.1	9.8	10.6	1.0	21.0	17.7	0.9
		1S-GSA	31.6	31.4	23.2	52.8	49.0	34.0	72.7	71.1	60.4
		2S-GSA	3.5	2.8	1.4	14.3	14.4	2.2	30.0	25.1	5.1
		STEM	0.0	0.1	0.0	0.1	0.0	0.9	0.0	0.0	3.8
		maSigFun	0.0	0.0	0.0	0.0	0.0	0.0	0.0	0.0	0.0
50	0.05	2T-GSA	20.7	18.8	5.7	67.6	66.7	9.0	93.9	91.4	20.9
		1S-GSA	40.5	41.9	32.9	87.0	84.7	73.5	99.3	99.6	99.6
		2S-GSA	23.1	21.5	8.8	74.0	71.9	19.8	96.4	94.4	62.0
		STEM	0.0	0.0	0.2	0.0	0.0	2.3	0.0	0.0	6.7
		maSigFun	0.0	0.0	0.0	0.0	0.0	0.0	0.0	0.0	0.0
	0.1	2T-GSA	22.7	18.9	6.5	68.3	67.0	8.5	92.3	90.4	21.5
		1S-GSA	40.8	39.6	29.5	85.1	84.5	71.2	99.2	99.2	98.7
		2S-GSA	26.3	22.4	7.7	74.9	71.8	19.9	94.3	93.3	60.5
		STEM	0.0	0.0	0.3	0.0	0.0	1.3	0.0	0.0	4.9
		maSigFun	0.0	0.0	0.0	0.0	0.0	0.0	0.0	0.0	0.0
	0.25	2T-GSA	19.1	16.3	5.7	52.7	48.8	5.4	84.8	82.9	13.2
		1S-GSA	47.8	46.7	34.0	85.2	82.7	62.9	98.7	99.0	95.6
		2S-GSA	31.0	28.9	11.0	62.0	60.0	12.9	93.2	92.0	66.3
		STEM	0.0	0.1	0.3	0.0	0.0	2.9	0.0	0.0	5.4
		maSigFun	0.0	0.0	0.0	0.0	0.0	0.0	0.0	0.0	0.0
	0.5	2T-GSA	8.1	8.3	3.7	35.3	29.8	2.1	57.5	53.7	6.6
		1S-GSA	40.5	36.0	29.6	74.5	78.5	59.4	92.4	92.8	89.9
		2S-GSA	17.7	16.8	9.4	54.6	48.8	12.1	73.4	71.8	50.5
		STEM	0.2	0.0	0.4	0.0	0.0	3.7	0.0	0.0	7.8
		maSigFun	0.0	0.0	0.0	0.0	0.0	0.0	0.0	0.0	0.0
100	0.05	2T-GSA	27.9	29.7	8.0	89.0	87.9	24.8	99.7	99.8	100.0
		1S-GSA	52.0	54.6	40.8	97.2	97.2	87.3	100.0	100.0	100.0
		2S-GSA	41.3	41.0	15.1	94.4	92.9	84.1	99.9	100.0	100.0
		STEM	0.0	0.0	0.2	0.0	0.0	1.6	0.1	0.1	35.8
		maSigFun	0.0	0.0	0.0	0.0	0.0	0.0	0.0	0.0	0.0
	0.1	2T-GSA	32.9	31.8	7.3	90.7	88.7	33.8	99.7	99.7	100.0
		1S-GSA	55.1	53.1	40.6	98.2	97.1	90.4	100.0	100.0	100.0
		2S-GSA	41.4	41.1	14.1	94.3	94.3	87.0	99.8	99.9	100.0
		STEM	0.0	0.0	0.4	0.0	0.0	1.4	0.2	0.0	38.3
		maSigFun	0.0	0.0	0.0	0.0	0.0	0.0	0.0	0.0	0.0
	0.25	2T-GSA	26.7	27.4	6.8	81.2	79.7	22.2	97.4	99.1	100.0
		1S-GSA	59.5	61.3	43.0	97.6	96.9	88.6	100.0	100.0	100.0
		2S-GSA	36.4	37.9	11.7	89.8	87.8	75.7	99.0	99.6	100.0
		STEM	0.0	0.1	0.4	0.0	0.0	1.7	0.5	0.1	47.0
		maSigFun	0.0	0.0	0.0	0.0	0.0	0.0	0.0	0.0	0.0
	0.5	2T-GSA	20.2	17.4	4.9	59.5	58.7	11.8	82.4	87.0	95.5
		1S-GSA	46.1	43.0	34.0	87.5	87.2	70.7	99.1	99.5	100.0
		2S-GSA	36.4	30.1	12.8	79.6	76.0	68.3	93.5	94.2	100.0
		STEM	0.0	0.0	0.7	0.0	0.0	2.2	1.0	0.5	60.8
		maSigFun	0.0	0.0	0.0	0.0	0.0	0.0	0.0	0.0	0.0

Table B.2: ACC and NPV for identifying spiked in activation profiles in the simulation to compare the five profile algorithms on basis of the AH data.

		Aldosterone heart data			ACC per p_{ag}						NPV per p_{ag}			
s_{SIM}	p_{as}	algorithm	0.1	0.2	0.4	0.6	0.8	1	0.1	0.2	0.4	0.6	0.8	1
10	0.05	2T-GSA	91.74	91.99	92.29	92.67	93.08	93.45	95.05	95.17	95.35	95.66	95.97	96.17
		1S-GSA	79.96	80.09	80.62	81.14	81.92	82.65	95.13	95.32	95.65	96.13	96.71	97.38
		2S-GSA	89.65	89.80	90.37	90.75	91.27	91.75	95.09	95.24	95.54	95.90	96.31	96.66
		STEM	94.99	95.00	94.99	94.99	94.99	95.00	95.00	95.00	95.00	95.00	95.00	95.01
		maSigFun	95.00	95.00	95.00	95.00	95.00	95.00	95.00	95.00	95.00	95.00	95.00	95.00
	0.1	2T-GSA	87.30	87.62	88.29	88.59	89.43	90.01	90.16	90.34	90.71	91.13	91.67	92.18
		1S-GSA	71.33	72.22	73.51	75.06	76.69	77.91	90.25	90.98	91.98	93.20	94.53	95.93
		2S-GSA	85.62	86.15	87.03	87.48	88.62	89.14	90.20	90.52	91.02	91.65	92.45	93.02
		STEM	89.99	90.00	89.99	89.99	90.00	90.01	90.00	90.00	90.00	90.00	90.01	90.02
		maSigFun	90.00	90.00	90.00	90.00	90.00	90.00	90.00	90.00	90.00	90.00	90.00	90.00
	0.25	2T-GSA	74.19	74.56	75.52	76.08	77.44	78.14	75.18	75.39	76.03	76.39	77.40	77.95
		1S-GSA	65.04	65.95	69.33	71.98	75.68	78.89	76.01	76.95	79.19	81.51	84.71	87.16
		2S-GSA	73.84	74.13	75.66	76.66	78.67	79.47	75.30	75.56	76.55	77.32	78.78	79.52
		STEM	74.98	74.98	75.00	75.02	75.01	75.01	75.00	75.00	75.00	75.01	75.01	75.02
		maSigFun	75.00	75.00	75.00	75.00	75.00	75.00	75.00	75.00	75.00	75.00	75.00	75.00
	0.5	2T-GSA	50.03	50.10	50.32	50.73	51.35	52.27	50.02	50.05	50.16	50.37	50.69	51.16
		1S-GSA	50.82	52.48	56.28	58.63	62.15	65.80	50.48	51.46	53.77	55.25	57.57	60.41
		2S-GSA	50.05	50.18	50.42	50.93	51.88	53.08	50.03	50.09	50.21	50.47	50.96	51.59
		STEM	50.02	49.98	50.02	50.03	50.02	50.13	50.01	49.99	50.01	50.02	50.01	50.07
		maSigFun	50.00	50.00	50.00	50.00	50.00	50.00	50.00	50.00	50.00	50.00	50.00	50.00
20	0.05	2T-GSA	91.22	91.41	91.98	92.61	93.23	93.74	95.12	95.26	95.67	96.16	96.68	97.06
		1S-GSA	78.45	78.76	80.09	80.91	81.90	82.94	95.15	95.49	96.28	97.14	98.02	98.75
		2S-GSA	90.78	90.96	91.54	92.24	92.85	93.33	95.13	95.27	95.69	96.21	96.77	97.14
		STEM	94.99	94.99	95.00	95.00	95.02	95.02	95.00	95.00	95.00	95.00	95.02	95.02
		maSigFun	95.00	95.00	95.00	95.00	95.00	95.00	95.00	95.00	95.00	95.00	95.00	95.00
	0.1	2T-GSA	86.69	86.86	87.92	89.06	90.35	91.19	90.28	90.49	91.23	92.13	92.97	93.68
		1S-GSA	72.05	73.57	74.69	77.38	79.02	80.98	90.50	91.25	92.54	94.62	96.43	97.92
		2S-GSA	86.29	86.51	87.57	88.77	90.17	91.09	90.30	90.55	91.31	92.27	93.26	94.11
		STEM	90.00	89.98	89.98	89.99	90.01	90.04	90.00	90.00	90.00	90.00	90.02	90.04
		maSigFun	90.00	90.00	90.00	90.00	90.00	90.00	90.00	90.00	90.00	90.00	90.00	90.00
	0.25	2T-GSA	73.87	74.30	75.71	77.93	79.72	81.88	75.39	75.66	76.65	78.25	79.59	81.40
		1S-GSA	66.51	67.88	72.01	76.96	80.39	85.05	75.71	76.59	79.68	83.63	87.06	91.61
		2S-GSA	73.62	74.23	75.87	78.61	80.65	82.67	75.40	75.84	76.98	79.03	80.63	82.44
		STEM	74.98	75.00	74.98	75.02	75.02	75.11	74.99	75.01	75.00	75.04	75.03	75.12
		maSigFun	75.00	75.00	75.00	75.00	75.00	75.00	75.00	75.00	75.00	75.00	75.00	75.00
	0.5	2T-GSA	50.05	50.65	51.32	53.28	54.90	56.55	50.03	50.33	50.67	51.71	52.58	53.51
		1S-GSA	52.42	55.23	60.08	66.02	71.98	79.13	51.56	53.42	56.80	61.32	66.50	73.85
		2S-GSA	50.07	50.95	52.02	54.68	57.20	59.93	50.03	50.48	51.04	52.48	53.90	55.53
		STEM	49.98	50.02	50.03	50.10	50.28	50.53	49.99	50.01	50.02	50.05	50.14	50.27
		maSigFun	50.00	50.00	50.00	50.00	50.00	50.00	50.00	50.00	50.00	50.00	50.00	50.00
50	0.05	2T-GSA	90.41	90.56	91.78	92.77	93.63	94.33	95.24	95.48	96.32	97.19	97.86	98.31
		1S-GSA	75.09	75.77	76.97	78.91	80.01	81.07	95.31	96.00	97.45	98.79	99.63	99.97
		2S-GSA	89.11	89.25	90.51	91.65	92.62	93.58	95.29	95.56	96.59	97.54	98.33	99.13
		STEM	95.00	95.00	95.00	95.03	95.09	95.10	95.00	95.00	95.01	95.04	95.09	95.11
		maSigFun	95.00	95.00	95.00	95.00	95.00	95.00	95.00	95.00	95.00	95.00	95.00	95.00
	0.1	2T-GSA	86.36	87.10	89.15	91.25	92.52	93.90	90.35	91.06	92.54	94.32	95.37	96.46
		1S-GSA	73.70	75.31	78.48	81.42	83.41	84.99	90.67	91.87	94.66	97.38	99.11	99.87
		2S-GSA	85.34	86.00	88.57	90.67	92.04	94.05	90.43	91.20	93.01	95.03	96.30	98.02
		STEM	89.97	89.99	90.01	90.03	90.15	90.12	90.00	90.01	90.03	90.04	90.15	90.14
		maSigFun	90.00	90.00	90.00	90.00	90.00	90.00	90.00	90.00	90.00	90.00	90.00	90.00
	0.25	2T-GSA	73.70	75.16	79.23	80.97	86.36	88.88	75.84	76.88	79.80	80.23	85.77	88.15
		1S-GSA	63.95	67.94	75.83	82.76	86.77	89.72	77.07	80.02	86.69	90.76	97.16	99.15
		2S-GSA	72.06	73.93	80.62	82.72	88.86	92.92	76.51	78.08	83.00	82.44	90.72	94.68
		STEM	74.97	74.99	75.05	72.94	75.35	75.40	75.00	75.01	75.07	72.90	75.30	75.33
		maSigFun	75.00	75.00	75.00	72.73	75.00	75.00	75.00	75.00	75.00	72.73	75.00	75.00
	0.5	2T-GSA	50.72	52.40	56.35	60.55	56.50	69.45	50.37	51.25	53.43	55.98	47.89	62.13
		1S-GSA	54.32	58.25	68.15	76.85	81.96	91.87	52.92	55.66	63.71	72.65	71.13	91.73
		2S-GSA	51.60	55.53	60.70	67.93	68.10	82.20	50.86	53.04	56.20	61.29	55.71	74.04
		STEM	50.03	50.03	50.10	50.52	41.52	51.02	50.02	50.02	50.05	50.26	40.56	50.52
		maSigFun	50.00	50.00	50.00	50.00	40.00	50.00	50.00	50.00	50.00	50.00	40.00	50.00
100	0.05	2T-GSA	89.36	90.11	91.88	92.99	94.91	95.43	95.28	95.80	97.10	98.20	99.91	99.99
		1S-GSA	71.95	72.89	74.90	76.72	78.12	78.66	95.60	96.51	98.41	99.78	99.99	100.00
		2S-GSA	85.16	86.13	88.05	90.02	91.15	91.51	95.42	96.16	97.74	99.45	99.96	100.00
		STEM	94.99	94.99	95.03	95.02	95.15	95.58	95.00	95.00	95.03	95.02	95.16	95.57
		maSigFun	95.00	95.00	95.00	95.00	95.00	95.00	95.00	95.00	95.00	95.00	95.00	95.00
	0.1	2T-GSA	85.88	87.53	90.50	93.30	96.46	97.19	90.64	91.80	94.36	96.75	99.80	99.98
		1S-GSA	71.61	73.35	77.97	80.88	83.19	84.70	91.13	93.14	96.91	99.34	99.98	100.00
		2S-GSA	83.22	85.28	88.84	92.76	94.18	95.10	90.91	92.37	95.48	99.04	99.91	99.93
		STEM	89.99	89.99	90.03	90.02	90.29	91.25	90.00	90.01	90.05	90.04	90.28	91.17
		maSigFun	90.00	90.00	90.00	90.00	90.00	90.00	90.00	90.00	90.00	90.00	90.00	90.00
	0.25	2T-GSA	73.69	76.59	83.09	88.05	97.27	98.98	76.08	78.21	83.03	87.04	97.89	99.61
		1S-GSA	64.28	69.49	79.33	86.89	89.65	92.04	78.31	83.11	92.57	97.55	99.96	100.00
		2S-GSA	72.88	77.10	85.06	93.22	97.38	98.58	76.52	79.68	86.08	94.30	99.15	99.45
		STEM	74.96	74.99	75.17	72.82	76.16	78.63	74.99	75.02	75.15	72.82	75.99	78.02
		maSigFun	75.00	75.00	75.00	72.73	75.00	75.00	75.00	75.00	75.00	72.73	75.00	75.00
	0.5	2T-GSA	52.32	55.70	63.00	71.28	76.78	93.97	51.23	53.11	57.64	63.65	63.29	89.49
		1S-GSA	55.47	61.55	76.22	86.67	89.84	97.35	53.73	58.19	71.66	83.42	84.18	96.20
		2S-GSA	54.25	60.17	71.48	86.23	89.22	97.60	52.43	56.07	64.39	79.41	78.95	96.03
		STEM	49.95	50.07	50.47	50.18	42.76	57.87	49.97	50.03	50.23	50.09	40.83	54.52
		maSigFun	50.00	50.00	50.00	50.00	40.00	50.00	50.00	50.00	50.00	50.00	40.00	50.00

Table B.3: ACC and NPV for identifying spiked in activation profiles in the simulation to compare the five profile algorithms on basis of the OD data.

Ovary development data			ACC per p_{ag}						NPV per p_{ag}					
$|s|_{SIM}$	p_{ag}	algorithm	0.1	0.2	0.4	0.6	0.8	1	0.1	0.2	0.4	0.6	0.8	1
10	0.05	2T-GSA	94.94	95.17	95.76	96.31	96.67	96.99	95.03	95.25	95.99	96.65	97.18	97.58
		1S-GSA	86.30	86.79	88.74	90.00	90.50	90.99	95.15	95.58	97.32	98.38	98.74	98.92
		2S-GSA	94.80	95.07	95.51	95.83	95.95	96.20	95.04	95.29	96.06	96.68	97.19	97.58
		STEM	95.00	95.00	95.00	95.00	95.00	95.01	95.00	95.00	95.00	95.00	95.00	95.01
		maSigFun	95.00	95.00	95.00	95.00	95.00	95.00	95.00	95.00	95.00	95.00	95.00	95.10
	0.1	2T-GSA	89.88	90.72	92.72	94.41	95.63	96.10	90.06	90.89	92.98	94.68	96.00	96.56
		1S-GSA	82.77	83.59	87.17	89.95	90.97	91.66	90.34	90.96	94.39	96.64	97.42	97.76
		2S-GSA	89.64	90.73	92.48	93.86	94.75	95.29	90.10	91.29	93.46	94.93	96.11	96.63
		STEM	90.00	90.00	89.99	90.00	90.00	90.03	90.00	90.00	90.00	90.00	90.00	90.03
		maSigFun	90.00	90.00	90.00	90.00	90.00	90.01	90.00	90.00	90.00	90.00	90.01	90.15
	0.25	2T-GSA	75.00	76.82	82.97	88.22	91.62	92.75	75.23	76.65	81.93	86.82	90.39	91.67
		1S-GSA	70.80	74.72	83.61	87.47	89.61	90.30	76.14	79.19	86.98	90.63	92.48	93.33
		2S-GSA	74.84	78.64	85.59	89.57	91.97	92.53	75.56	78.60	85.08	88.88	91.36	92.04
		STEM	75.00	75.00	75.00	75.02	75.04	75.13	75.00	75.00	75.00	75.01	75.04	75.11
		maSigFun	75.00	75.00	75.00	75.00	75.02	75.48	75.00	75.00	75.00	75.00	75.01	75.37
	0.5	2T-GSA	50.77	53.36	61.82	72.82	81.24	85.35	50.39	51.75	56.76	64.89	72.88	77.50
		1S-GSA	52.85	57.99	72.32	78.91	82.82	84.01	51.63	54.68	65.42	72.13	76.88	79.00
		2S-GSA	51.93	57.81	71.27	79.23	84.75	86.54	51.01	54.34	63.77	71.02	77.14	79.23
		STEM	49.97	50.00	50.01	49.99	50.11	50.30	49.98	50.00	50.01	49.99	50.06	50.15
		maSigFun	50.00	50.00	50.00	50.00	50.03	51.08	50.00	50.00	50.00	50.00	50.02	50.55
20	0.05	2T-GSA	94.23	94.53	95.64	96.26	96.65	96.99	95.06	95.52	96.83	97.68	98.08	98.32
		1S-GSA	85.63	86.98	88.88	89.77	90.17	90.60	95.39	96.59	98.33	98.83	98.95	99.00
		2S-GSA	92.58	92.80	93.47	93.69	93.98	94.20	95.15	95.73	96.96	97.68	98.06	98.31
		STEM	94.99	95.00	95.00	95.00	95.02	95.05	95.00	95.00	95.00	95.00	95.02	95.05
		maSigFun	95.00	95.00	95.00	95.00	95.00	95.12	95.00	95.00	95.00	95.00	95.00	95.11
	0.1	2T-GSA	90.29	91.57	93.88	95.39	96.04	96.40	90.37	91.66	94.06	95.76	96.50	96.85
		1S-GSA	82.62	84.90	89.39	90.52	91.02	91.59	90.93	93.01	96.80	97.68	97.89	98.02
		2S-GSA	90.22	91.43	93.49	94.58	95.17	95.39	90.42	91.84	94.29	95.83	96.51	96.88
		STEM	89.99	90.00	89.99	90.01	90.05	90.13	90.00	90.00	90.00	90.01	90.05	90.13
		maSigFun	90.00	90.00	90.00	90.00	90.00	90.27	90.00	90.00	90.00	90.00	90.00	90.24
	0.25	2T-GSA	76.22	80.16	88.84	92.07	93.51	93.70	76.09	79.37	87.47	90.94	92.37	92.64
		1S-GSA	71.68	78.54	87.23	89.31	90.29	90.34	77.04	82.63	91.08	93.02	93.73	93.92
		2S-GSA	76.41	82.19	89.92	92.07	93.22	93.38	76.50	81.55	89.22	91.47	92.46	92.75
		STEM	74.98	74.97	74.94	75.06	75.15	75.42	74.99	74.99	74.99	75.06	75.15	75.36
		maSigFun	75.00	75.00	75.00	75.00	75.00	75.75	75.00	75.00	75.00	75.00	75.00	75.57
	0.5	2T-GSA	52.38	58.40	74.16	82.43	85.77	87.81	51.23	54.62	66.01	74.14	78.04	80.62
		1S-GSA	54.33	62.39	79.25	82.67	83.65	83.08	52.46	57.93	73.13	78.41	80.47	81.56
		2S-GSA	56.67	67.30	80.90	85.29	86.99	88.11	53.66	60.73	72.80	77.80	79.85	81.47
		STEM	49.98	50.00	50.04	50.13	50.55	50.97	49.99	50.00	50.02	50.07	50.28	50.50
		maSigFun	50.00	50.00	50.00	50.00	50.01	51.66	50.00	50.00	50.00	50.00	50.01	50.84
50	0.05	2T-GSA	90.55	92.32	94.05	94.41	94.62	95.01	95.64	97.18	98.68	98.79	98.88	99.01
		1S-GSA	83.36	85.20	86.86	87.32	87.94	88.61	96.09	97.81	98.91	99.02	99.12	99.25
		2S-GSA	80.81	82.70	84.62	85.24	85.89	86.66	96.38	97.85	98.65	98.71	98.80	98.95
		STEM	95.00	95.00	95.03	95.13	95.29	95.45	95.00	95.00	95.04	95.13	95.28	95.44
		maSigFun	95.00	95.00	95.00	95.00	95.00	95.16	95.00	95.00	95.00	95.00	95.00	95.15
	0.1	2T-GSA	88.03	91.34	94.45	94.62	96.13	95.47	94.89	94.16	97.14	97.43	97.58	97.77
		1S-GSA	82.48	86.01	88.69	89.54	90.29	90.89	92.25	95.76	97.63	97.77	97.94	98.12
		2S-GSA	83.27	86.71	88.94	90.74	91.72	93.10	92.84	95.62	97.08	97.36	97.55	97.80
		STEM	89.99	89.99	90.02	90.11	90.32	90.48	90.00	90.00	90.03	90.12	90.30	90.46
		maSigFun	90.00	90.00	90.00	90.00	90.00	90.41	90.00	90.00	90.00	90.00	90.00	90.37
	0.25	2T-GSA	78.34	84.77	91.88	93.05	93.44	93.98	77.73	83.50	90.62	91.85	92.30	92.92
		1S-GSA	76.31	84.75	88.88	89.32	89.74	89.99	81.99	89.92	93.44	93.72	94.29	94.61
		2S-GSA	79.09	85.77	91.51	92.63	93.17	93.72	78.53	84.96	90.77	91.91	92.46	93.10
		STEM	74.98	74.99	75.07	75.29	75.82	76.02	75.00	75.01	75.08	75.30	75.75	76.05
		maSigFun	75.00	75.00	75.00	75.00	75.00	76.05	75.00	75.00	75.00	75.00	75.00	75.80
	0.5	2T-GSA	57.83	70.24	84.01	87.13	88.18	88.65	54.26	62.76	75.93	79.67	81.12	81.76
		1S-GSA	63.04	76.77	84.94	84.42	83.92	80.45	58.77	71.20	81.53	82.42	83.15	82.14
		2S-GSA	63.40	77.07	86.15	87.39	88.10	88.52	57.83	68.90	78.79	80.28	81.35	82.03
		STEM	49.99	49.94	50.30	50.82	50.17	47.96	49.99	49.97	50.15	50.42	50.09	48.83
		maSigFun	50.00	50.00	50.00	50.00	50.00	52.10	50.00	50.00	50.00	50.00	50.00	51.07
100	0.05	2T-GSA	95.00	95.00	95.00	95.00	95.00	95.00	95.00	95.00	95.00	95.00	95.00	95.00
		1S-GSA	79.68	81.54	82.72	83.70	84.74	85.69	96.99	98.49	98.79	98.86	99.01	99.18
		2S-GSA	95.00	95.00	95.00	95.00	95.00	95.00	95.00	95.00	95.00	95.00	95.00	95.00
		STEM	95.00	95.01	95.30	95.88	96.29	96.62	95.01	95.01	95.29	95.85	96.25	96.58
		maSigFun	95.00	95.00	95.00	95.00	95.00	95.21	95.00	95.00	95.00	95.00	95.00	95.20
	0.1	2T-GSA	88.04	91.55	93.33	93.99	94.57	94.89	92.46	95.89	97.46	97.70	98.13	98.39
		1S-GSA	81.47	84.64	86.50	87.53	88.53	89.29	93.79	97.06	97.91	98.16	98.59	98.99
		2S-GSA	79.22	82.06	84.04	85.13	86.09	87.17	94.20	96.59	97.26	97.55	98.01	98.31
		STEM	90.00	90.02	90.28	90.77	91.20	91.66	90.00	90.03	90.25	90.71	91.11	91.54
		maSigFun	90.00	90.00	90.00	90.00	90.00	90.54	90.00	90.00	90.00	90.00	90.00	90.48
	0.25	2T-GSA	79.36	88.70	92.42	93.15	94.26	95.32	79.21	87.75	91.58	92.27	93.49	94.53
		1S-GSA	77.50	85.00	87.19	87.53	88.72	88.00	85.04	92.26	93.64	94.06	95.34	96.32
		2S-GSA	80.72	88.39	91.27	92.37	93.81	95.10	81.79	88.81	91.58	92.31	93.78	95.01
		STEM	74.98	75.02	75.28	75.64	75.91	75.18	75.00	75.02	75.26	75.66	76.08	76.15
		maSigFun	75.00	75.00	75.00	75.00	75.00	76.44	75.00	75.00	75.00	75.00	75.00	76.09
	0.5	2T-GSA	61.82	78.26	86.04	87.27	88.40	89.25	56.72	69.77	78.33	79.95	81.36	82.64
		1S-GSA	69.96	82.65	84.88	83.66	83.39	64.50	71.34	78.13	82.33	82.82	84.05	85.01
		2S-GSA	66.81	82.01	86.24	87.32	88.45	89.58	60.20	73.78	78.82	80.24	81.78	83.43
		STEM	49.97	50.00	50.56	50.01	45.37	41.60	49.98	50.00	50.29	50.01	47.23	44.41
		maSigFun	50.00	50.00	50.00	50.00	50.00	52.89	50.00	50.00	50.00	50.00	50.00	51.49

Table B.4: ACC and NPV for identifying spiked in activation profiles in the simulation to compare the five profile algorithms on basis of the SH data.

	Skin healing data				ACC per p_{ag}					NPV per p_{ag}				
$\|s\|_{SIM}$	p_{as}	algorithm	0.1	0.2	0.4	0.6	0.8	1	0.1	0.2	0.4	0.6	0.8	1
10	0.05	2T-GSA	94.94	95.59	96.47	96.83	97.12	97.37	95.46	96.10	96.88	97.19	97.42	97.62
		1S-GSA	82.22	83.16	85.41	86.52	87.15	87.74	95.23	95.91	97.95	98.90	99.18	99.42
		2S-GSA	94.91	95.58	96.45	96.84	97.16	97.43	95.47	96.15	96.99	97.29	97.55	97.76
		STEM	95.00	95.00	95.00	95.00	95.00	95.01	95.00	95.00	95.00	95.00	95.00	95.01
		maSigFun	95.00	95.00	95.00	95.00	95.01	95.79	95.00	95.00	95.00	95.00	95.01	95.76
	0.1	2T-GSA	89.99	91.34	93.18	93.99	94.48	94.92	90.22	91.49	93.20	93.90	94.34	94.73
		1S-GSA	79.72	80.84	84.95	87.58	88.58	89.20	90.61	91.45	95.44	97.60	98.23	98.54
		2S-GSA	89.91	91.51	93.54	94.33	94.81	95.22	90.27	91.80	93.68	94.36	94.73	95.09
		STEM	90.00	90.00	90.00	90.00	90.01	90.03	90.00	90.00	90.00	90.00	90.01	90.03
		maSigFun	90.00	90.00	90.00	90.00	90.03	91.50	90.00	90.00	90.00	90.00	90.02	91.37
	0.25	2T-GSA	75.09	77.03	81.05	82.87	84.14	85.56	75.21	76.66	79.89	81.45	82.56	83.86
		1S-GSA	68.94	73.19	81.60	85.12	87.32	87.50	76.83	80.01	88.14	92.35	94.42	95.24
		2S-GSA	75.12	78.65	82.64	84.60	85.94	87.02	75.37	78.01	81.32	83.06	84.24	85.27
		STEM	74.99	75.00	75.00	75.00	75.03	75.12	75.00	75.00	75.00	75.00	75.03	75.09
		maSigFun	75.00	75.00	75.00	75.00	75.09	78.87	75.00	75.00	75.00	75.00	75.06	78.02
	0.5	2T-GSA	50.40	52.04	55.96	59.89	64.09	68.46	50.20	51.04	53.17	55.49	58.20	61.32
		1S-GSA	52.23	57.03	68.32	74.77	78.04	78.68	51.35	54.37	62.88	69.74	74.09	76.20
		2S-GSA	50.71	54.57	60.13	64.47	68.64	72.13	50.36	52.40	55.64	58.46	61.46	64.22
		STEM	50.00	49.96	50.02	50.01	50.16	50.30	50.00	49.98	50.01	50.01	50.08	50.15
		maSigFun	50.00	50.00	50.00	50.00	50.16	57.37	50.00	50.00	50.00	50.00	50.08	53.98
20	0.05	2T-GSA	95.42	96.26	96.88	97.38	97.58	97.75	95.54	96.42	97.05	97.49	97.68	97.83
		1S-GSA	81.64	82.75	85.14	86.06	86.69	87.04	95.71	97.06	98.91	99.28	99.49	99.58
		2S-GSA	95.43	96.29	97.02	97.48	97.68	97.84	95.55	96.45	97.19	97.61	97.78	97.91
		STEM	95.00	95.00	95.00	95.01	95.02	95.05	95.00	95.00	95.00	95.01	95.02	95.05
		maSigFun	95.00	95.00	95.00	95.00	95.01	95.92	95.00	95.00	95.00	95.00	95.01	95.89
	0.1	2T-GSA	91.08	92.84	94.34	94.95	95.38	95.69	91.27	92.91	94.29	94.83	95.24	95.49
		1S-GSA	79.42	82.26	86.42	87.59	88.63	89.22	91.17	94.00	97.62	98.52	98.81	99.14
		2S-GSA	91.12	92.99	94.61	95.19	95.56	95.90	91.31	93.09	94.59	95.08	95.43	95.68
		STEM	90.00	89.99	90.00	90.01	90.04	90.09	90.00	90.00	90.00	90.01	90.04	90.09
		maSigFun	90.00	90.00	90.00	90.00	90.00	91.85	90.00	90.00	90.00	90.00	90.00	91.70
	0.25	2T-GSA	76.68	80.22	83.29	85.71	87.30	88.51	76.38	79.22	81.81	84.01	85.52	86.73
		1S-GSA	70.57	76.66	85.02	87.06	87.71	87.90	78.08	83.50	92.68	95.00	96.27	97.04
		2S-GSA	77.00	81.55	84.81	87.02	88.22	88.97	76.66	80.38	83.19	85.26	86.44	87.19
		STEM	74.98	75.00	75.00	75.00	75.08	75.10	75.22	74.99	75.01	75.06	75.08	75.18
		maSigFun	75.00	75.00	75.00	75.00	75.02	79.59	75.00	75.00	75.00	75.00	75.02	78.61
	0.5	2T-GSA	52.09	55.58	61.70	66.25	71.03	75.38	51.07	52.96	56.63	59.71	63.32	67.02
		1S-GSA	54.23	61.42	76.74	81.19	82.75	80.74	52.64	57.75	72.68	80.39	85.45	87.58
		2S-GSA	54.65	59.83	66.41	70.88	74.19	76.98	52.46	55.48	59.84	63.21	65.96	68.48
		STEM	50.00	49.97	50.03	50.08	49.81	47.08	50.00	49.98	50.02	50.04	49.90	48.43
		maSigFun	50.00	50.00	50.00	50.00	50.05	59.09	50.00	50.00	50.00	50.00	50.03	55.00
50	0.05	2T-GSA	95.43	96.21	97.11	97.35	97.47	97.56	95.43	96.19	97.11	97.34	97.47	97.54
		1S-GSA	79.29	81.17	82.55	83.59	83.89	84.67	96.53	98.39	99.13	99.34	99.51	99.64
		2S-GSA	95.45	96.23	97.22	97.48	97.64	97.73	95.45	96.21	97.20	97.45	97.60	97.69
		STEM	95.00	95.00	95.02	95.10	95.18	95.28	95.00	95.00	95.02	95.10	95.17	95.27
		maSigFun	95.00	95.00	95.00	95.00	95.00	95.99	95.00	95.00	95.00	95.00	95.00	95.95
	0.1	2T-GSA	92.48	93.77	95.02	95.36	95.73	95.92	92.46	93.67	94.85	95.34	95.51	95.69
		1S-GSA	78.75	82.54	85.51	86.51	87.33	87.85	93.37	97.17	98.48	98.97	99.33	99.44
		2S-GSA	92.53	94.01	95.22	95.74	95.89	96.04	92.51	93.91	95.06	95.53	95.67	95.81
		STEM	90.00	90.00	90.03	90.11	90.23	90.37	90.00	90.00	90.03	90.11	90.22	90.36
		maSigFun	90.00	90.00	90.00	90.00	90.00	91.98	90.00	90.00	90.00	90.00	90.00	91.82
	0.25	2T-GSA	80.99	84.61	87.27	88.48	89.23	90.08	79.93	83.07	85.50	86.69	87.46	88.31
		1S-GSA	74.98	83.08	86.50	87.01	87.67	87.44	83.72	92.23	95.42	97.08	97.87	98.37
		2S-GSA	81.92	85.52	88.12	89.10	89.66	90.01	80.76	83.94	86.35	87.33	87.89	88.25
		STEM	74.97	74.99	75.04	75.02	74.47	71.30	74.99	75.00	75.05	75.09	75.23	74.87
		maSigFun	75.00	75.00	75.00	75.00	75.00	79.95	75.00	75.00	75.00	75.00	75.00	78.91
	0.5	2T-GSA	57.27	62.96	69.19	72.53	76.60	78.87	53.93	57.45	61.88	64.55	68.12	70.30
		1S-GSA	63.35	75.87	83.09	83.96	82.33	81.01	59.79	72.66	84.45	89.84	92.04	93.71
		2S-GSA	60.96	67.09	72.70	75.31	77.61	79.22	56.19	60.32	64.69	66.95	69.07	70.65
		STEM	49.97	49.98	48.77	44.98	39.92	32.86	49.98	49.99	49.37	47.12	43.26	36.04
		maSigFun	50.00	50.00	50.00	50.00	50.00	59.62	50.00	50.00	50.00	50.00	50.00	55.32
100	0.05	2T-GSA	95.11	95.16	95.24	95.38	95.70	95.94	95.39	95.83	96.34	96.53	96.83	97.04
		1S-GSA	76.94	78.75	80.21	80.76	81.56	82.06	97.28	98.93	99.53	99.68	99.78	99.80
		2S-GSA	95.14	95.27	95.34	95.53	95.78	96.05	95.42	95.93	96.42	96.64	96.87	97.08
		STEM	94.99	95.03	95.28	95.58	95.90	96.05	95.00	95.03	95.28	95.66	95.87	96.02
		maSigFun	95.00	95.00	95.00	95.00	95.00	95.96	95.00	95.00	95.00	95.00	95.00	95.92
	0.1	2T-GSA	91.75	93.38	94.54	95.11	95.48	95.69	91.63	93.18	94.34	94.87	95.25	95.46
		1S-GSA	77.53	80.79	82.42	83.70	84.86	85.46	94.89	97.65	98.56	98.89	99.32	99.42
		2S-GSA	91.84	93.63	94.86	95.40	95.68	95.82	91.71	93.41	94.61	95.15	95.44	95.58
		STEM	89.99	90.03	90.19	90.35	90.47	90.49	90.00	90.03	90.19	90.35	90.49	90.58
		maSigFun	90.00	90.00	90.00	90.00	90.00	91.94	90.00	90.00	90.00	90.00	90.00	91.83
	0.25	2T-GSA	82.91	86.00	88.11	88.84	89.73	90.76	81.53	84.29	86.34	87.06	87.97	89.05
		1S-GSA	76.33	82.40	84.87	85.99	86.33	85.95	88.32	94.17	96.92	97.78	98.42	98.69
		2S-GSA	83.71	86.91	88.69	89.12	89.81	90.33	82.25	85.17	86.92	87.34	88.05	88.58
		STEM	74.99	74.96	74.92	71.99	62.98	48.58	75.00	75.01	75.17	74.82	72.95	68.01
		maSigFun	75.00	75.00	75.00	75.00	75.00	79.97	75.00	75.00	75.00	75.00	75.00	78.93
	0.5	2T-GSA	61.40	66.46	71.55	74.14	77.10	80.14	56.45	59.86	63.75	65.91	68.59	71.57
		1S-GSA	62.69	81.60	85.13	86.17	84.12	82.65	65.59	80.56	88.58	92.62	94.13	96.10
		2S-GSA	63.91	69.72	73.70	75.53	77.50	80.19	58.10	62.29	65.54	67.14	68.97	71.62
		STEM	49.89	49.47	44.25	33.49	23.87	20.18	49.94	49.73	46.65	37.12	23.05	14.43
		maSigFun	50.00	50.00	50.00	50.00	50.00	59.73	50.00	50.00	50.00	50.00	50.00	55.39

Table B.5: ACC and NPV for identifying spiked in activation profiles in the simulation to compare the five profile algorithms on basis of the TH data.

			ACC per p_{ag}						NPV per p_{ag}					
Tongue healing data														
$|s|_{SIM}$	p_{as}	algorithm	0.1	0.2	0.4	0.6	0.8	1	0.1	0.2	0.4	0.6	0.8	1
10	0.05	2T-GSA	95.45	96.19	97.23	97.82	98.21	98.41	95.63	96.42	97.52	98.00	98.38	98.56
		1S-GSA	82.13	82.84	85.00	86.76	87.32	88.04	95.33	95.84	97.97	99.25	99.67	99.85
		2S-GSA	95.45	96.19	97.23	97.84	98.22	98.42	95.63	96.42	97.53	98.06	98.43	98.59
		STEM	95.00	95.00	95.00	95.00	95.00	95.00	95.00	95.00	95.00	95.00	95.00	95.00
		maSigFun	95.00	95.00	95.00	95.00	95.01	96.03	95.00	95.00	95.00	95.00	95.01	95.99
	0.1	2T-GSA	90.07	92.23	94.79	96.02	96.83	97.26	90.10	92.11	94.60	95.86	96.72	97.15
		1S-GSA	79.41	80.78	85.34	88.16	89.57	90.37	90.53	91.62	95.62	98.34	99.33	99.64
		2S-GSA	90.07	92.25	95.10	96.23	96.93	97.29	90.11	92.17	94.94	96.15	96.90	97.27
		STEM	90.00	90.00	90.00	90.00	90.00	90.00	90.00	90.00	90.00	90.00	90.00	90.00
		maSigFun	90.00	90.00	90.00	90.00	90.03	92.03	90.00	90.00	90.00	90.00	90.02	91.87
	0.25	2T-GSA	75.26	78.12	85.04	88.88	91.38	92.85	75.22	77.48	83.43	87.11	89.73	91.32
		1S-GSA	68.49	73.97	84.15	89.09	91.03	92.10	76.81	80.80	90.65	95.29	97.76	98.90
		2S-GSA	75.37	80.20	87.06	89.95	92.31	93.38	75.31	79.21	85.39	88.26	90.78	91.94
		STEM	75.00	75.00	75.00	75.00	75.00	75.00	75.00	75.00	75.00	75.00	75.00	75.01
		maSigFun	75.00	75.00	75.00	75.00	75.06	79.97	75.00	75.00	75.00	75.00	75.05	78.92
	0.5	2T-GSA	50.63	53.15	60.32	70.45	76.50	81.18	50.32	51.63	55.76	62.85	68.03	72.66
		1S-GSA	52.96	58.68	73.99	82.67	87.54	88.89	51.78	55.44	67.87	77.75	85.06	88.65
		2S-GSA	50.99	57.20	68.18	75.70	80.14	83.71	50.50	53.89	61.13	67.30	71.59	75.43
		STEM	50.00	50.00	50.02	50.01	49.97	50.14	50.00	50.00	50.01	50.01	49.98	50.07
		maSigFun	50.00	50.00	50.00	50.01	50.04	59.91	50.00	50.00	50.00	50.01	50.02	55.50
20	0.05	2T-GSA	95.61	96.88	97.90	98.28	98.45	98.54	95.61	96.89	97.93	98.37	98.58	98.66
		1S-GSA	81.23	82.75	85.35	86.22	87.01	87.54	95.67	97.14	99.28	99.81	99.90	99.94
		2S-GSA	95.61	96.88	97.90	98.28	98.43	98.53	95.61	96.89	97.93	98.37	98.58	98.66
		STEM	95.00	95.00	95.00	95.00	95.01	95.01	95.00	95.00	95.00	95.00	95.01	95.01
		maSigFun	95.00	95.00	95.00	95.00	95.00	96.00	95.00	95.00	95.00	95.00	95.00	95.96
	0.1	2T-GSA	91.86	94.24	96.13	97.05	97.51	97.64	91.76	94.06	96.01	96.96	97.45	97.57
		1S-GSA	78.79	82.04	87.06	88.91	89.46	90.03	91.14	93.94	98.42	99.49	99.78	99.91
		2S-GSA	91.86	94.24	96.16	97.11	97.54	97.63	91.76	94.06	96.06	97.03	97.49	97.59
		STEM	90.00	90.00	90.00	90.00	90.00	90.03	90.00	90.00	90.00	90.00	90.00	90.03
		maSigFun	90.00	90.00	90.00	90.00	90.01	92.00	90.00	90.00	90.00	90.00	90.01	91.84
	0.25	2T-GSA	77.85	83.75	89.78	93.01	94.08	94.77	77.24	82.22	88.04	91.48	92.71	93.48
		1S-GSA	70.52	77.19	87.84	91.06	91.53	92.10	78.28	84.03	95.37	98.60	99.46	99.81
		2S-GSA	77.96	84.84	90.53	93.42	94.31	94.86	77.33	83.23	88.84	91.96	92.99	93.59
		STEM	74.98	75.00	75.00	75.00	75.00	75.12	75.00	75.00	75.00	75.00	75.02	75.10
		maSigFun	75.00	75.00	75.00	75.00	75.02	80.03	75.00	75.00	75.00	75.00	75.01	78.97
	0.5	2T-GSA	53.14	59.88	73.56	79.90	85.26	87.61	51.62	55.49	65.41	71.34	77.23	80.14
		1S-GSA	55.00	64.08	83.25	89.16	90.95	90.12	53.14	59.67	80.50	91.26	96.50	98.31
		2S-GSA	56.75	67.67	77.71	83.37	87.12	88.69	53.63	60.76	69.19	75.08	79.52	81.55
		STEM	49.99	50.00	49.95	50.02	49.75	48.09	49.99	50.00	49.97	50.01	49.87	49.01
		maSigFun	50.00	50.00	50.00	50.00	50.04	59.98	50.00	50.00	50.00	50.00	50.02	55.54
50	0.05	2T-GSA	96.70	97.50	97.97	98.23	98.37	98.43	96.77	97.72	98.34	98.52	98.63	98.67
		1S-GSA	79.59	81.62	83.50	84.34	85.10	85.76	96.70	98.67	99.71	99.89	99.91	99.95
		2S-GSA	96.70	97.50	97.93	98.19	98.33	98.39	96.77	97.72	98.35	98.53	98.63	98.68
		STEM	95.00	95.00	95.00	95.01	95.01	95.04	95.06	95.00	95.00	95.01	95.01	95.04	95.06
		maSigFun	95.00	95.00	95.00	95.00	95.00	96.00	95.00	95.00	95.00	95.00	95.00	95.96
	0.1	2T-GSA	93.75	95.67	96.99	97.62	97.75	93.56	95.50	96.87	97.22	97.49	97.63	
		1S-GSA	78.77	83.51	86.58	87.82	88.72	89.28	93.45	97.83	99.63	99.85	99.95	99.99
		2S-GSA	93.75	95.67	97.00	97.33	97.60	97.73	93.56	95.52	96.88	97.24	97.50	97.65
		STEM	90.00	90.00	89.99	90.03	90.06	90.09	90.00	90.00	90.00	90.00	90.03	90.06	90.09
		maSigFun	90.00	90.00	90.00	90.00	90.00	92.01	90.00	90.00	90.00	90.00	90.00	91.85
	0.25	2T-GSA	84.02	88.85	93.38	94.58	94.99	95.15	82.46	87.11	91.93	93.30	93.74	93.93
		1S-GSA	75.20	85.10	89.79	91.30	91.25	91.00	84.27	94.50	99.14	99.78	99.92	99.98
		2S-GSA	84.15	89.48	93.66	94.64	95.01	95.17	82.59	88.75	92.26	93.37	93.78	93.96
		STEM	74.98	75.00	75.02	75.00	74.53	69.73	74.99	75.00	75.03	75.07	75.04	73.92
		maSigFun	75.00	75.00	75.00	75.00	75.00	80.03	75.00	75.00	75.00	75.00	75.00	78.97
	0.5	2T-GSA	61.71	72.77	83.61	87.28	89.28	89.84	56.64	64.75	75.32	79.72	82.35	83.11
		1S-GSA	64.06	81.92	90.22	90.65	89.52	87.47	60.31	79.81	95.60	98.82	99.35	99.55
		2S-GSA	66.72	76.53	85.88	88.53	89.69	90.27	60.05	68.07	77.99	81.34	82.92	83.71
		STEM	49.94	49.93	49.17	42.16	33.24	25.72	49.97	49.96	49.58	45.54	38.81	31.01
		maSigFun	50.00	50.00	50.00	50.00	50.01	60.06	50.00	50.00	50.00	50.00	50.01	55.59
100	0.05	2T-GSA	94.90	94.91	95.08	95.60	96.05	96.34	95.02	95.09	95.46	96.09	96.70	97.17
		1S-GSA	77.51	79.50	81.25	82.12	82.97	83.32	97.83	99.47	99.93	99.99	99.99	100.00
		2S-GSA	94.90	94.94	95.24	95.89	96.34	96.56	95.02	95.10	95.51	96.22	96.85	97.29
		STEM	94.99	95.02	95.33	95.74	96.08	96.47	95.00	95.02	95.32	95.72	96.05	96.43
		maSigFun	95.00	95.00	95.00	95.00	95.00	96.01	95.00	95.00	95.00	95.00	95.00	95.97
	0.1	2T-GSA	94.45	96.01	97.00	97.31	97.49	97.75	94.45	96.11	96.99	97.27	97.40	97.66
		1S-GSA	77.74	81.59	83.90	85.14	85.70	86.01	95.15	98.33	99.59	99.71	99.68	99.72
		2S-GSA	94.45	96.11	96.99	97.24	97.43	97.80	94.45	96.16	97.01	97.28	97.46	97.80
		STEM	89.99	90.00	90.02	90.09	90.11	90.06	90.00	90.01	90.03	90.11	90.15	90.22
		maSigFun	90.00	90.00	90.00	90.00	90.00	92.01	90.00	90.00	90.00	90.00	90.00	91.84
	0.25	2T-GSA	86.09	91.14	94.12	94.62	94.83	94.94	84.41	89.57	92.74	93.33	93.56	93.70
		1S-GSA	77.38	84.25	87.11	87.44	86.76	85.95	89.56	96.94	99.50	99.80	99.96	99.98
		2S-GSA	86.28	91.53	94.17	94.66	94.84	95.02	84.59	89.93	92.81	93.39	93.60	93.80
		STEM	74.98	74.95	74.67	69.20	52.91	39.69	75.00	74.99	74.97	73.85	68.72	62.29
		maSigFun	75.00	75.00	75.00	75.00	75.00	80.03	75.00	75.00	75.00	75.00	75.00	78.97
	0.5	2T-GSA	68.59	78.23	87.28	88.72	88.85	89.39	61.42	69.68	79.72	81.59	81.77	82.49
		1S-GSA	72.58	86.26	90.37	88.99	85.05	81.72	68.61	87.32	97.78	99.12	99.43	99.69
		2S-GSA	70.24	80.72	88.21	89.23	89.14	89.67	62.69	72.19	80.92	82.29	82.16	82.88
		STEM	49.93	49.55	40.13	27.88	22.11	17.43	49.96	49.77	44.13	33.16	25.10	16.62
		maSigFun	50.00	50.00	50.00	50.00	50.00	60.05	50.00	50.00	50.00	50.00	50.00	55.59

Hexbin Plots for the *fill* direction of smoothing

Hexbin Plots use shaded hexagons to represent the number of observations on the covered area. In the following plots, this kind of figure is used to visualize the pattern and deviation of the functional relation of the smoothing parameter λ_{fill} and the general smoothing proportion for the proposed smoothing methods and the four data sets, where the smoothing was applied. The general smoothing proportion (GSP) is a weighted mean of the smoothing proportions yielded by smoothing three samples of 8000 smoothing constellations triples for a fixed group size and a fixed smoothing parameter while the number of differential expressed genes is fixed to the median of differential expressed genes (both up and down) in each data set. The weights are chosen according to the number of the fixed group size in the corresponding data set. Details are given in section 5.4.

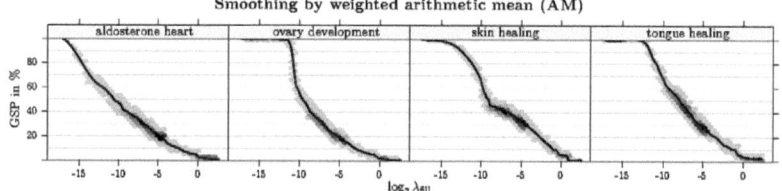

Figure B.1: Hexbin Plot of the general smoothing proportion (GSP) in dependence of the smoothing parameter λ_{fill} for the arithmetic mean smoothing and the four used data sets..

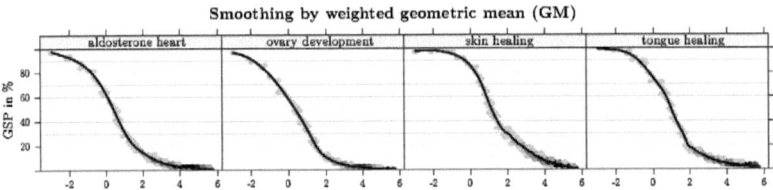

Figure B.2: Hexbin Plot of the general smoothing proportion (GSP) in dependence of the smoothing parameter λ_fill for the geometric mean smoothing and the four used data sets.

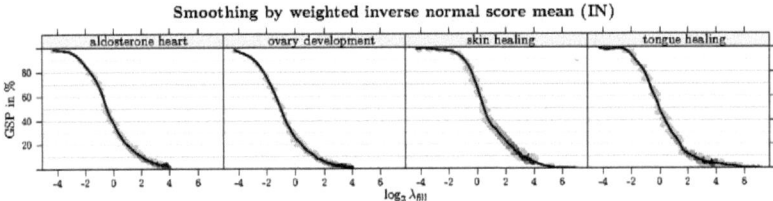

Figure B.3: Hexbin Plot of the general smoothing proportion (GSP) in dependence of the smoothing parameter λ_fill for the inverse normal score smoothing and the four used data sets.

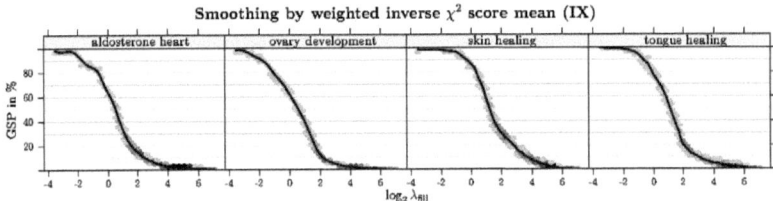

Figure B.4: Hexbin Plot of the general smoothing proportion (GSP) in dependence of the smoothing parameter λ_fill for the inverse χ^2 score smoothing and the four used data sets.

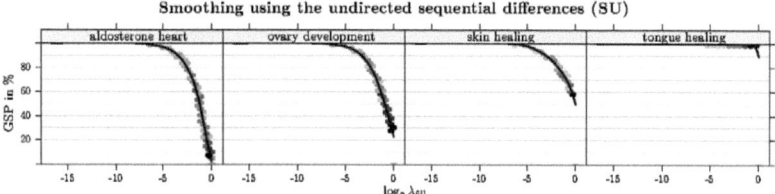

Figure B.5: Hexbin Plot of the general smoothing proportion (GSP) in dependence of the smoothing parameter λ_{fill} for the undirected sequential difference smoothing and the four used data sets.

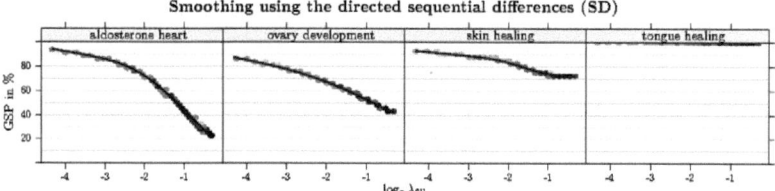

Figure B.6: Hexbin Plot of the general smoothing proportion (GSP) in dependence of the smoothing parameter λ_{fill} for the directed sequential difference smoothing and the four used data sets.

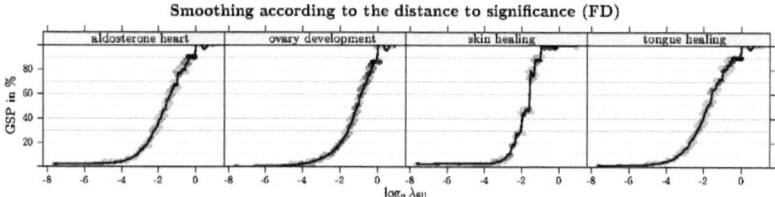

Figure B.7: Hexbin Plot of the general smoothing proportion (GSP) in dependence of the smoothing parameter λ_{fill} for the distance to significance smoothing and the four used data sets.

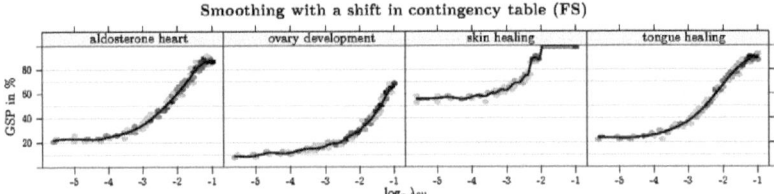

Figure B.8: Hexbin Plot of the general smoothing proportion (GSP) in dependence of the smoothing parameter λ_{fill} for the shift in contingency table smoothing and the four used data sets.

Hexbin Plots for the *wipe* direction of smoothing

Hexbin Plots use shaded hexagons to represent the number of observations on the covered area. In the following plots, this kind of figure is used to visualize the pattern and deviation of the functional relation of the smoothing parameter λ_{wipe} and the general smoothing proportion for the proposed smoothing methods and the four data sets, where the smoothing was applied. The general smoothing proportion (GSP) is a weighted mean of the smoothing proportions yielded by smoothing three samples of 8000 smoothing constellations triples for a fixed group size and a fixed smoothing parameter while the number of differential expressed genes is fixed to the median of differential expressed genes (both up and down) in each data set. The weights are chosen according to the number of the fixed group size in the corresponding data set as described in section 5.4 (pp. 104).

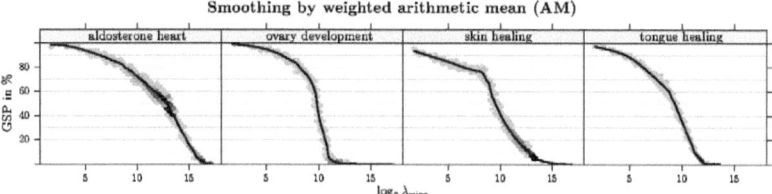

Figure B.9: Hexbin Plot of the general smoothing proportion (GSP) in dependence of the smoothing parameter λ_{wipe} for the arithmetic mean smoothing and the four used data sets.

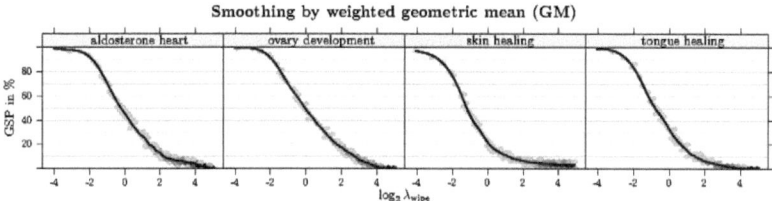

Figure B.10: Hexbin Plot of the general smoothing proportion (GSP) in dependence of the smoothing parameter λ_{wipe} for the geometric mean smoothing and the four used data sets.

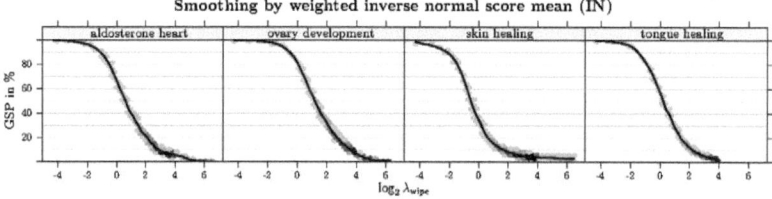

Figure B.11: Hexbin Plot of the general smoothing proportion (GSP) in dependence of the smoothing parameter λ_{wipe} for the inverse normal score smoothing and the four used data sets.

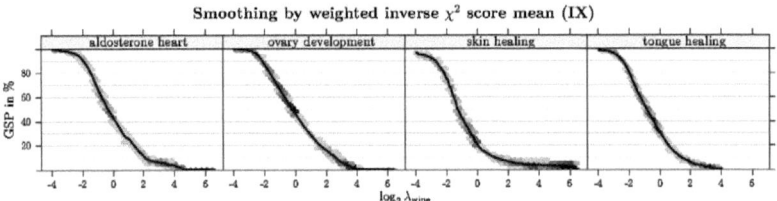

Figure B.12: Hexbin Plot of the general smoothing proportion (GSP) in dependence of the smoothing parameter λ_{wipe} for the inverse χ^2 score smoothing and the four used data sets.

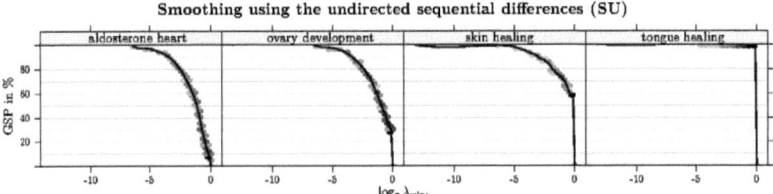

Figure B.13: Hexbin Plot of the general smoothing proportion (GSP) in dependence of the smoothing parameter λ_{wipe} for the undirected sequential difference smoothing and the four used data sets.

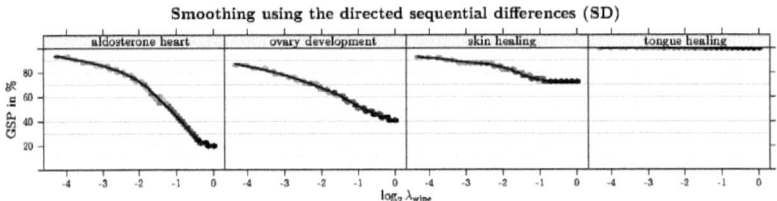

Figure B.14: Hexbin Plot of the general smoothing proportion (GSP) in dependence of the smoothing parameter λ_{wipe} for the directed sequential difference smoothing and the four used data sets.

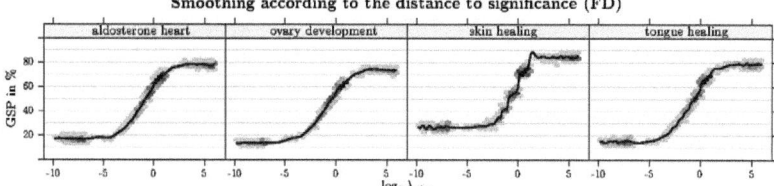

Figure B.15: Hexbin Plot of the general smoothing proportion (GSP) in dependence of the smoothing parameter λ_{wipe} for the distance to significance smoothing and the four used data sets.

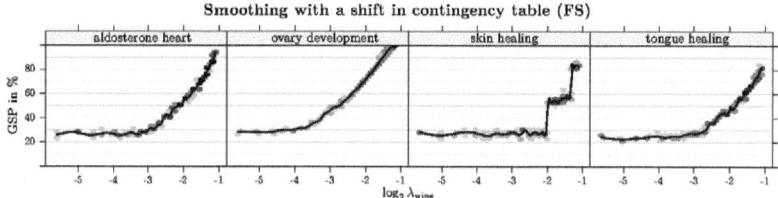

Figure B.16: Hexbin Plot of the general smoothing proportion (GSP) in dependence of the smoothing parameter λ_{wipe} for the shift in contingency table smoothing and the four used data sets.

Table B.6: All smoothing paramters used in the simulation study for the validation of smoothing methods. The smoothing methods in first column are abbreviated with AM (arithmetic mean), GM (geometric mean), IN (inverse normal score mean), IX (invers χ^2 score mean), FD (fisher test distance to significance), FS (Fisher test shift in contingency table), SU (sequential undirected difference) and SD (sequential directed difference). The second column gives the corresponding general smoothing proportion (GSP). Missing values could not be achieved by a reasonable paramter and the combination of smoothing method and data set.

	%	aldosterone heart λ_{fill}	λ_{wipe}	ovary development λ_{fill}	λ_{wipe}	skin healing λ_{fill}	λ_{wipe}	tongue healing λ_{fill}	λ_{wipe}
AM	2.5	1.3454	83,835.1573	0.9775	2846.6160	1.5919	22,336.6244	3.1852	4215.2213
	5	0.6760	58,986.8823	0.5536	2106.6650	1.3244	10,938.2568	0.9497	3492.7848
	7.5	0.3383	49,913.5248	0.2698	1866.5432	0.8175	7475.5569	0.5831	2835.2904
	10	0.1813	45,904.5923	0.1407	1828.4005	0.5946	6742.8316	0.4257	2433.1448
	15	0.0548	34,668.4615	0.0468	1744.9925	0.2971	4683.1338	0.1981	2083.2975
	20	0.0365	27,894.9612	0.0298	1564.3751	0.1660	3122.0351	0.0900	1750.1108
	25	0.0181	22,147.1388	0.0132	1408.6396	0.0890	2302.8079	0.0428	1447.5645
	30	0.0086	17,360.3853	0.0078	1251.3156	0.0427	1736.9771	0.0290	1238.6130
	35	0.0046	14,440.7175	0.0047	1124.7171	0.0257	1351.4548	0.0169	1038.5850
	40	0.0021	10,616.5831	0.0027	1014.6518	0.0096	1088.4165	0.0098	869.5550
	45	0.0011	9131.3370	0.0018	948.2326	0.0040	884.1354	0.0066	751.9674
	50	0.0007	7460.4425	0.0013	906.3096	0.0018	732.9309	0.0044	622.1785
	60	0.0002	3380.4277	0.0008	792.4283	0.0011	540.4841	0.0017	398.9328
	70	0.0001	1313.9135	0.0006	558.6815	0.0008	405.5668	0.0009	195.4759
	80	$3.8 \cdot 10^{-5}$	498.1275	0.0006	287.8479	0.0004	100.0780	0.0005	78.6816
	90	$2.0 \cdot 10^{-5}$	80.0904	0.0005	86.3824	0.0002	8.8592	0.0003	25.3714
GM	2.5	22.4920	20.1212	13.6310	14.1976	35.1406	20.6318	36.1962	8.9723
	5	10.5241	12.4811	7.1750	10.4269	22.5748	12.7372	16.7157	5.6013

continued on next page . . .

... continued from previous page

	%	ovary development λ_{fill}	ovary development λ_{wipe}	aldosterone heart λ_{fill}	aldosterone heart λ_{wipe}	skin healing λ_{fill}	skin healing λ_{wipe}	tongue healing λ_{fill}	tongue healing λ_{wipe}
	7.5	7.3347	7.2833	5.1663	8.0421	16.4815	4.2800	10.2918	3.8214
	10	5.6376	4.9257	4.0877	6.6056	12.6741	2.2898	7.6259	3.0386
	15	3.8591	3.5633	3.1660	4.7280	8.7224	1.4583	4.9331	2.0624
	20	2.8946	2.7207	2.7121	3.3704	6.3765	1.0769	3.6844	1.6145
	25	2.3825	2.1584	2.3979	2.5291	5.0284	0.9000	3.2836	1.2773
	30	2.0313	1.7057	2.1563	2.0996	3.8425	0.7766	3.0371	1.0903
	35	1.7792	1.3989	1.8985	1.6847	3.1820	0.6596	2.6619	0.9443
	40	1.6080	1.1979	1.6706	1.3990	2.7085	0.5657	2.3406	0.8031
	45	1.4529	0.9884	1.4536	1.1576	2.4243	0.4936	2.1178	0.6880
	50	1.3001	0.8521	1.2893	0.9672	2.1841	0.4377	1.9609	0.5936
	60	1.0395	0.6367	0.9359	0.6835	1.8212	0.3572	1.5980	0.4492
	70	0.8148	0.4893	0.6776	0.5016	1.5332	0.2950	1.1806	0.3451
	80	0.5778	0.3899	0.4772	0.3633	1.1768	0.2350	0.8459	0.2719
	90	0.3295	0.2770	0.2618	0.2635	0.8438	0.1428	0.5755	0.2147
IN	2.5	13.6444	27.2342	9.4303	29.0392	19.6587	30.0564	25.9509	13.8784
	5	8.1826	15.2342	4.9361	19.5376	13.9967	11.2096	11.2899	8.8694
	7.5	5.6146	10.8076	3.3381	14.1738	12.0991	5.5596	7.0388	6.1633
	10	4.1948	7.6029	2.6036	12.1302	9.7433	3.4642	5.0545	4.8777
	15	2.8088	5.3526	1.7765	8.5713	6.5846	2.0444	3.2544	3.4189
	20	1.9941	4.2333	1.3229	6.6401	4.9413	1.5467	2.6696	2.6619
	25	1.5450	3.3588	1.0428	5.2963	3.5591	1.2829	2.2598	2.2747
	30	1.2513	2.7573	0.8378	4.3437	2.7674	1.1261	1.7584	1.9252
	35	1.0778	2.3469	0.7148	3.5331	2.1752	0.9805	1.4770	1.6965

continued on next page ...

... continued from previous page

%	ovary development λ_{fill}	λ_{wipe}	aldosterone heart λ_{fill}	λ_{wipe}	skin healing λ_{fill}	λ_{wipe}	tongue healing λ_{fill}	λ_{wipe}
40	0.9300	1.9550	0.6185	3.0668	1.7548	0.8507	1.2826	1.4755
45	0.7955	1.6672	0.5495	2.6059	1.5286	0.7539	1.0862	1.2616
50	0.7030	1.5027	0.4870	2.2789	1.3877	0.6832	0.9588	1.1390
60	0.5907	1.1615	0.3874	1.7369	1.1489	0.5727	0.7887	0.9156
70	0.4608	0.9060	0.2925	1.3620	0.9488	0.4644	0.5702	0.7219
80	0.3166	0.6988	0.2149	1.0454	0.7436	0.3712	0.4691	0.5553
90	0.2046	0.4647	0.1302	0.6924	0.5093	0.2135	0.2984	0.3634
IX 2.5	23.0178	18.7772	14.2455	12.8154	35.8762	43.5362	37.5894	8.5979
5	10.7923	11.7925	7.5755	9.6881	23.7096	17.6538	16.8636	5.1199
7.5	7.4287	7.0715	5.3045	7.2692	16.7050	3.8835	10.4687	3.4886
10	5.4515	4.5783	4.3873	6.0302	12.8007	2.4523	7.7612	2.6472
15	4.0124	3.3553	3.5418	4.4580	8.8589	1.3577	5.0000	1.8542
20	3.0437	2.6484	3.0501	2.9867	6.7034	0.9598	3.9864	1.4678
25	2.4600	2.0497	2.7374	2.3487	4.8689	0.8611	3.4570	1.2062
30	2.1533	1.5823	2.4458	1.8509	3.9351	0.7690	3.2153	0.9743
35	1.9039	1.3192	2.1662	1.5031	3.2350	0.6312	2.8884	0.8605
40	1.7067	1.0849	1.9332	1.2410	2.8065	0.5419	2.5989	0.7459
45	1.5382	0.9284	1.6826	0.9968	2.5310	0.4738	2.3638	0.6441
50	1.4006	0.7890	1.4563	0.8568	2.3098	0.4177	2.1294	0.5467
60	1.1082	0.5984	1.0611	0.6156	1.9018	0.3436	1.7111	0.4077
70	0.8230	0.4634	0.7519	0.4451	1.5765	0.2817	1.3019	0.3183
80	0.6364	0.3625	0.4818	0.3195	1.2669	0.2225	0.9039	0.2536
90	0.3183	0.2625	0.2874	0.2289	0.8103	0.1504	0.6211	0.1865

continued on next page ...

... continued from previous page

	%	ovary development λ_{fill}	λ_{wipe}	aldosterone heart λ_{fill}	λ_{wipe}	skin healing λ_{fill}	λ_{wipe}	tongue healing λ_{fill}	λ_{wipe}
FD	2.5	0.0195		0.0615		0.0112		0.0347	
	5	0.0688		0.0997		0.1154		0.0646	
	7.5	0.0911		0.1300		0.1395		0.0838	
	10	0.1090		0.1589		0.1458		0.0975	
	15	0.1387	0.0049	0.2105	0.0057	0.1369		0.1263	0.0071
	20	0.1700	0.0095	0.2615	0.0988	0.2106		0.1466	0.0469
	25	0.1925	0.1349	0.2960	0.1618	0.2068	0.0058	0.1734	0.1536
	30	0.2149	0.1891	0.3360	0.2428	0.2468	0.0297	0.1946	0.2074
	35	0.2484	0.2462	0.3666	0.3245	0.2565	0.2855	0.2232	0.2788
	40	0.2699	0.3345	0.4045	0.4291	0.2474	0.4013	0.2387	0.3824
	45	0.2940	0.4329	0.4323	0.5529	0.2776	0.4351	0.2588	0.5105
	50	0.3230	0.5646	0.4711	0.7524	0.3352	0.5523	0.2908	0.6463
	60	0.3903	0.9368	0.5408	1.2798	0.3972	0.8759	0.3483	1.0461
	70	0.4766	1.7365	0.6421	3.6535	0.3361	1.2511	0.4197	1.7647
	80	0.6292	27.9746	0.7758	16.9952	0.3493	4.5754	0.5413	30.2826
	90	0.7936	2.8297		45.0709	0.4954	16.3412	0.7170	35.9035
FS	7.5			0.0242					
	10			0.0389					
	15			0.0923					
	20	0.0412		0.1565					
	25	0.0637	0.0717	0.2064		0.1375		0.0457	
	30	0.0960	0.1089	0.2489	0.0468	0.1608		0.0761	0.1364
	35	0.1269	0.1489	0.2853	0.0895	0.1872		0.1106	0.1795

continued on next page ...

... continued from previous page

	%	ovary development		aldosterone heart		skin healing		tongue healing	
		λ_{fill}	λ_{wipe}	λ_{fill}	λ_{wipe}	λ_{fill}	λ_{wipe}	λ_{fill}	λ_{wipe}
	40	0.1515	0.1777	0.3174	0.1065		0.2163	0.1390	0.2032
	45	0.1744	0.2001	0.3471	0.1190		0.2477	0.1602	0.2337
	50	0.1990	0.2261	0.3784	0.1434		0.2805	0.1786	0.2717
	60	0.2480	0.2886	0.4343	0.1899		0.3445	0.2167	0.3355
	70	0.3027	0.3493	0.5023	0.2391	0.0872	0.3943	0.2611	0.3985
	80	0.3624	0.4036	0.5704	0.2914	0.1578	0.4312	0.3250	0.4522
	90	0.4815	0.4502	0.6792	0.3467	0.2017	0.4651	0.4542	0.5012
SU	7.5	0.9203	0.9265			0.2267		0.9203	0.9265
	10	0.8582	0.8589					0.8582	0.8589
	15	0.7752	0.7773					0.7752	0.7773
	20	0.7070	0.7074					0.7070	0.7074
	25	0.6514	0.6549					0.6514	0.6549
	30	0.6091	0.6104	0.8779	0.8747			0.6091	0.6104
	35	0.5553	0.5569	0.7620	0.7632			0.5553	0.5569
	40	0.5268	0.5220	0.6821	0.6841			0.5268	0.5220
	45	0.4863	0.4851	0.5995	0.6011			0.4863	0.4851
	50	0.4502	0.4433	0.5149	0.5121			0.4502	0.4433
	60	0.3585	0.3561	0.3958	0.3928	0.7496	0.7484	0.3585	0.3561
	70	0.2592	0.2555	0.2893	0.2905	0.5434	0.5372	0.2592	0.2555
	80	0.1679	0.1715	0.1882	0.1891	0.3023	0.2812	0.1679	0.1715
	90	0.0930	0.0950	0.0977	0.0965	0.1129	0.1147	0.0930	0.0950
SD	20	1.0059	0.9802			1.0059	0.9802	1.0059	0.9802

continued on next page ...

... continued from previous page

%	ovary development		aldosterone heart		skin healing		tongue healing	
	λ_{fill}	λ_{wipe}	λ_{fill}	λ_{wipe}	λ_{fill}	λ_{wipe}	λ_{fill}	λ_{wipe}
25	0.7460	0.7437			0.7460	0.7437	0.7460	0.7437
30	0.6619	0.6650			0.6619	0.6650	0.6619	0.6650
35	0.6077	0.6058			0.6077	0.6058	0.6077	0.6058
40	0.5443	0.5444	1.0369	1.0371	0.5443	0.5444	0.5443	0.5444
45	0.4938	0.4902	0.7027	0.7009	0.4938	0.4902	0.4938	0.4902
50	0.4387	0.4425	0.5562	0.5540	0.4387	0.4425	0.4387	0.4425
60	0.3520	0.3549	0.3663	0.3658	0.3520	0.3549	0.3520	0.3549
70	0.2720	0.2729	0.2152	0.2156	0.2720	0.2729	0.2720	0.2729
80	0.1834	0.1830	0.1059	0.1059	0.1834	0.1830	0.1834	0.1830
90	0.0827	0.0826	0.0332		0.0827	0.0826	0.0827	0.0826

Figures of smoothing evaluation simulation results

This appendix section includes the 3D surface accuracy plots, which show the sensitivity (TPR), specificity (1-FPR), PPV_{p90} and accuracy (ACC) for the smoothing parameter combinations resulting from Table B.6. The section is divided in subsections regarding the data set and activation profile algorithm used for the simulation. Each figure consists of four subplots, i.e. one for each accuracy measure. The maximum and minimum in each plot are marked in gray scale on the smoothing value plain and the corresponding values are labeled at the north-eastern y-axis together with the value without smoothing (medium gray). The basement axes give the extend of smoothing within the current simulation setting, i.e. they are linked to the smoothing parameters λ_{wipe} (symbolized as $\downarrow^o_o \,^+_o \,^o_o \downarrow$) and λ_{fill} (symbolized as $\downarrow ^+_+ \,^o_o \,^+_+ \downarrow$) by the corresponding general smoothing proportion (GSP). The additional value of the best possible sensitivity value is added for the sensitivity plots. This value will be reached if all smoothing positions would be smoothed towards the preset pattern if possible, i.e. this would be the sensitivity of a oracle smoothing method, which knows the true profile in the current simulation setting. The shown surfaces cover only the smoothing parameter area, which is available for the current smoothing method. On the vertical limiting plot sides the smoothing parameters are shown corresponding their GSP as dashed vertical lines. The dashed horizontal lines display the minimum, the maximum, the 25%, 50% and the 75% quantiles of the current accuracy measure in the plot. This auxiliary lines help to better assess the height value of the 3D surface. The marked maximum for the ACC surface plot is chosen as best smoothing parameter combination for the current smoothing method and data set listed in Tables 5.11 to 5.14.

Aldosterone effect on heart data set

Two-threshold GSA-type activation profile algorithm (2T-GSA)

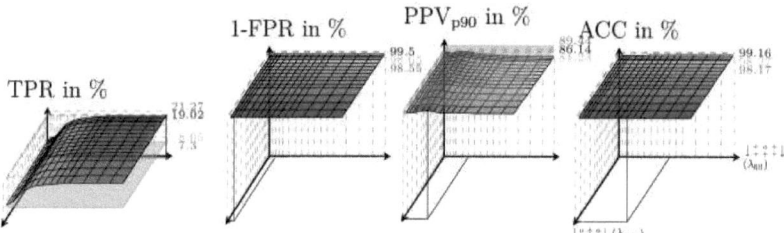

Figure B.17: Accuracy Plots with data: AH, profile algorithm: 2T-GSA, smoothing algorithm: AM. Detailed figure description at section start on page 228.

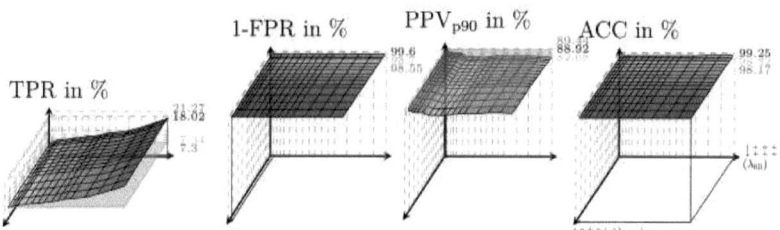

Figure B.18: Accuracy Plots with data: AH, profile algorithm: 2T-GSA, smoothing algorithm: GM. Detailed figure description at section start on page 228.

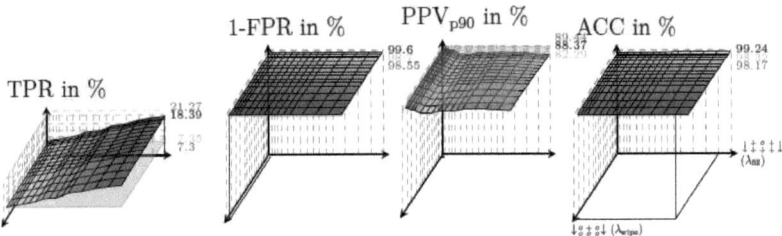

Figure B.19: Accuracy Plots with data: AH, profile algorithm: 2T-GSA, smoothing algorithm: IN. Detailed figure description at section start on page 228.

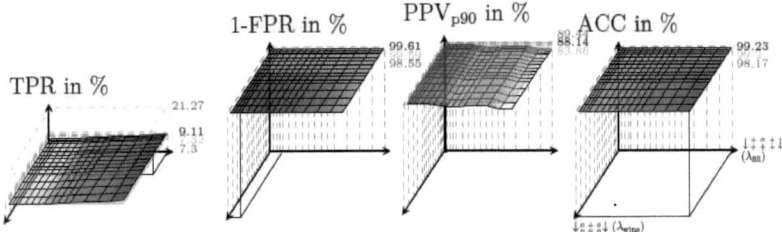

Figure B.20: Accuracy Plots with data: AH, profile algorithm: 2T-GSA, smoothing algorithm: IX. Detailed figure description at section start on page 228.

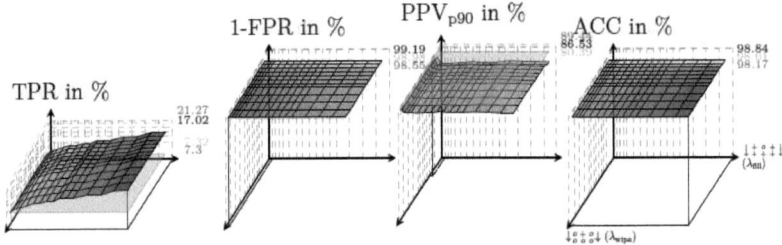

Figure B.21: Accuracy Plots with data: AH, profile algorithm: 2T-GSA, smoothing algorithm: FD. Detailed figure description at section start on page 228.

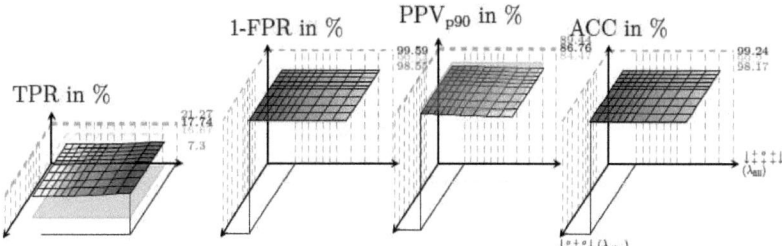

Figure B.22: Accuracy Plots with data: AH, profile algorithm: 2T-GSA, smoothing algorithm: FS. Detailed figure description at section start on page 228.

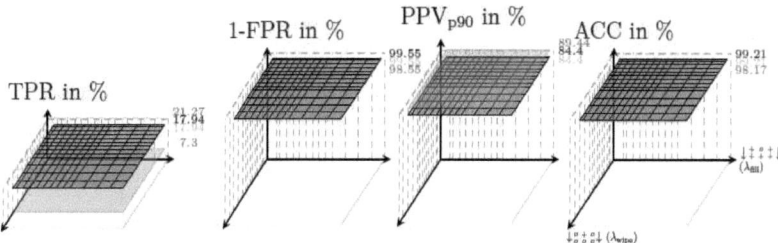

Figure B.23: Accuracy Plots with data: AH, profile algorithm: 2T-GSA, smoothing algorithm: SU. Detailed figure description at section start on page 228.

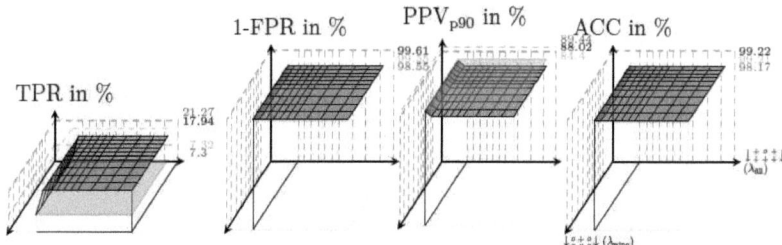

Figure B.24: Accuracy Plots with data: AH, profile algorithm: 1S-GSA, smoothing algorithm: SD. Detailed figure description at section start on page 228.

One-threshold segmentation GSA-type activation profile algorithm (1S-GSA)

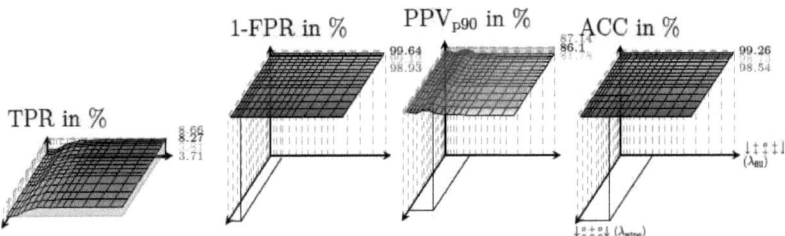

Figure B.25: Accuracy Plots with data: AH, profile algorithm: 1S-GSA, smoothing algorithm: AM. Detailed figure description at section start on page 228.

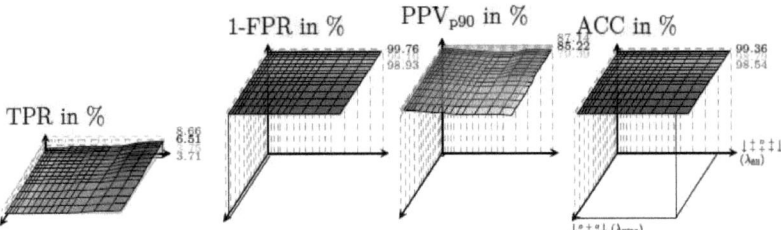

Figure B.26: Accuracy Plots with data: AH, profile algorithm: 1S-GSA, smoothing algorithm: GM. Detailed figure description at section start on page 228.

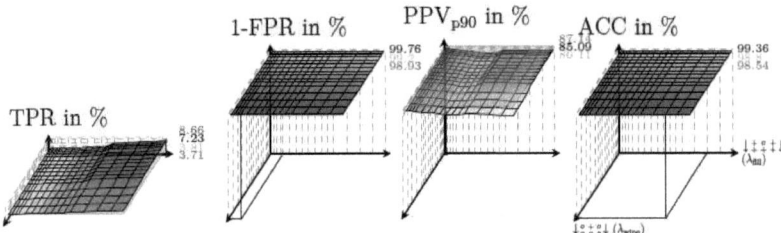

Figure B.27: Accuracy Plots with data: AH, profile algorithm: 1S-GSA, smoothing algorithm: IN. Detailed figure description at section start on page 228.

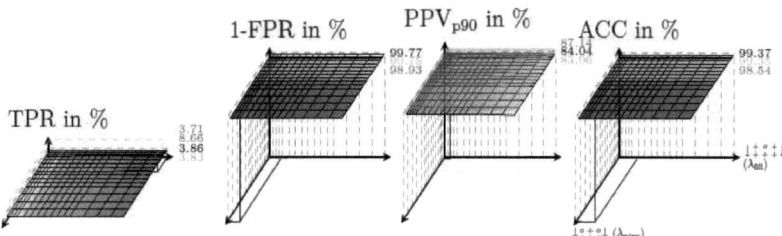

Figure B.28: Accuracy Plots with data: AH, profile algorithm: 1S-GSA, smoothing algorithm: IX. Detailed figure description at section start on page 228.

Two-threshold segmentation GSA-type activation profile algorithm (2S-GSA)

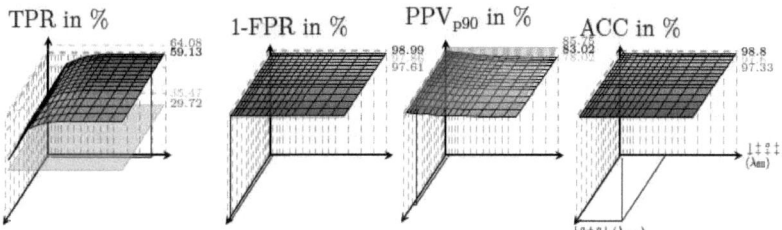

Figure B.29: Accuracy Plots with data: AH, profile algorithm: 2S-GSA, smoothing algorithm: AM. Detailed figure description at section start on page 228.

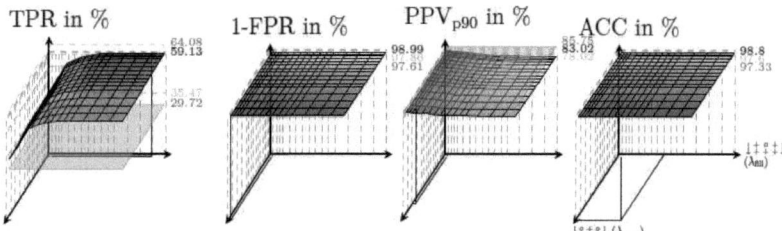

Figure B.30: Accuracy Plots with data: AH, profile algorithm: 2S-GSA, smoothing algorithm: GM. Detailed figure description at section start on page 228.

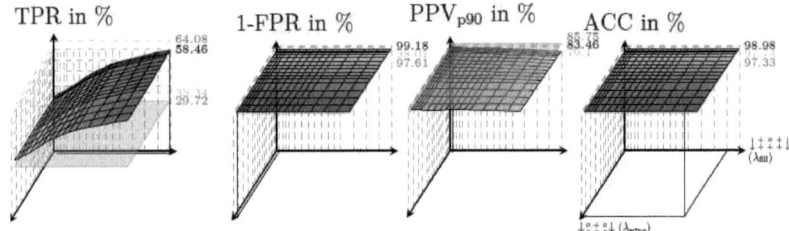

Figure B.31: Accuracy Plots with data: AH, profile algorithm: 2S-GSA, smoothing algorithm: IN. Detailed figure description at section start on page 228.

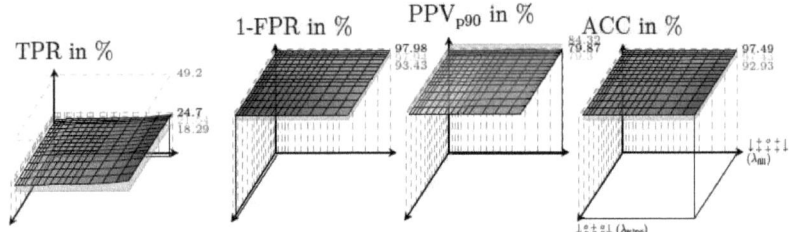

Figure B.32: Accuracy Plots with data: AH, profile algorithm: 2S-GSA, smoothing algorithm: IX. Detailed figure description at section start on page 228.

Embryonic ovary development data set

Two-threshold GSA-type activation profile algorithm (2T-GSA)

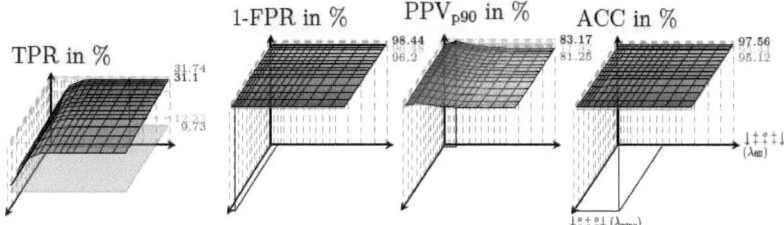

Figure B.33: Accuracy Plots with data: OD, profile algorithm: 2T-GSA, smoothing algorithm: AM. Detailed figure description at section start on page 228.

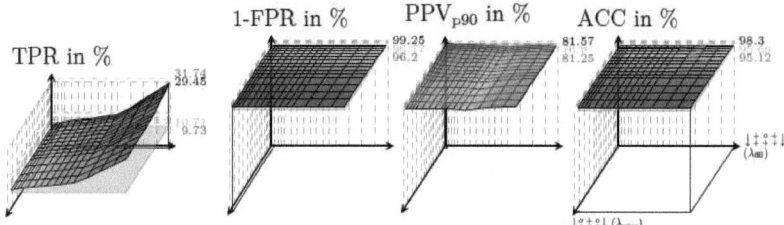

Figure B.34: Accuracy Plots with data: OD, profile algorithm: 2T-GSA, smoothing algorithm: GM. Detailed figure description at section start on page 228.

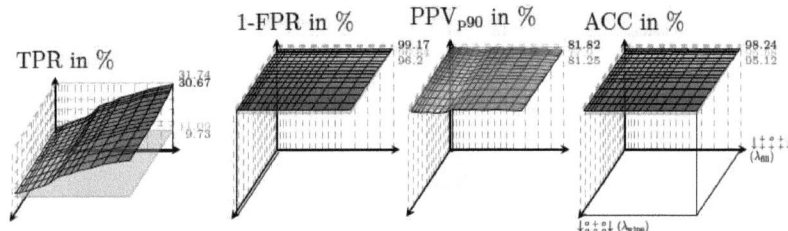

Figure B.35: Accuracy Plots with data: OD, profile algorithm: 2T-GSA, smoothing algorithm: IN. Detailed figure description at section start on page 228.

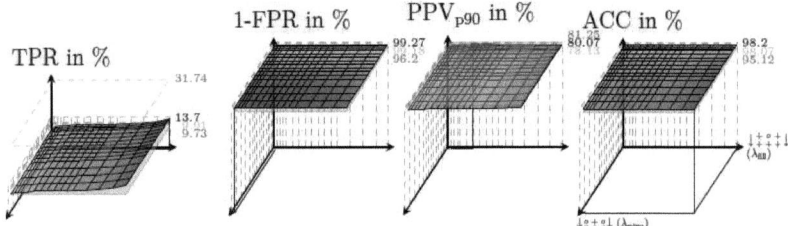

Figure B.36: Accuracy Plots with data: OD, profile algorithm: 2T-GSA, smoothing algorithm: IX. Detailed figure description at section start on page 228.

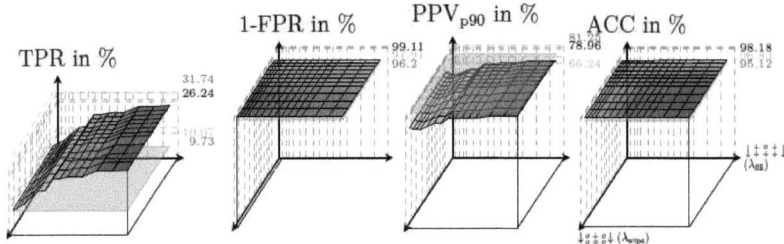

Figure B.37: Accuracy Plots with data: OD, profile algorithm: 2T-GSA, smoothing algorithm: FD. Detailed figure description at section start on page 228.

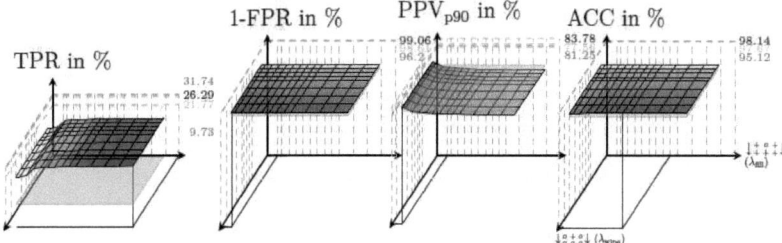

Figure B.38: Accuracy Plots with data: OD, profile algorithm: 2T-GSA, smoothing algorithm: FS. Detailed figure description at section start on page 228.

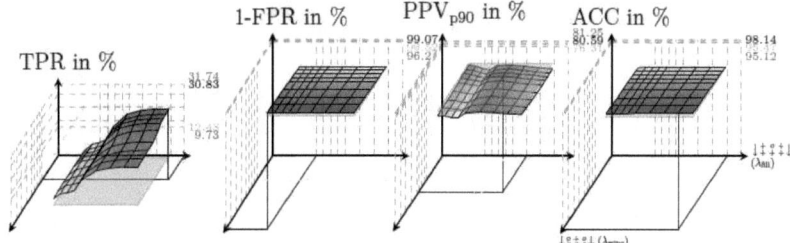

Figure B.39: Accuracy Plots with data: OD, profile algorithm: 2T-GSA, smoothing algorithm: SU. Detailed figure description at section start on page 228.

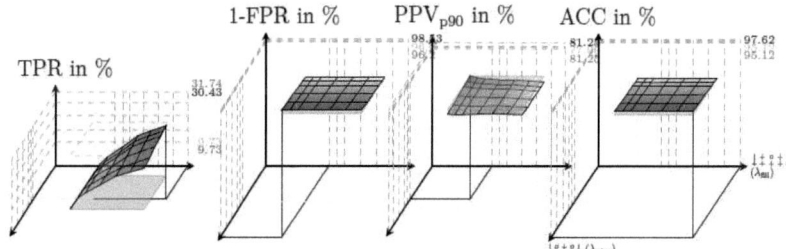

Figure B.40: Accuracy Plots with data: OD, profile algorithm: 1S-GSA, smoothing algorithm: SD. Detailed figure description at section start on page 228.

One-threshold segmentation GSA-type activation profile algorithm (1S-GSA)

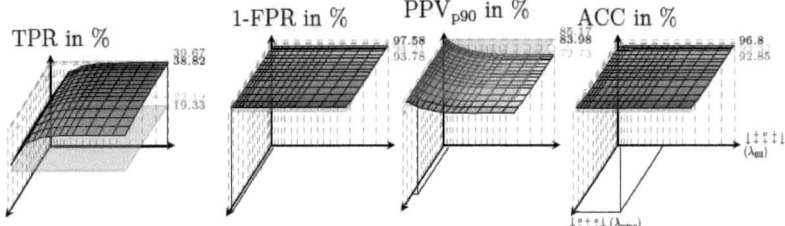

Figure B.41: Accuracy Plots with data: OD, profile algorithm: 1S-GSA, smoothing algorithm: AM. Detailed figure description at section start on page 228.

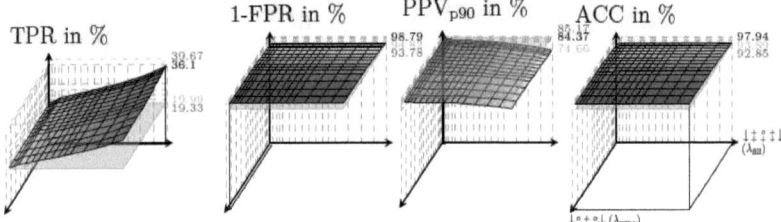

Figure B.42: Accuracy Plots with data: OD, profile algorithm: 1S-GSA, smoothing algorithm: GM. Detailed figure description at section start on page 228.

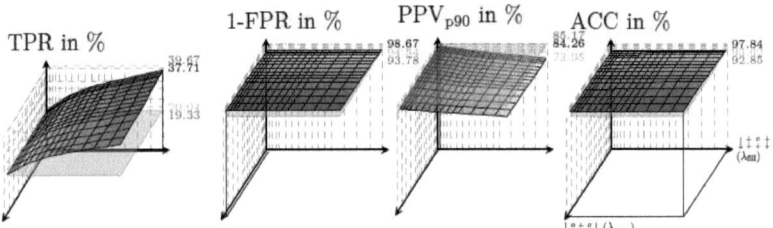

Figure B.43: Accuracy Plots with data: OD, profile algorithm: 1S-GSA, smoothing algorithm: IN. Detailed figure description at section start on page 228.

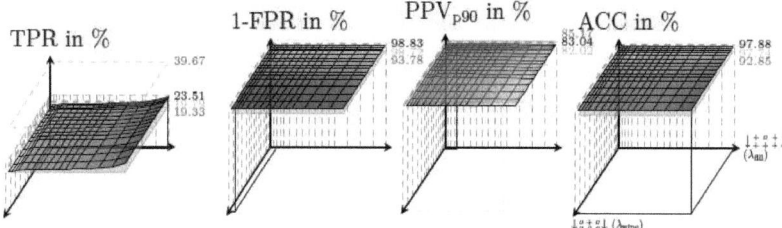

Figure B.44: Accuracy Plots with data: OD, profile algorithm: 1S-GSA, smoothing algorithm: IX. Detailed figure description at section start on page 228.

Two-threshold segmentation GSA-type activation profile algorithm (2S-GSA)

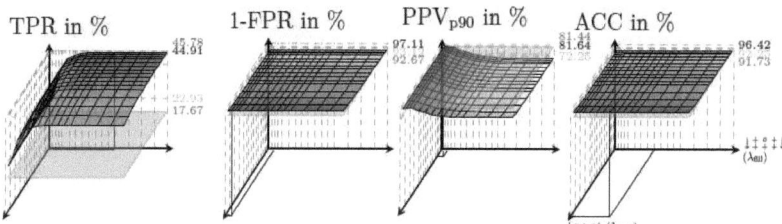

Figure B.45: Accuracy Plots with data: OD, profile algorithm: 2S-GSA, smoothing algorithm: AM. Detailed figure description at section start on page 228.

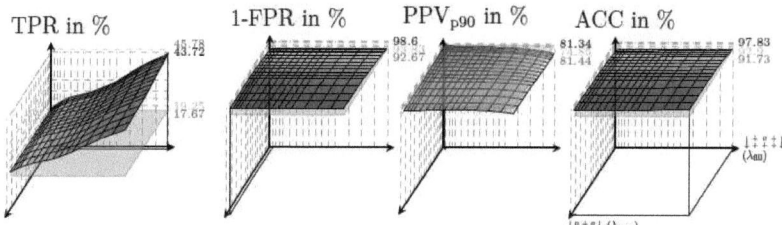

Figure B.46: Accuracy Plots with data: OD, profile algorithm: 2S-GSA, smoothing algorithm: GM. Detailed figure description at section start on page 228.

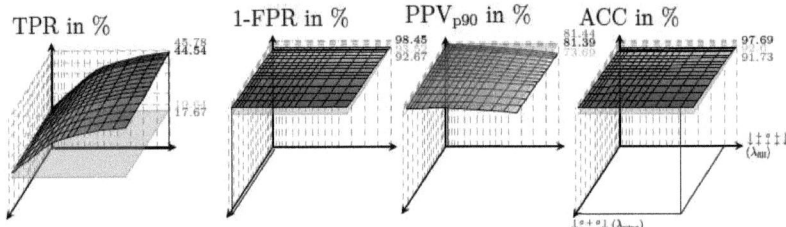

Figure B.47: Accuracy Plots with data: OD, profile algorithm: 2S-GSA, smoothing algorithm: IN. Detailed figure description at section start on page 228.

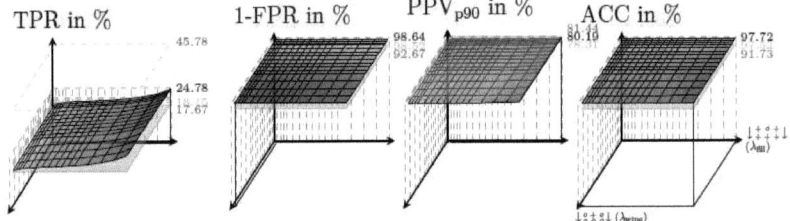

Figure B.48: Accuracy Plots with data: OD, profile algorithm: 2S-GSA, smoothing algorithm: IX. Detailed figure description at section start on page 228.

Skin healing data set

Two-threshold GSA-type activation profile algorithm (2T-GSA)

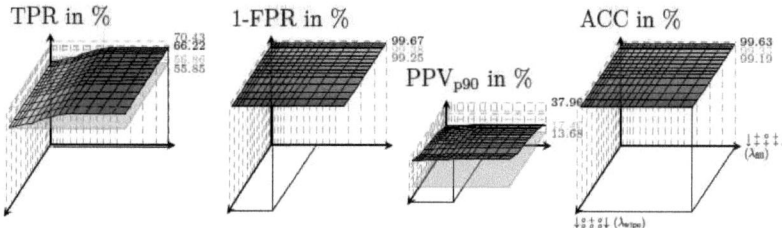

Figure B.49: Accuracy Plots with data: SH, profile algorithm: 2T-GSA, smoothing algorithm: AM. Detailed figure description at section start on page 228.

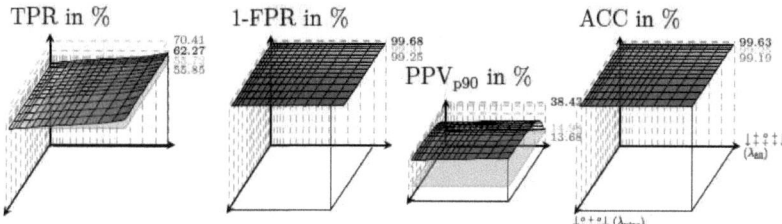

Figure B.50: Accuracy Plots with data: SH, profile algorithm: 2T-GSA, smoothing algorithm: GM. Detailed figure description at section start on page 228.

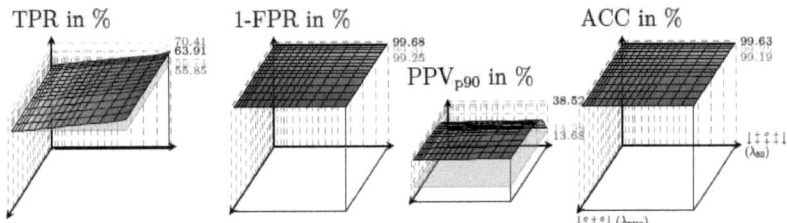

Figure B.51: Accuracy Plots with data: SH, profile algorithm: 2T-GSA, smoothing algorithm: IN. Detailed figure description at section start on page 228.

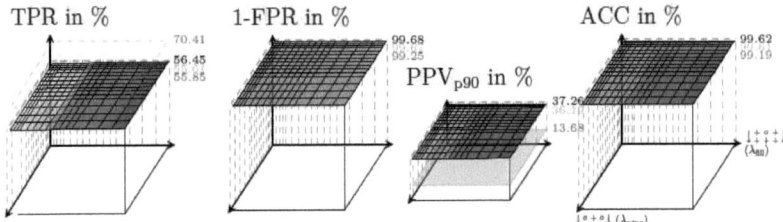

Figure B.52: Accuracy Plots with data: SH, profile algorithm: 2T-GSA, smoothing algorithm: IX. Detailed figure description at section start on page 228.

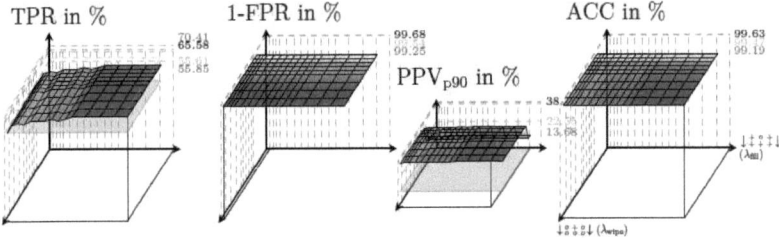

Figure B.53: Accuracy Plots with data: SH, profile algorithm: 2T-GSA, smoothing algorithm: FD. Detailed figure description at section start on page 228.

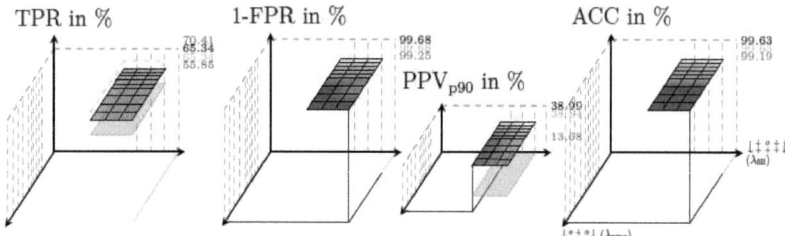

Figure B.54: Accuracy Plots with data: SH, profile algorithm: 2T-GSA, smoothing algorithm: FS. Detailed figure description at section start on page 228.

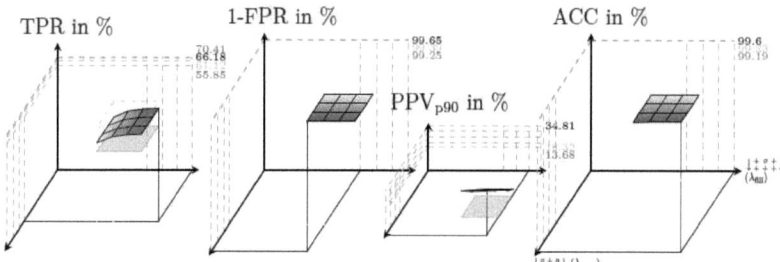

Figure B.55: Accuracy Plots with data: SH, profile algorithm: 2T-GSA, smoothing algorithm: SU. Detailed figure description at section start on page 228.

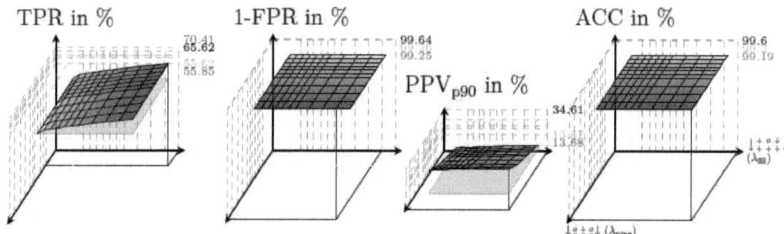

Figure B.56: Accuracy Plots with data: SH, profile algorithm: 1S-GSA, smoothing algorithm: SD. Detailed figure description at section start on page 228.

One-threshold segmentation GSA-type activation profile algorithm (1S-GSA)

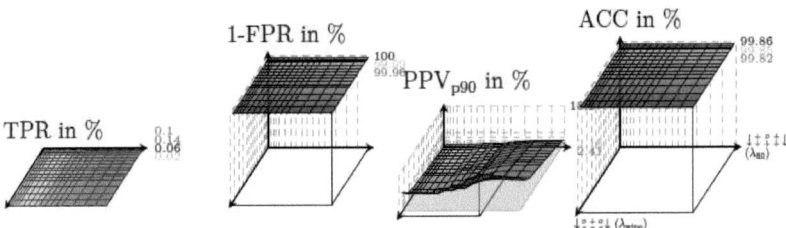

Figure B.57: Accuracy Plots with data: SH, profile algorithm: 1S-GSA, smoothing algorithm: AM. Detailed figure description at section start on page 228.

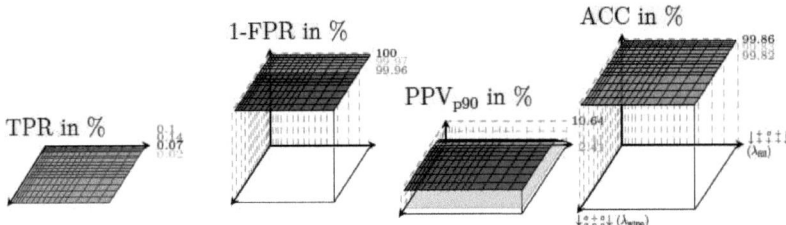

Figure B.58: Accuracy Plots with data: SH, profile algorithm: 1S-GSA, smoothing algorithm: GM. Detailed figure description at section start on page 228.

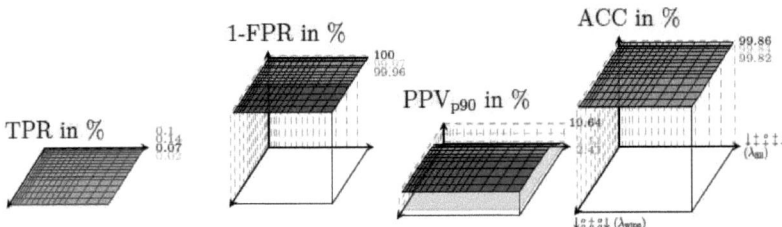

Figure B.59: Accuracy Plots with data: SH, profile algorithm: 1S-GSA, smoothing algorithm: IN. Detailed figure description at section start on page 228.

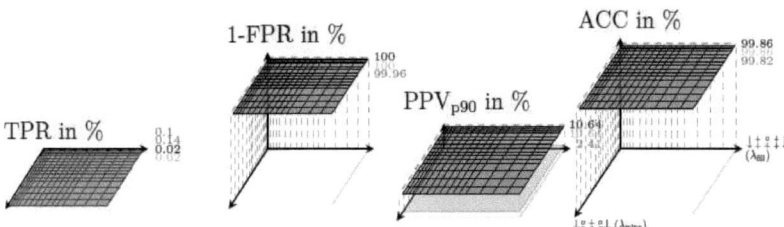

Figure B.60: Accuracy Plots with data: SH, profile algorithm: 1S-GSA, smoothing algorithm: IX. Detailed figure description at section start on page 228.

Two-threshold segmentation GSA-type activation profile algorithm (2S-GSA)

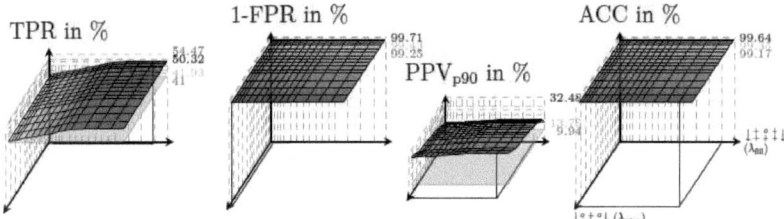

Figure B.61: Accuracy Plots with data: SH, profile algorithm: 2S-GSA, smoothing algorithm: AM. Detailed figure description at section start on page 228.

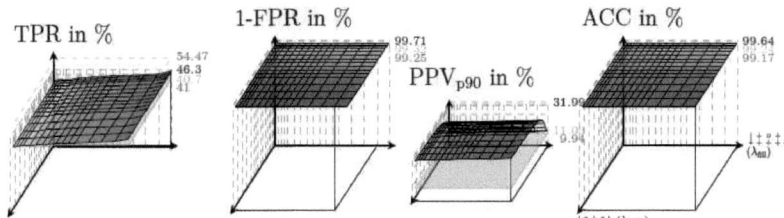

Figure B.62: Accuracy Plots with data: SH, profile algorithm: 2S-GSA, smoothing algorithm: GM. Detailed figure description at section start on page 228.

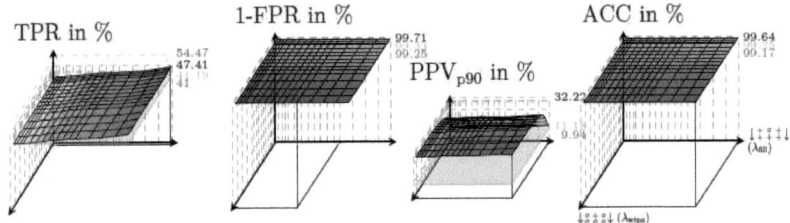

Figure B.63: Accuracy Plots with data: SH, profile algorithm: 2S-GSA, smoothing algorithm: IN. Detailed figure description at section start on page 228.

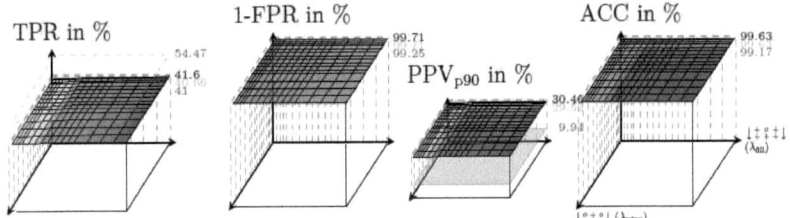

Figure B.64: Accuracy Plots with data: SH, profile algorithm: 2S-GSA, smoothing algorithm: IX. Detailed figure description at section start on page 228.

Tongue healing data set

Two-threshold GSA-type activation profile algorithm (2T-GSA)

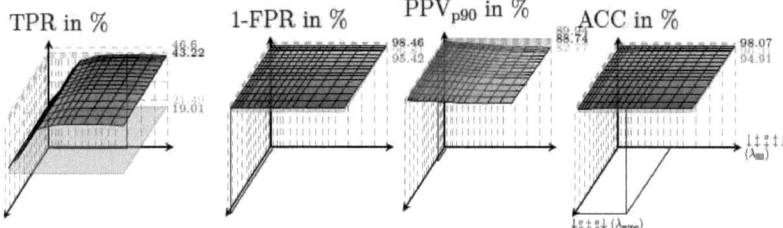

Figure B.65: Accuracy Plots with data: TH, profile algorithm: 2T-GSA, smoothing algorithm: AM. Detailed figure description at section start on page 228.

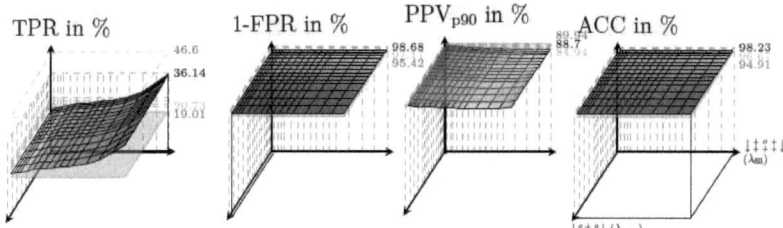

Figure B.66: Accuracy Plots with data: TH, profile algorithm: 2T-GSA, smoothing algorithm: GM. Detailed figure description at section start on page 228.

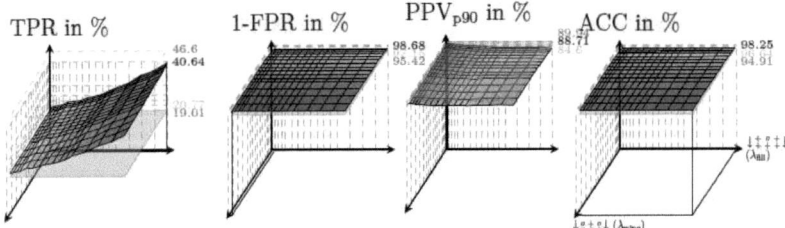

Figure B.67: Accuracy Plots with data: TH, profile algorithm: 2T-GSA, smoothing algorithm: IN. Detailed figure description at section start on page 228.

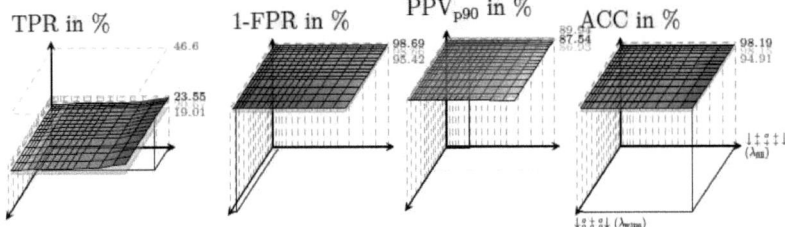

Figure B.68: Accuracy Plots with data: TH, profile algorithm: 2T-GSA, smoothing algorithm: IX. Detailed figure description at section start on page 228.

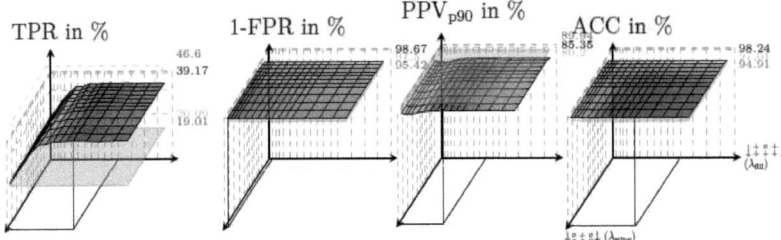

Figure B.69: Accuracy Plots with data: TH, profile algorithm: 2T-GSA, smoothing algorithm: FD. Detailed figure description at section start on page 228.

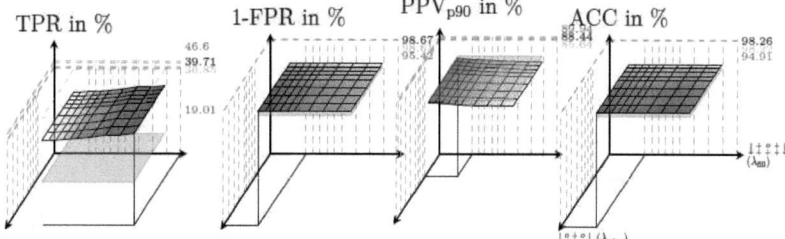

Figure B.70: Accuracy Plots with data: TH, profile algorithm: 2T-GSA, smoothing algorithm: FS. Detailed figure description at section start on page 228.

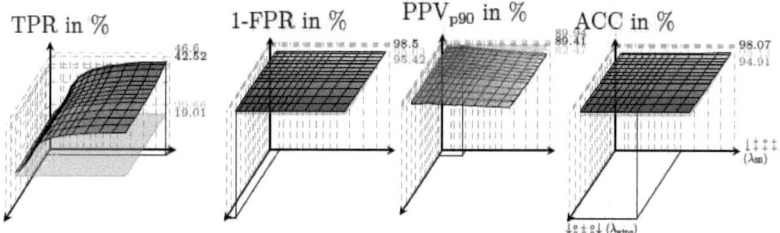

Figure B.71: Accuracy Plots with data: TH, profile algorithm: 2T-GSA, smoothing algorithm: SU. Detailed figure description at section start on page 228.

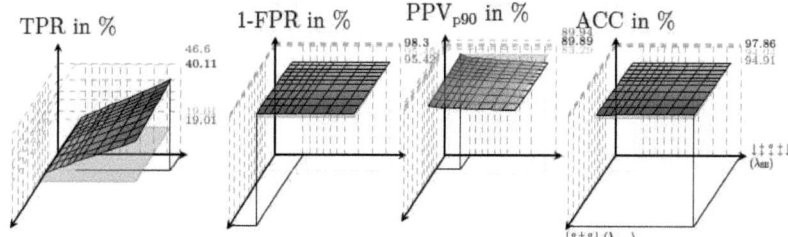

Figure B.72: Accuracy Plots with data: TH, profile algorithm: 1S-GSA, smoothing algorithm: SD. Detailed figure description at section start on page 228.

One-threshold segmentation GSA-type activation profile algorithm (1S-GSA)

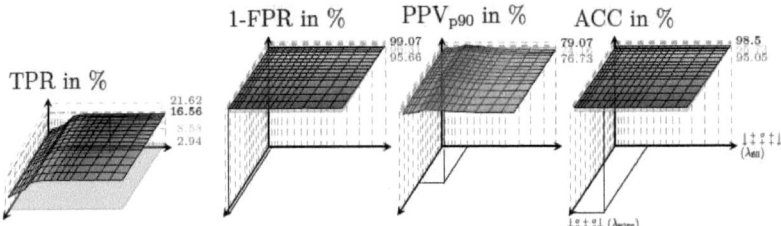

Figure B.73: Accuracy Plots with data: TH, profile algorithm: 1S-GSA, smoothing algorithm: AM. Detailed figure description at section start on page 228.

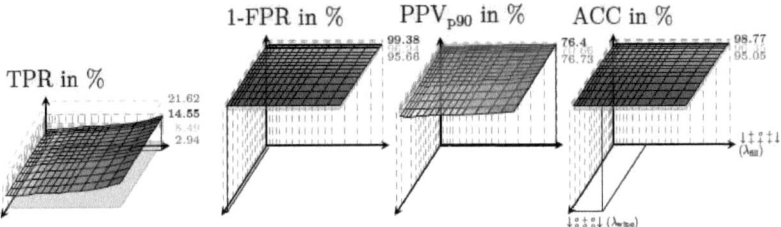

Figure B.74: Accuracy Plots with data: TH, profile algorithm: 1S-GSA, smoothing algorithm: GM. Detailed figure description at section start on page 228.

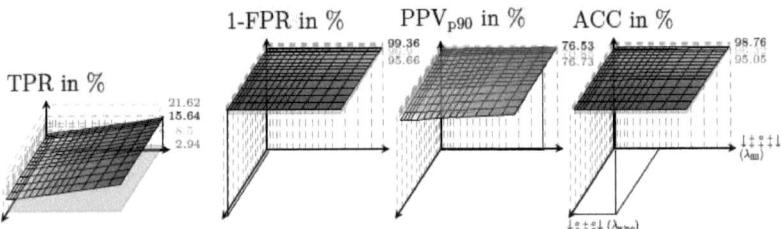

Figure B.75: Accuracy Plots with data: TH, profile algorithm: 1S-GSA, smoothing algorithm: IN. Detailed figure description at section start on page 228.

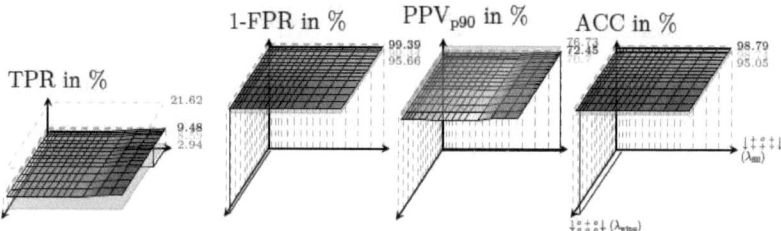

Figure B.76: Accuracy Plots with data: TH, profile algorithm: 1S-GSA, smoothing algorithm: IX. Detailed figure description at section start on page 228.

Two-threshold segmentation GSA-type activation profile algorithm (2S-GSA)

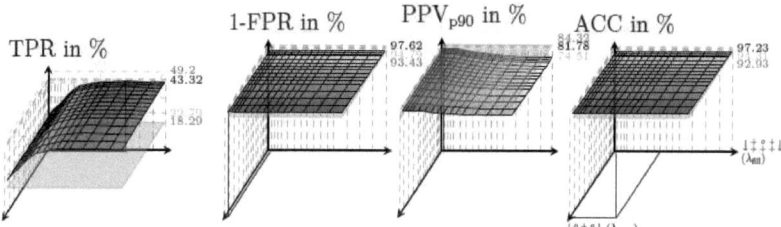

Figure B.77: Accuracy Plots with data: TH, profile algorithm: 2S-GSA, smoothing algorithm: AM. Detailed figure description at section start on page 228.

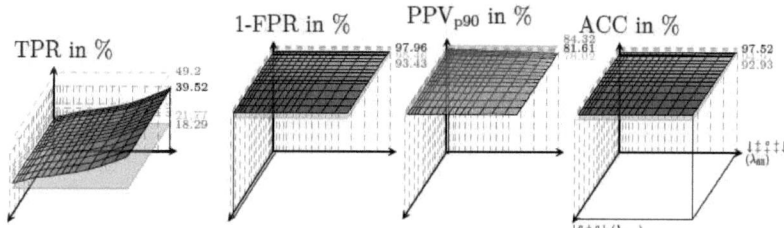

Figure B.78: Accuracy Plots with data: TH, profile algorithm: 2S-GSA, smoothing algorithm: GM. Detailed figure description at section start on page 228.

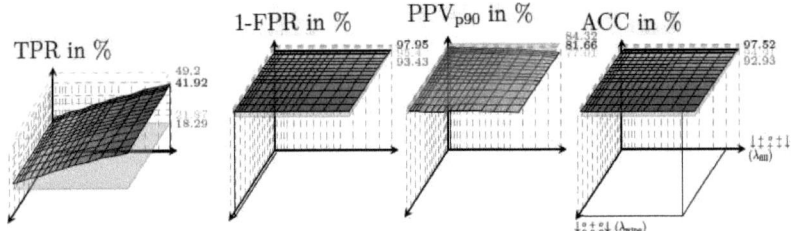

Figure B.79: Accuracy Plots with data: TH, profile algorithm: 2S-GSA, smoothing algorithm: IN. Detailed figure description at section start on page 228.

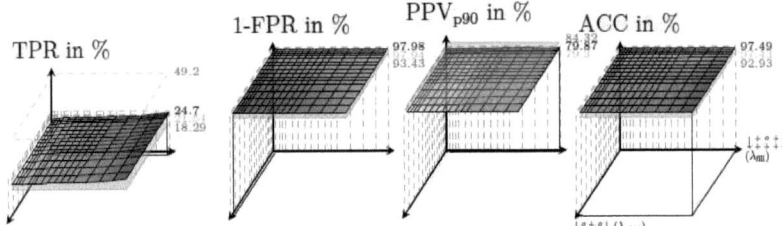

Figure B.80: Accuracy Plots with data: TH, profile algorithm: 2S-GSA, smoothing algorithm: IX. Detailed figure description at section start on page 228.

Tables of significant smoothed profiles

This appendix section includes the significant gene set activation profiles resulting from applying the simulation based selected profile algorithms and smoothing procedures for each of the four application data sets. All tables are ordered according to a decreasing $D_s^{|\text{med}|}$ score. Only the 100 most differentially expressed significant sets are printed for OD and TH data sets due to the large number of significant gene sets. The complete table is embedded in the electronic version of the document only.

Aldosterone heart data set

Table B.7: All 105 significant gene set activation profiles resulting from a 2S-GSA profile algorithm combined with AM smoothing on the AH data.

ID	Profile	Description	$\|s\|$	2T-GSA	1S-GSA	STEM	maSigFun
GO:0035456	ooo+ooo	response to interferon-beta	15	✓	✓	✗	✗
GO:0035458	ooo+ooo	cellular response to interferon-beta	14	✓	✓	✗	✗
GO:0051856	ooo+ooo	adhesion to symbiont	8	✓	✓	✗	✗
GO:0044403	ooo+ooo	symbiosis, encompassing mutualism through parasitism	87	✗	✓	✗	✗
GO:0051825	ooo+ooo	adhesion to other organism involved in symbiotic interaction	8	✓	✓	✗	✗
GO:0030595	o-oooo	leukocyte chemotaxis	94	✗	✓	✗	✗
GO:0060326	o-oooo	cell chemotaxis	120	✗	✓	✗	✗
GO:0001562	ooo+ooo	response to protozoan	17	✓	✓	✗	✗
GO:0042832	ooo+ooo	defense response to protozoan	15	✓	✓	✗	✗
GO:0044419	ooo+ooo	interspecies interaction between organisms	104	✓	✓	✗	✗
GO:0051702	ooo+ooo	interaction with symbiont	37	✗	✓	✗	✗
REACT:GRB2	ooooo+o	genes involved in GRB2	10	✓	✓	✗	✗

continued on next page ...

... continued from previous page

ID	Profile	Description	$\|s\|$	2T-GSA	1S-GSA	STEM	maSigFun
REACT: P130CAS	ooooo+o	genes involved in p130Cas linkage to MAPK signaling for integrins	10	✓	✓	✗	✗
GO:0035457	ooo+o-o	cellular response to interferon-alpha	4	✗	✓	✗	✗
GO:0006935	o-ooooo	chemotaxis	324	✗	✓	✗	✗
GO:0042330	o-ooooo	taxis	325	✗	✓	✗	✗
GO:0009607	ooo+ooo	response to biotic stimulus	369	✓	✓	✗	✗
GO:0042742	ooo+ooo	defense response to bacterium	102	✓	✓	✗	✗
GO:0050830	ooo+ooo	defense response to Gram-positive bacterium	37	✓	✓	✗	✗
GO:0051707	ooo+ooo	response to other organism	338	✓	✓	✗	✗
GO:0071346	ooo+ooo	cellular response to interferon-gamma	22	✗	✓	✗	✗
GO:0034341	ooo+ooo	response to interferon-gamma	37	✓	✓	✗	✗
GO:0051704	ooo+ooo	multi-organism process	512	✓	✓	✗	✗
KEGG:04610	ooooo+o	complement and coagulation cascades	69	✓	✓	✗	✗
BioCarta: ACTINY	oo+oooo	Y branching of actin filaments	700	✓	✓	✗	✗
GO:0050900	o-ooooo	leukocyte migration	144	✗	✓	✗	✗
GO:0002523	o-ooooo	leukocyte migration involved in inflammatory response	7	✗	✗	✗	✗
GO:0040011	o-ooooo	locomotion	898	✓	✓	✗	✗
GO:0048870	o-ooooo	cell motility	750	✓	✓	✗	✗
GO:0051674	o-ooooo	localization of cell	750	✓	✓	✗	✗
GO:0009605	o-ooooo	response to external stimulus	796	✗	✓	✗	✗
GO:0032101	o-ooooo	regulation of response to external stimulus	301	✗	✗	✗	✗
GO:0030198	o-ooooo	extracellular matrix organization	137	✓	✓	✗	✗

continued on next page ...

... continued from previous page

ID	Profile	Description	$\|s\|$	2T-GSA	1S-GSA	STEM	maSigFun
GO:0043062	o-ooooo	extracellular structure organization	137	✓	✓	✗	✗
GO:0009611	o-ooooo	response to wounding	549	✓	✓	✗	✗
GO:0050727	o-ooooo	regulation of inflammatory response	145	✗	✗	✗	✗
GO:0007155	o-ooooo	cell adhesion	726	✗	✗	✗	✗
GO:0022610	o-ooooo	biological adhesion	731	✗	✓	✗	✗
GO:0018149	o-ooooo	peptide cross-linking	25	✗	✗	✗	✗
GO:0006631	o-ooooo	fatty acid metabolic process	264	✗	✗	✗	✗
GO:0044255	o-ooooo	cellular lipid metabolic process	634	✗	✗	✗	✗
GO:0009617	ooo+ooo	response to bacterium	228	✓	✓	✗	✗
GO:0048871	o-ooooo	multicellular organismal homeostasis	137	✓	✓	✗	✗
GO:0016337	o-ooooo	cell-cell adhesion	261	✗	✗	✗	✗
GO:0045087	ooo+ooo	innate immune response	237	✓	✓	✗	✗
GO:0050873	oo+oooo	brown fat cell differentiation	28	✗	✗	✗	✗
GO:0006953	ooooo+o	acute-phase response	38	✓	✓	✗	✗
GO:0009615	ooo+ooo	response to virus	114	✗	✗	✗	✗
GO:0019216	oo+oooo	regulation of lipid metabolic process	178	✗	✓	✗	✗
GO:0043200	o-ooooo	response to amino acid stimulus	39	✗	✗	✗	✗
GO:0006952	o-o+ooo	defense response	639	✓	✓	✗	✗
GO:0002376	o-o+ooo	immune system process	1140	✓	✓	✗	✗
GO:0044283	o-+oooo	small molecule biosynthetic process	337	✓	✓	✗	✗
GO:0042221	o-o+ooo	response to chemical stimulus	1821	✓	✓	✗	✗
GO:0070887	o-o+ooo	cellular response to chemical stimulus	913	✓	✓	✗	✗
GO:0034097	ooo+ooo	response to cytokine stimulus	278	✓	✓	✗	✗
GO:0001101	o-ooooo	response to acid	54	✗	✗	✗	✗

continued on next page ...

... continued from previous page

ID	Profile	Description	$|s|$	2T-GSA	1S-GSA	STEM	maSigFun
KEGG:05144	ooo+ooo	Malaria	43	✓	✓	✗	✗
GO:0071345	ooo+ooo	cellular response to cytokine stimulus	210	✓	✓	✗	✗
GO:0006955	o-o+ooo	immune response	572	✓	✓	✗	✗
GO:0016477	o-ooooo	cell migration	695	✓	✓	✗	✗
GO:0006928	o-ooooo	cellular component movement	876	✓	✓	✗	✗
GO:0009888	ooo-ooo	tissue development	1040	✗	✓	✗	✗
GO:0010033	ooo+ooo	response to organic substance	1107	✓	✓	✗	✗
GO:0071310	ooo+ooo	cellular response to organic substance	675	✓	✓	✗	✗
GO:0007507	ooo-ooo	heart development	336	✗	✗	✗	✗
GO:0046364	oo+oooo	monosaccharide biosynthetic process	54	✓	✗	✗	✗
GO:0072358	ooo-ooo	cardiovascular system development	662	✓	✓	✗	✗
GO:0072359	ooo-ooo	circulatory system development	662	✓	✓	✗	✗
KEGG:05150	o-ooooo	staphylococcus aureus infection	46	✓	✓	✗	✗
GO:0006954	o-o+ooo	inflammatory response	341	✓	✓	✗	✗
GO:0006897	oo+oooo	endocytosis	358	✓	✓	✗	✗
GO:0010324	oo+oooo	membrane invagination	358	✓	✓	✗	✗
GO:0001568	ooo-ooo	blood vessel development	443	✓	✓	✗	✗
GO:0050896	o-o+ooo	response to stimulus	4884	✓	✓	✗	✗
GO:0032374	oo+oooo	regulation of cholesterol transport	27	✓	✓	✗	✗
GO:0032371	oo+oooo	regulation of sterol transport	27	✓	✓	✗	✗
GO:0016044	oo+oooo	cellular membrane organization	551	✓	✓	✗	✗
GO:0061024	oo+oooo	membrane organization	553	✓	✓	✗	✗
GO:0045444	oo+oooo	fat cell differentiation	106	✓	✓	✗	✗

continued on next page ...

... continued from previous page

ID	Profile	Description	$\|s\|$	2T-GSA	1S-GSA	STEM	maSigFun
GO:0006950	o-o+ooo	response to stress	1801	✓	✓	✗	✗
GO:0060255	ooo-ooo	regulation of macromolecule metabolic process	3416	✓	✓	✗	✗
GO:0031532	o-ooooo	actin cytoskeleton reorganization	40	✗	✗	✗	✗
GO:0048514	ooo-ooo	blood vessel morphogenesis	385	✗	✓	✗	✗
GO:0001944	o-o-ooo	vasculature development	468	✓	✓	✗	✗
GO:0048583	o-ooooo	regulation of response to stimulus	1787	✓	✓	✗	✗
GO:0080090	ooo-ooo	regulation of primary metabolic process	3638	✓	✓	✗	✗
GO:0031323	ooo-ooo	regulation of cellular metabolic process	3684	✓	✓	✗	✗
GO:0019222	ooo-ooo	regulation of metabolic process	4207	✓	✓	✗	✗
GO:0002682	ooo+ooo	regulation of immune system process	590	✓	✓	✗	✗
GO:0050789	ooo-ooo	regulation of biological process	6841	✗	✓	✗	✗
GO:0032268	ooo-ooo	regulation of cellular protein metabolic process	977	✗	✗	✗	✗
GO:0048523	ooo-ooo	negative regulation of cellular process	2382	✓	✓	✗	✗
GO:0048519	ooo-ooo	negative regulation of biological process	2612	✓	✓	✗	✗
GO:0048585	ooo-ooo	negative regulation of response to stimulus	618	✓	✓	✗	✗
GO:0008283	o-ooooo	cell proliferation	1207	✓	✓	✗	✗
GO:0044237	-ooooo	cellular metabolic process	7240	✓	✓	✗	✗
GO:0044260	-ooooo	cellular macromolecule metabolic process	5235	✓	✓	✗	✗

continued on next page ...

... continued from previous page

ID	Profile	Description	$|s|$	2T-GSA	1S-GSA	STEM	maSigFun
GO:0043170	-oooooo	macromolecule metabolic process	5877	✓	✓	✗	✗
GO:0006139	-oooooo	nucleobase-containing compound metabolic process	4241	✓	✓	✗	✗
GO:0071843	oooo+oo	cellular component biogenesis at cellular level	189	✓	✗	✗	✗
GO:0090304	-oooooo	nucleic acid metabolic process	3396	✓	✓	✗	✗
GO:0034641	-oooooo	cellular nitrogen compound metabolic process	4560	✓	✓	✗	✗
GO:0006807	-oooooo	nitrogen compound metabolic process	4644	✓	✓	✗	✗
GO:0010467	-o-oooo	gene expression	3374	✓	✓	✗	✗

Ovary development data set

Table B.8: Top 100 significant gene set activation profiles according to the $D_s^{|\text{med}|}$ score resulting from a 2S-GSA profile algorithm combined with AM smoothing on the OD data. The complete list is available (in electronic version only).

ID	Profile	Description	$|s|$	2T-GSA	1S-GSA	STEM	maSigFun
GO:0070301	-ooooo	cellular response to hydrogen peroxide	30	✗	✓	✗	✗
REACT:GRB2	--oooo	genes involved in GRB2	11	✗	✓	✗	✗
REACT:P130CAS	--oooo	genes involved in p130Cas linkage to MAPK signaling for integrins	11	✗	✓	✗	✗
GO:0042542	--oooo	response to hydrogen peroxide	52	✗	✓	✗	✗

continued on next page ...

... continued from previous page

ID	Profile	Description	$\|s\|$	2T-GSA	1S-GSA	STEM	maSigFun
BioCarta: FIBRINOLYSIS	------	fibrinolysis pathway	10	✗	✓	✗	✗
GO:0050790	o-oooo	regulation of catalytic activity	1427	✗	✓	✗	✗
BioCarta: AMI	------	acute myocardial infarction	15	✓	✓	✗	✗
BioCarta: EXTRINSIC	------	extrinsic prothrombin activation pathway	13	✗	✓	✗	✗
BioCarta: INTRINSIC	------	intrinsic prothrombin	16	✓	✓	✗	✗
GO:0007066	oo+ooo	female meiosis sister chromatid cohesion	4	✗	✓	✗	✗
REACT: COMMON	------	genes involved in common pathway	13	✓	✓	✗	✗
GO:0044236	oooo--	multicellular organismal metabolic process	60	✗	✓	✗	✗
GO:0010916	-ooooo	negative regulation of very-low-density lipoprotein particle clearance	4	✗	✓	✗	✗
GO:0010915	-ooooo	regulation of very-low-density lipoprotein particle clearance	4	✗	✓	✗	✗
GO:0032374	------	regulation of cholesterol transport	27	✓	✓	✗	✗
GO:0032371	------	regulation of sterol transport	27	✓	✓	✗	✗
GO:0032963	ooooo-	collagen metabolic process	49	✗	✓	✗	✗
GO:0050881	------	musculoskeletal movement	24	✓	✓	✗	✗
GO:0050879	------	multicellular organismal movement	24	✓	✓	✗	✗
GO:0003009	------	skeletal muscle contraction	14	✓	✓	✗	✗
GO:0010896	------	regulation of triglyceride catabolic process	7	✗	✓	✗	✗

continued on next page ...

... continued from previous page

ID	Profile	Description	$|s|$	2T-GSA	1S-GSA	STEM	maSigFun
GO:0010901	------	regulation of very-low-density lipoprotein particle remodeling	4	✗	✓	✗	✗
GO:0006094	------	gluconeogenesis	35	✓	✓	✗	✗
GO:0019319	------	hexose biosynthetic process	43	✓	✓	✗	✗
REACT: MUSCLE	------	genes involved in muscle contraction	30	✓	✓	✗	✗
GO:0030299	------	intestinal cholesterol absorption	9	✓	✓	✗	✗
GO:0051336	------	regulation of hydrolase activity	738	✓	✓	✗	✗
GO:0006111	ooo-oo	regulation of gluconeogenesis	15	✗	✓	✗	✗
GO:0045103	ooo---	intermediate filament-based process	28	✗	✓	✗	✗
BioCarta: UCALPAIN	ooo--o	uCalpain and friends in Cell spread	13	✗	✓	✗	✗
GO:0030194	------	positive regulation of blood coagulation	16	✓	✓	✗	✗
GO:1900048	------	positive regulation of hemostasis	16	✓	✓	✗	✗
REACT: STRIATED	------	genes involved in striated muscle contraction	20	✓	✓	✗	✗
GO:0006942	o-oooo	regulation of striated muscle contraction	24	✗	✓	✗	✗
GO:0050892	------	intestinal absorption	18	✓	✓	✗	✗
GO:0051346	------	negative regulation of hydrolase activity	275	✓	✓	✗	✗
GO:0050996	------	positive regulation of lipid catabolic process	18	✓	✓	✗	✗
GO:0052548	------	regulation of endopeptidase activity	261	✓	✓	✗	✗

continued on next page ...

... continued from previous page

ID	Profile	Description	$|s|$	2T-GSA	1S-GSA	STEM	maSigFun
GO:0052547	------	regulation of peptidase activity	280	✓	✓	✗	✗
GO:0046503	------	glycerolipid catabolic process	22	✓	✓	✗	✗
GO:0006941	------	striated muscle contraction	55	✓	✓	✗	✗
GO:0051918	------	negative regulation of fibrinolysis	7	✗	✓	✗	✗
GO:0019433	------	triglyceride catabolic process	13	✓	✓	✗	✗
GO:0034372	------	very-low-density lipoprotein particle remodeling	9	✓	✓	✗	✗
REACT: RECYCLING	-oo---	Genes involved in Recycling of bile acids and salts	8	✗	✓	✗	✗
GO:0050820	------	positive regulation of coagulation	19	✓	✓	✗	✗
GO:0033344	------	cholesterol efflux	33	✓	✓	✗	✗
GO:0010466	------	negative regulation of peptidase activity	184	✓	✓	✗	✗
GO:0046461	------	neutral lipid catabolic process	15	✓	✓	✗	✗
GO:0046464	------	acylglycerol catabolic process	15	✓	✓	✗	✗
GO:0044269	------	glycerol ether catabolic process	15	✓	✓	✗	✗
GO:0010898	------	positive regulation of triglyceride catabolic process	6	✗	✓	✗	✗
GO:0010951	------	negative regulation of endopeptidase activity	171	✓	✓	✗	✗
GO:0006642	-ooooo	triglyceride mobilization	6	✗	✓	✗	✗
GO:0030300	------	regulation of intestinal cholesterol absorption	6	✓	✓	✗	✗
GO:0033700	------	phospholipid efflux	11	✗	✓	✗	✗
REACT: CHYLOMICRON	------	genes involved in chylomicron-mediated lipid transport	15	✓	✓	✗	✗

continued on next page ...

... continued from previous page

ID	Profile	Description	$\|s\|$	2T-GSA	1S-GSA	STEM	maSigFun
GO:0043086	------	negative regulation of catalytic activity	472	✓	✓	✗	✗
KEGG:04972	------	Pancreatic secretion	100	✓	✓	✗	✗
GO:0015671	o+oooo	oxygen transport	6	✗	✓	✗	✗
GO:0030195	------	negative regulation of blood coagulation	25	✓	✓	✗	✗
GO:1900047	------	negative regulation of hemostasis	25	✓	✓	✗	✗
GO:0033275	ooo---	actin-myosin filament sliding	8	✗	✓	✗	✗
GO:0070252	ooo---	actin-mediated cell contraction	10	✗	✓	✗	✗
KEGG:04975	------	fat digestion and absorption	45	✓	✓	✗	✗
GO:0006936	------	muscle contraction	148	✓	✓	✗	✗
GO:0034370	------	triglyceride-rich lipoprotein particle remodeling	10	✓	✓	✗	✗
GO:0046364	------	monosaccharide biosynthetic process	54	✓	✓	✗	✗
GO:0045723	oo----	positive regulation of fatty acid biosynthetic process	15	✗	✓	✗	✗
GO:0065005	------	protein-lipid complex assembly	12	✗	✓	✗	✗
GO:0048608	ooooo+	reproductive structure development	202	✗	✓	✗	✗
GO:0051917	------	regulation of fibrinolysis	10	✓	✓	✗	✗
MetaCyc: PWY-4984	o----o	urea cycle	4	✗	✓	✗	✗
GO:0009566	oooo++	fertilization	82	✗	✓	✗	✗
GO:0022602	ooooo+	ovulation cycle process	74	✗	✓	✗	✗
GO:0042698	ooooo+	ovulation cycle	78	✗	✓	✗	✗
GO:0006508	------	proteolysis	819	✓	✓	✗	✗

continued on next page ...

... continued from previous page

ID	Profile	Description	$|s|$	2T-GSA	1S-GSA	STEM	maSigFun
GO:0045104	------	intermediate filament cytoskeleton organization	27	✗	✓	✗	✗
GO:0048511	ooooo+	rhythmic process	160	✗	✓	✗	✗
GO:0010872	------	regulation of cholesterol esterification	9	✓	✓	✗	✗
GO:0010873	------	positive regulation of cholesterol esterification	7	✓	✓	✗	✗
GO:0042730	------	fibrinolysis	14	✓	✓	✗	✗
GO:2000194	ooooo+	regulation of female gonad development	6	✗	✓	✗	✗
GO:0006559	o-oo--	L-phenylalanine catabolic process	6	✗	✓	✗	✗
GO:0032368	------	regulation of lipid transport	49	✓	✓	✗	✗
GO:0034435	------	cholesterol esterification	12	✓	✓	✗	✗
GO:0034434	------	sterol esterification	12	✓	✓	✗	✗
GO:0034433	------	steroid esterification	12	✓	✓	✗	✗
GO:0030162	------	regulation of proteolysis	156	✓	✓	✗	✗
GO:0019751	oooo-o	polyol metabolic process	37	✓	✓	✗	✗
GO:0007586	------	digestion	68	✓	✓	✗	✗
GO:0034442	oooo-o	regulation of lipoprotein oxidation	4	✗	✓	✗	✗
GO:0045780	oooo-o	positive regulation of bone resorption	12	✗	✓	✗	✗
GO:0046852	oooo-o	positive regulation of bone remodeling	12	✗	✓	✗	✗
GO:0015918	------	sterol transport	53	✓	✓	✗	✗
GO:0030301	------	cholesterol transport	53	✓	✓	✗	✗
GO:0046486	------	glycerolipid metabolic process	219	✓	✓	✗	✗
GO:0097006	------	regulation of plasma lipoprotein particle levels	41	✓	✓	✗	✗
GO:0044241	------	lipid digestion	11	✓	✓	✗	✗

continued on next page ...

ID	Profile	Description	$\|s\|$	2T-GSA	1S-GSA	STEM	maSigFun
		... continued from previous page					
GO:0051919	------	positive regulation of fibrinolysis	4	✗	✓	✗	✗

Skin healing data set

Table B.9: All 50 significant gene set activation profiles according to the $D_s^{|\text{med}|}$ score resulting from a 2S-GSA profile algorithm combined with AM smoothing on the OD data.

ID	Profile	Description	$\|s\|$	1S-GSA	2T-GSA	STEM	maSigFun
KEGG:04620	oo+oooo	Toll-like receptor signaling pathway	93	✓	✓	✗	✗
GO:0031424	ooooo+o	keratinization	25	✓	✓	✗	✗
REACT:CHEMOKINE	+++oooo	genes involved in Chemokine receptors bind chemokines	36	✓	✓	✗	✗
REACT:PEPTIDE	+++oooo	genes involved in Peptide ligand-binding receptors	115	✓	✓	✗	✗
REACT:GPCR	o++oooo	genes involved in GPCR ligand binding	250	✓	✓	✗	✗
GO:0040011	o++oooo	locomotion	898	✓	✓	✗	✗
GO:0050715	o++oooo	positive regulation of cytokine secretion	46	✓	✓	✗	✗
GO:0050714	o++oooo	positive regulation of protein secretion	71	✓	✓	✗	✗
KEGG:04060	+++oooo	cytokine-cytokine receptor interaction	219	✓	✓	✗	✗
GO:0006954	o++oooo	inflammatory response	341	✓	✓	✗	✗

continued on next page ...

... continued from previous page

ID	Profile	Description	$|s|$	1S-GSA	2T-GSA	STEM	maSigFun
GO:0009605	o++oooo	response to external stimulus	796	✓	✓	✗	✗
GO:0030593	++++ooo	neutrophil chemotaxis	44	✓	✓	✗	✗
GO:0050663	o++oooo	cytokine secretion	72	✓	✓	✗	✗
GO:0050707	o++oooo	regulation of cytokine secretion	61	✓	✓	✗	✗
GO:0050708	o++oooo	regulation of protein secretion	103	✓	✓	✗	✗
GO:0051222	o++oooo	positive regulation of protein transport	138	✓	✓	✗	✗
GO:0030595	++++ooo	leukocyte chemotaxis	94	✓	✓	✗	✗
GO:0060326	++++ooo	cell chemotaxis	120	✓	✓	✗	✗
REACT: STRIATED	ooooo+o	genes involved in striated muscle contraction	20	✓	✓	✗	✗
GO:0009611	o++oooo	response to wounding	549	✓	✓	✗	✗
GO:0050900	++++ooo	leukocyte migration	144	✓	✓	✗	✗
GO:0007249	oo+oooo	I-kappaB kinase/NF-kappaB cascade	155	✓	✓	✗	✗
GO:0006952	+++oooo	defense response	639	✓	✓	✗	✗
GO:0002690	oo++ooo	positive regulation of leukocyte chemotaxis	41	✓	✓	✗	✗
GO:0002688	oo++ooo	regulation of leukocyte chemotaxis	47	✓	✓	✗	✗
GO:0002376	+++oooo	immune system process	1140	✓	✓	✗	✗
GO:0006935	++++ooo	chemotaxis	324	✓	✓	✗	✗
GO:0042330	++++ooo	taxis	325	✓	✓	✗	✗
GO:0006955	+++oooo	immune response	572	✓	✓	✗	✗
GO:0001816	+++oooo	cytokine production	333	✓	✓	✗	✗
GO:0001817	+++oooo	regulation of cytokine production	295	✓	✓	✗	✗
GO:0010627	oo+oooo	regulation of intracellular protein kinase cascade	493	✓	✓	✗	✗

continued on next page ...

... continued from previous page

ID	Profile	Description	$|s|$	1S-GSA	2T-GSA	STEM	maSigFun
GO:0043122	oo+oooo	regulation of I-kappaB kinase/NF-kappaB cascade	124	✓	✓	✗	✗
GO:0001775	oo+oooo	cell activation	521	✓	✓	✗	✗
GO:0045321	oo+oooo	leukocyte activation	461	✓	✓	✗	✗
GO:0002274	oo+oooo	myeloid leukocyte activation	102	✓	✗	✗	✗
GO:0002684	oo+oooo	positive regulation of immune system process	383	✓	✓	✗	✗
GO:0006950	o+ooooo	response to stress	1801	✓	✓	✗	✗
GO:0002682	oo+oooo	regulation of immune system process	590	✓	✓	✗	✗
GO:0050896	+++oooo	response to stimulus	4884	✓	✓	✗	✗
GO:0002443	o+ooooo	leukocyte mediated immunity	181	✓	✗	✗	✗
GO:0032940	o+ooooo	secretion by cell	527	✓	✓	✗	✗
GO:0051223	o+ooooo	regulation of protein transport	233	✓	✓	✗	✗
GO:0009306	o+ooooo	protein secretion	133	✓	✓	✗	✗
GO:0002252	o+ooooo	immune effector process	310	✓	✓	✗	✗
GO:0050921	oo++ooo	positive regulation of chemotaxis	66	✓	✓	✗	✗
GO:0002687	ooo+ooo	positive regulation of leukocyte migration	59	✓	✓	✗	✗
GO:0033205	o-ooooo	cell cycle cytokinesis	27	✓	✓	✗	✗
GO:0046654	oo+oooo	tetrahydrofolate biosynthetic process	5	✗	✗	✗	✗
GO:0000910	o-ooooo	cytokinesis	77	✓	✓	✗	✗

Tongue healing data set

Table B.10: Top 100 significant gene set activation profiles according to the $D_s^{|\text{med}|}$ score resulting from a 2S-GSA profile algorithm combined with AM smoothing on the OD data. The complete list is available (in electronic version only)

ID	Profile	Description	$\|s\|$	2T-GSA	1S-GSA	STEM	maSigFun
KEGG:04621	o++oooo	NOD-like receptor signaling pathway	51	✗	✓	✗	✗
GO:0035457	+++oooo	cellular response to interferon-alpha	4	✗	✓	✗	✗
GO:0010573	++++ooo	vascular endothelial growth factor production	16	✓	✓	✗	✗
GO:0010574	++++ooo	regulation of vascular endothelial growth factor production	16	✓	✓	✗	✗
GO:0001660	o+++ooo	fever generation	12	✓	✓	✗	✗
GO:0031649	o+++ooo	heat generation	17	✓	✓	✗	✗
GO:0071354	++ooooo	cellular response to interleukin-6	10	✗	✓	✗	✗
GO:0031622	o+++ooo	positive regulation of fever generation	8	✗	✓	✗	✗
GO:0031652	o+++ooo	positive regulation of heat generation	10	✓	✓	✗	✗
REACT: STRIATED	ooo++++	genes involved in striated muscle contraction	20	✓	✓	✗	✗
GO:0032846	oo+oooo	positive regulation of homeostatic process	59	✓	✓	✗	✗
REACT: MUSCLE	ooo++++	genes involved in muscle contraction	31	✓	✓	✗	✗
GO:0001659	o+++ooo	temperature homeostasis	29	✗	✓	✗	✗
GO:0002673	++++ooo	regulation of acute inflammatory response	46	✓	✓	✗	✗
GO:0002526	++++ooo	acute inflammatory response	88	✓	✓	✗	✗
GO:0055074	o++oooo	calcium ion homeostasis	242	✓	✓	✗	✗
GO:0035455	+++++oo	response to interferon-alpha	6	✓	✓	✗	✗
GO:0006953	++++ooo	acute-phase response	38	✓	✓	✗	✗

continued on next page ...

... continued from previous page

ID	Profile	Description	$|s|$	2T-GSA	1S-GSA	STEM	maSigFun
GO:2000242	o++oooo	negative regulation of reproductive process	42	✗	✗	✗	✗
GO:0050792	o++oooo	regulation of viral reproduction	42	✗	✗	✗	✗
GO:2000503	o++oooo	positive regulation of natural killer cell chemotaxis	5	✓	✓	✗	✗
GO:0008347	o+ooooo	glial cell migration	13	✗	✓	✗	✗
GO:0048524	o++oooo	positive regulation of viral reproduction	32	✗	✗	✗	✗
GO:0019058	o++oooo	viral infectious cycle	51	✗	✗	✗	✗
GO:0031620	++++ooo	regulation of fever generation	9	✓	✓	✗	✗
GO:0031650	++++ooo	regulation of heat generation	11	✓	✓	✗	✗
GO:0071346	+++oooo	cellular response to interferon-gamma	22	✓	✓	✗	✗
GO:0051856	+++oooo	adhesion to symbiont	8	✓	✓	✗	✗
GO:0051825	+++oooo	adhesion to other organism involved in symbiotic interaction	8	✓	✓	✗	✗
GO:0051702	+++oooo	interaction with symbiont	37	✓	✓	✗	✗
GO:0034340	+++oooo	response to type I interferon	8	✓	✓	✗	✗
GO:0007610	+++oooo	behavior	485	✗	✓	✗	✗
BioCarta: GRANULOCYTES	o++oooo	adhesion and diapedesis of granulocytes	9	✓	✓	✗	✗
GO:0002675	+++++oo	positive regulation of acute inflammatory response	23	✓	✓	✗	✗
GO:0071621	o++oooo	granulocyte chemotaxis	11	✓	✓	✗	✗
KEGG:05160	+++oooo	hepatitis C	127	✓	✓	✗	✗
GO:0032897	o++oooo	negative regulation of viral transcription	14	✗	✓	✗	✗
GO:0002831	+++oooo	regulation of response to biotic stimulus	44	✓	✓	✗	✗

continued on next page ...

... continued from previous page

ID	Profile	Description	$|s|$	2T-GSA	1S-GSA	STEM	maSigFun
GO:0035458	++++ooo	cellular response to interferon-beta	14	✓	✓	✗	✗
GO:0043491	+++oooo	protein kinase B signaling cascade	86	✓	✓	✗	✗
GO:0010759	o++oooo	positive regulation of macrophage chemotaxis	8	✓	✓	✗	✗
GO:0071622	o++oooo	regulation of granulocyte chemotaxis	10	✗	✓	✗	✗
GO:0071674	o++oooo	mononuclear cell migration	10	✗	✓	✗	✗
GO:0010758	o++oooo	regulation of macrophage chemotaxis	9	✓	✓	✗	✗
GO:0071675	o++oooo	regulation of mononuclear cell migration	9	✓	✓	✗	✗
GO:0006910	ooo++oo	phagocytosis, recognition	12	✗	✓	✗	✗
KEGG:04623	+++oooo	cytosolic DNA-sensing pathway	47	✓	✓	✗	✗
GO:0022415	o++oooo	viral reproductive process	76	✓	✓	✗	✗
GO:0042832	++++ooo	defense response to protozoan	15	✓	✓	✗	✗
KEGG:05323	+++oooo	rheumatoid arthritis	78	✓	✓	✗	✗
BioCarta: EGF	+oooooo	EGF signaling pathway	10	✗	✓	✗	✗
BioCarta: INSULIN	+oooooo	insulin signaling pathway	11	✗	✓	✗	✗
BioCarta: PDGF	+oooooo	PDGF signaling pathway	11	✗	✓	✗	✗
BioCarta: TPO	+oooooo	thrombopoietin signaling pathway	11	✗	✓	✗	✗
GO:0001562	++++ooo	response to protozoan	17	✓	✓	✗	✗
GO:0016032	o++oooo	viral reproduction	94	✓	✓	✗	✗
GO:0048246	o++oooo	macrophage chemotaxis	16	✓	✓	✗	✗
GO:0034341	++++ooo	response to interferon-gamma	37	✓	✓	✗	✗

continued on next page ...

... continued from previous page

ID	Profile	Description	$\|s\|$	2T-GSA	1S-GSA	STEM	maSigFun
GO:0002691	++++ooo	regulation of cellular extravasation	6	✓	✓	✗	✗
GO:0006873	oo++ooo	cellular ion homeostasis	498	✗	✓	✗	✗
GO:0035747	+++oooo	natural killer cell chemotaxis	6	✓	✓	✗	✓
GO:2000501	+++oooo	regulation of natural killer cell chemotaxis	6	✓	✓	✗	✓
GO:0045123	++++ooo	cellular extravasation	19	✓	✓	✗	✗
GO:2000108	o++oooo	positive regulation of leukocyte apoptosis	15	✓	✓	✗	✗
REACT: CHEMOKINE	++++ooo	genes involved in chemokine receptors bind chemokines	36	✓	✓	✗	✗
GO:0030003	+++oooo	cellular cation homeostasis	312	✓	✓	✗	✗
GO:0055080	+++oooo	cation homeostasis	364	✗	✓	✗	✗
GO:0046903	o++oooo	secretion	600	✓	✓	✗	✗
GO:0030574	oo+oooo	collagen catabolic process	17	✓	✓	✗	✗
GO:0044243	oo+oooo	multicellular organismal catabolic process	20	✗	✓	✗	✗
GO:0043922	o+ooooo	negative regulation by host of viral transcription	12	✗	✗	✗	✗
GO:2000403	o++oooo	positive regulation of lymphocyte migration	12	✓	✓	✗	✗
GO:0048247	+++oooo	lymphocyte chemotaxis	17	✓	✓	✗	✗
GO:0006875	+++oooo	cellular metal ion homeostasis	276	✓	✓	✗	✗
GO:0055065	+++oooo	metal ion homeostasis	293	✓	✓	✗	✗
GO:0006911	o++++oo	phagocytosis, engulfment	18	✗	✓	✗	✗
GO:0050691	+++oooo	regulation of defense response to virus by host	15	✗	✓	✗	✗
GO:0055082	oo+oooo	cellular chemical homeostasis	533	✗	✗	✗	✗
GO:0016525	++++ooo	negative regulation of angiogenesis	51	✗	✓	✗	✗
GO:0035456	+++++oo	response to interferon-beta	15	✓	✓	✗	✗

continued on next page ...

... continued from previous page

ID	Profile	Description	$|s|$	2T-GSA	1S-GSA	STEM	maSigFun
GO:0042119	o++oooo	neutrophil activation	15	✓	✓	✗	✗
GO:0045429	++++ooo	positive regulation of nitric oxide biosynthetic process	30	✓	✓	✗	✗
GO:0032966	o+ooooo	negative regulation of collagen biosynthetic process	4	✗	✓	✗	✗
GO:0010713	o+ooooo	negative regulation of collagen metabolic process	4	✗	✓	✗	✗
GO:0044252	o+ooooo	negative regulation of multicellular organismal metabolic process	4	✗	✓	✗	✗
GO:0045807	o+++ooo	positive regulation of endocytosis	68	✓	✓	✗	✗
GO:0090026	o+++ooo	positive regulation of monocyte chemotaxis	6	✓	✓	✗	✗
GO:0051607	+++oooo	defense response to virus	62	✓	✓	✗	✗
GO:0043032	+oooooo	positive regulation of macrophage activation	11	✗	✓	✗	✗
GO:0006874	+++oooo	cellular calcium ion homeostasis	234	✓	✓	✗	✗
GO:0032940	o+++ooo	secretion by cell	527	✓	✓	✗	✗
GO:0072503	++++ooo	cellular divalent inorganic cation homeostasis	250	✓	✓	✗	✗
GO:0043615	oo++ooo	astrocyte cell migration	6	✗	✗	✗	✗
GO:0032677	++++ooo	regulation of interleukin-8 production	32	✓	✓	✗	✗
KEGG:04062	++++ooo	chemokine signaling pathway	172	✓	✓	✗	✗
GO:0007204	+++oooo	elevation of cytosolic calcium ion concentration	161	✓	✓	✗	✗
GO:0051480	+++oooo	cytosolic calcium ion homeostasis	176	✓	✓	✗	✗

continued on next page ...

... continued from previous page

ID	Profile	Description	$|s|$	2T-GSA	1S-GSA	STEM	maSigFun
GO:0051897	+++oooo	positive regulation of protein kinase B signaling cascade	50	✓	✓	✗	✗
GO:0051896	++ooooo	regulation of protein kinase B signaling cascade	67	✓	✓	✗	✗
GO:0072507	+++oooo	divalent inorganic cation homeostasis	258	✓	✓	✗	✗

Acronyms

2T-GSA	Two-threshold GSA-type activation profile algorithm
1S-GSA	One-threshold segmentation GSA-type activation profile algorithm
2S-GSA	Two-threshold segmentation GSA-type activation profile algorithm
ACC	accuracy
AH	data set investigating the effect of aldosterone on the mouse heart gene expression
AM	weighted arithmetic mean smoothing
ANOVA	analysis of variance
BATS	*Bayesian Analyisis of Time Series*
CAGED	Cluster analysis of gene expression dynamics
DAVID	Database for annotation, visualization and integrated discovery
DE	differentially expressed
DNA	Deoxyribonucleic acid
EDGE	*Extraction of Differential Gene Expression*
EM	Expectation Maximization algorithm
EPIG	Extracting microarray gene expression Patterns and Identifying co-expressed Genes (clustering method)
FC	fold change
FD	smoothing on basis of the (relative) distance to significance in the enrichment test
FDA	functional data analysis
FDR	false discovery rate
FN	false negative
FP	false positive
FPR	false positive rate
FS	smoothing using a shift of (not) differentially expressed genes in the Fisher enrichment test
FWER	family wise error rate
GD	gestational day

GE	smoothing by forcing continous differential expression status for all genes in the gene set
GEO	Gene Expression Omnibus
GM	weighted geometric mean smoothing
GO	Gene Ontology
GQL	Graphical Query Language
GSA	gene set analysis
GSEA	Gene Set Enrichment Analysis
GSP	general smoothing proportion
GRN	gene regulatory networks
HMM	hidden Markov model
IN	weighted inverse normal score mean smoothing
INSM	inverse normal score mean
IX	weighted inverse χ^2 score mean smoothing
KEGG	Kyoto Encyclopedia of Genes and Genomes
MARD	mean absolute rank difference
maSigPro	microarray significant profiles
maSigFun	(regression) functional microarray significant profiles
MM	mismatch
NPV	negative predictive value
OD	data set investigating the gene expression in the embryonic development of the mouse ovary
ORICC	order-restricted information criterion clustering
PACE	principal components analysis through conditional expectation
PAGE	parametric analysis of gene enrichment
PCA	principal componant analysis
piRNA	Piwi-interacting RNA: small non-coding RNA molecules with a role in RNA silencing during embryonic devolpment and spermatogenesis in particular
PM	perfect match
PN	postnatal day
POSGODA	POsitive artificial GO DAta generator
PPV_{p90}	positive predictive value restricted to the preset profile types and a maximum similarity of 90% with preset gene sets
RMA	Robust Multiarray Analysis
RNA	Ribonucleic acid
RSS	residual sum of squares
RT-PCR	Real-time polymerase chain reaction
SAM	Significance Analysis of Microarrays

SD	smoothing using sequential tests for enrichment with directed differential expression
SH	data set investigating the gene expression during skin healing in mice
SNP	single nucleotide polymorphism
SOM	self organizing map
SVD	singular value decomposition
SVM	support vector machine
STEM	Short Time-series Expression Miner
SU	smoothing using sequential tests for enrichment with undirected differential expression
TF	transcription factor
TH	data set investigating the gene expression during tongue healing in mice
TN	true negative
TP	true positive
TPR	true positive rate
TS-ABH	two-stage adaptive linear step-up procedure of Benjamini, Krieger, and Yekutieli (2006)

List of Symbols

α_{genes} 46
significance threshold on the gene level (e.g. FDR, fold-change, FWER).

α_{sets} 46
significance threshold for enrichment tests (e.g. controlling the FDR or FWER).

$A\!P_i^\star$ 82
"mutated" preset profile in simulation.

$A\!P_s$ 47
activation profile of gene set s obtained by the threshold variant of the GSA approach.

$A\!P_s^{\text{rot}}$ 59
activation profile of gene set s obtained by rotation test variant of the GSEA approach.

$\widetilde{A\!P}_{s_\star}^{\text{2S-GSA}}$ 77
numeric vector representation of the 2S-GSA activation profile of gene set s ($A\!P_{s_\star}^{\text{ts}}$).

$A\!P_s(t)$ 47
t^{th} position of the threshold activation profile of gene set s.

$A\!P_s^{\text{tl}}$ 49
activation profile of gene set s obtained by the non-threshold variant of the GSA approach.

$A\!P_s^{\text{tl}}(t)$ 49
t^{th} position of the non-threshold activation profile of gene set s.

$I\!P_s^{rot}(t)$ 59
t^{th} position of the rotation test activation profile of gene set s.

$I\!P_s^{ts}$ 50
activation profile of gene set s obtained by the threshold-segmentation variant of the GSA approach.

$^{shift}I\!P_s(t)$ 54
new profile symbol after applying smoothing by shifting genes in the contingency table.

$\beta_{fill}^{AM,\alpha_{sets}}$ 137
Regression coefficient for the (logarithmized) enrichment significance level in the regression model to determine the AM smoothing parameters..

$\beta_{fill}^{AM,G}$ 137
Regression coefficient for the (logarithmized) total gene number G in the regression model to determine the AM smoothing parameters..

$\beta_{fill}^{AM,GSP}$ 137
Regression coefficient for the (logarithmized) general smoothing proportion in the regression model to determine the AM smoothing parameters..

$\beta_{fill}^{AM,\tilde{p}_{sg}}$ 137
Regression coefficient for the (logarithmized) proportion of significant differentially expressed genes in the regression model to determine the AM smoothing parameters..

$Bin(n,p)$ 62
Binomial distribution with parameters n and p.

$C_+^{(t)}$ 84
candidate set of genes, whose original test statistic values are used to designate up expressed genes in the simulation study to evaluate the smoothing methods.

$C_-^{(t)}$ 84
candidate set of genes, whose original test statistic values are used to designate down expressed genes in the simulation study to evaluate the smoothing methods.

$C_\circ^{(t)}$ 84
candidate set of genes, whose original test statistic values are used to designate

not differentially expressed genes in the preset gene sets at '+'/'−' positions in the simulation study to evaluate the smoothing methods.

$cor(.,.)$ 61
Pearson's empirical correlation coefficient between two vectors.

ct^{conti} 53
correction term for the determination of gene continuity smoothing.

ct^{shift} 54
correction term for the determination of gene shift smoothing.

$DE_+^{(t)}$ 46
gene set of the significantly up expressed genes at time point t.

$DE_-^{(t)}$ 46
gene set of the significantly down expressed genes at time point t.

d_g 77
vector of the difference of gene expression to reference for gene g.

D_g^{SAM} 39
test statistic of Significance Analysis of Microarrays to detect differential gene expression.

D_g^{shrink} 40
shrinkage t-test statistic for gene g to detect differential gene expression.

D_g^{t-test} 39
test statistic of the standard t-test for gene g.

$\mathbf{D}^{shrink}$ 58
Matrix of all shrinkage t-statistics resulting from the gene expression time course experiment.

D_s^{med} 57
median t-statistic score for ranking the non-constant-'o'-profiles of the GSA-type algorithms.

$D_s^{|med|}$ 57
median absolute t-statistic score for ranking the non-constant-'o'-profiles of the GSA-type algorithms.

d_{STEM} 77
    distance metric used in STEM to cluster.

$d_{\text{STEM}}(.,.)$ 61
    distance function between two profiles in the STEM algorithm.

$\text{ES}_{s+}^{(t)}$ 58
    enrichment score for overrepresentation of up regulated genes in gene set s at time point t.

$\text{ES}_{s+}^{(t),k}$ 58
    enrichment score for overrepresentation of up regulated genes in gene set s at time point t resulting from a random rotation in the rotation test GSEA type algorithm.

$\text{ES}_{s-}^{(t)}$ 58
    enrichment score for overrepresentation of down regulated genes in gene set s at time point t.

$\text{ES}_{s-}^{(t),k}$ 58
    enrichment score for overrepresentation of down regulated genes in gene set s at time point t resulting from a random rotation in the rotation test GSEA type algorithm.

F 52
    cumulative distribution function of a standard normal distribution.

F^{-1} 52
    inverse of the cumulative distribution function of a standard normal distribution.

FC_g 38
    fold change (logarithmic relation to base 2) of gene g of two measurements.

FN 43
    number of accepted null hypotheses while alternatives are true in a multiple testing setting.

FP 43
    number of rejected true null hypotheses in a multiple testing setting.

G 23
    total number of genes in the gene universe $\check{G}$.

g	....	23

gene of the universe $\check{G} = \{g_1, \ldots, g_G\}$.

$\check{g}$ 73
inconspicuous gene in simulation study to evaluate the competing profile algorithms.

$\check{G}$ 23
the gene universe $\check{G}$ (usually filtered according to gene set annotation or data quality).

$g_i^{\text{SIM}-1}$ 76
i^{th} sampled gene from the maSigFun prototype set in the simulation study for comparing the five profile algorithms.

G_{SIM}^o 74
number of inconspicuous (non-differential) genes in the simulation study to compare the competing profile algorithms.

GSP 137
Term of general smoothing proportion to put into the function to determine the smoothing parameters based on a linear regression estimation.

$\text{INS}_{s+}^{(t)}$ 52
inverse normal score applied in the *inverse normal score mean smoothing*.

K 58
total number of random rotations used in the rotation test GSEA type algorithm.

k 43
number of rejected tests in the Benjamini-Hochberg step-up procedure.

k^{I} 44
number of rejected hypotheses in the first stage of the TS-ABH procedure.

k^{II} 44
number of rejected hypotheses in the second stage of the TS-ABH procedure.

λ_{fill} 51
parameter for smoothing algorithms in the smoothing directions, which tries to fill a gap of significance in the activation profile (e.g. '++o+++' $\to$ '++++++').

$\lambda_{\text{fill}}^{\text{AM}}()$ 137
Function to determine the smoothing parameters based on a linear regression estimation for the AM smoothing in the *fill* direction.

$\hat{\lambda}$ 40
data driven shrinkage parameter for the shrinkage t-statistic $D^{\text{shrinkage}-t}$.

λ_i^{SIM} 76
ith sampled parameter for the convex combination to create an artificial active gene for an active maSigFun set in the simulation study for comparing the five profile algorithms.

λ_{wipe} 51
parameter for smoothing algorithms in the smoothing directions, which tries to wipe out a single significant position in the activation profile (e.g. 'oo-ooo' → 'oooooo').

$L_{\text{DE}}^{(t)}$ 48
ordered gene list according to the shrinkage t-statistics at time point t.

M 23
total number of replicates / multiple measurements.

m 23
index of replicates $(1, \ldots, M)$.

$\hat{m}_\star^{\text{STEM}}$ 77
model profile of the STEM prototype gene set in the simulation study to compare the five competing profile algorithms.

$\mu_{\text{fill}}^{\text{AM}}$ 137
Intercept in the regression model to determine the AM smoothing parameters based on a linear regression estimation.

$\check{N}$ 73
number of inconspicuous genes in simulation study to evaluate the competing profile algorithms.

$n_{\text{DE}+}$ 46
number of significantly up expressed genes at time point t.

$\tilde{n}_{\text{DE}_+^{(t)} \in s}^{(t,s)}$ 82
  minimum of genes needed to obtain a significant enrichment with up expressed genes in gene set s at time point t (equivalent for down expression).

$n_{\text{DE}_+^{(t)} \in s}^{\text{conti}}$ 53
  corrected number of differential expressed genes in gene set s, occuring in the gene expression continuity smoothing approach.

$n_{\text{DE}_+^{(t)} \in s}^{\text{shift}}$ 54
  corrected number of differential expressed genes in gene set s, occuring in the gene shift smoothing approach.

$n_{\text{DE}_+^{(t)} \in s}$ 46
  number of significantly up expressed genes in gene set s at time point t.

n_{DE_-} 46
  number of significantly down expressed genes at time point t.

$n_{\text{DE}_-^{(t)} \in s}$ 46
  number of significantly down expressed genes in gene set s at time point t.

$\text{NES}_{s+}^{(t)}$ 59
  normalized enrichment score in the rotation test GSEA type algorithm for overrepresentation of up regulated genes in gene set s at time point t.

$\text{NES}_{s-}^{(t)}$ 59
  normalized enrichment score in the rotation test GSEA type algorithm for overrepresentation of down regulated genes in gene set s at time point t.

n_{H_0} 43
  number of true null hypotheses in a multiple testing setting.

$\hat{n}_{H_0}$ 44
  estimated upper bound of the number of true null hypotheses.

$n_{\hat{H}_0}$ 43
  number of accepted null hypotheses in a multiple testing setting.

n_{H_1} 43
  number of true null alternatives in a multiple testing setting.

$n_{\hat{H}_1}$ 43
number of rejected null hypotheses in a multiple testing setting.

p 43
p-value of a statistical test.

p_{ag} 74
proportion of (*active*) differential expressed genes in *active* gene sets in simulation study for the comparison of profile algorithm.

$\tilde{p}_{\text{ag}}$ 137
Term of the proportion of differentially expressed genes to put into the function to determine the AM smoothing parameters based on a linear regression estimation.

p_{as} 74
proportion of (*active*) gene sets with spiked in differentially expressed genes in simulation study for the comparison of profile algorithm.

p^D 42
one-sided p-value for depletion in the Fisher's exact test.

p_{def} 147
proportion of gene sets according to the set of gene sets of the same definition type (e.g. KEGG)..

P_{diff} 84
the probability of assigning an up or down expressed test statistic (with respect to α_{genes}) to a gene selected for up or down expression in the preset profile (in simulation to evaluate smoothing method).

p^E 42
one-sided p-value for enrichment in the Fisher's exact test.

$p_{s+}^{E(t)}$ 46
enrichment test p-value for enrichment of up regulated genes in gene set s at time point t.

$p_{s-}^{E(t)}$ 46
enrichment test p-value for enrichment of down regulated genes in gene set s at time point t.

$p_{\check{S}}$ 147
proportion of gene sets according to the whole gene set universe $\check{S}$..

Q 43
unknown proportion of erroneously rejected true null hypotheses in a multiple testing setting.

q 43
threshold for the control of the FDR or q-value adjusted p-value for a single test in a multiple experiment setting., ... 44
FDR-q-value: smallest FDR, which can be controlled with the TS-ABH procedure while rejecting the corresponding hypothesis and all others with a more extreme test statistic.

Q 58
random $(J \times J)$ rotation matrix.

$q_g^{(t)}$ 46
q-value of the one-sided shrinkage t-test on the gene level.

q^{II} 44
critical threshold in the second stage of the TS-ABH procedure.

$q_{s+}^{E(t)}$ 46
enrichment test q-value for enrichment of up regulated genes in gene set s at time point t.

$\check{q}_{s+}^{E(t)}$ 49
q-value of the enrichment test with up expressed genes in the non-threshold GSA variant for gene set s at time point t.

$\tilde{q}_{s+}^{E(t)}$ 59
enrichment q-value in the rotation test GSEA type algorithm for overrepresentation of up regulated genes in gene set s at time point t.

$^{\text{AM}}q_{s+}^{(t)}$ 52
weighted arithmetic mean of q-values in a neighborhood of a activation profile position considered for smoothing.

$^{\text{conti}}q_{s+}^{(t)}$ 53
recalculated q-value of the smoothing approach, which forces continuity in significant gene expression differences., ... 54
recalculated q-value of the gene shift smoothing approach.

$^{\text{GM}}q_{s+}^{(t)}$ 52
  weighted geometric mean of q-values in a neighborhood of a activation profile position considered for smoothing.

$^{\text{INS}}q_{s+}^{(t)}$ 52
  weighted inverse normal score mean of q-values in a neighborhood of a activation profile position considered for smoothing.

$^{\text{IXS}}q_{s+}^{(t)}$ 52
  weighted inverse χ_1^2 score mean of q-values in a neighborhood of a activation profile position considered for smoothing.

$q_{s-}^{E(t)}$ 46
  enrichment test q-value for enrichment of down regulated genes in gene set s at time point t.

$\breve{q}_{s-}^{E(t)}$ 49
  q-value of the enrichment test with down expressed genes in the non-threshold GSA variant for gene set s at time point t.

$\check{q}_{s-}^{E(t)}$ 59
  enrichment q-value in the rotation test GSEA type algorithm for overrepresentation of down regulated genes in gene set s at time point t.

$q_{s,d}^{E(t)}$ 49
  q-value of the segmentation enrichment test with partitioning according to d in the non-threshold GSA variant for gene set s at time point t.

R^2 63
  unadjusted measure of determination in the regression model.

$r_g^{\text{2S-GSA}}$ 77
  rank for determining the sample probability for the spike-in genes from the 2S-GSA prototype gene set in the simulation study to compare the five competing profile algorithms.

r_g^{maSigFun} 76
  rank for determining the sample probability for the spike-in genes from the maSigFun prototype gene set in the simulation study to compare the five competing profile algorithms.

r_g^o 74
   rank of summarized gene expression difference to reference for an *inconspicuous* gene in simulation to compare the five competing profile algorithms.

r_g^{STEM} 77
   rank for determining the sample probability for the spike-in genes from the STEM prototype gene set in the simulation study to compare the five competing profile algorithms.

$r_{s^{\text{maSigFun}}}^{\text{maSigFun}}$ 75
   rank for determining the sample probability for the maSigFun prototype gene set s^{maSigFun} in the simulation study to compare the five competing profile algorithms.

RSS_g 76
   residual sum of squares for a single gene g and the corresponding maSigFun model used in the simulation study to compare the five competing profile algorithms.

S 23
   total number of gene sets in the gene set universe $\check{S}$.

s 23
   gene set of the gene set universe $\check{S} = \{s_1, \ldots, s_S\}$.

$S_+^{(t)}$ 80
   gene set, containing those genes, which are designated to be up expressed in a simulation turn (evaluation of smoothing algorithms).

$S_-^{(t)}$ 80
   gene set, containing those genes, which are designated to be down expressed in a simulation turn (evaluation of smoothing algorithms).

$S^{\text{1S-GSA}}$ 73
   number of 1S-GSA prototype gene sets in the profile algorithm comparing simulation.

$\tilde{s}^{\text{1S-GSA}}$ 77
   active gene set with spiked-in genes from a gene set with a significant 1S-GSA profile in the simulation study to compare the five competing profile algorithms.

S^{2S-GSA} 73
  number of 2S-GSA prototype gene sets in the profile algorithm comparing simulation.

$\tilde{s}^{2S-GSA}$ 77
  active gene set with spiked-in genes from a gene set with a significant 2S-GSA profile in the simulation study to compare the five competing profile algorithms.

S^{2T-GSA} 73
  number of 2T-GSA prototype gene sets in the profile algorithm comparing simulation.

$\tilde{s}^{2T-GSA}$ 77
  active gene set with spiked-in genes from a gene set with a significant 2T-GSA profile in the simulation study to compare the five competing profile algorithms.

$\check{S}$ 23
  the gene set universe on $\check{G}$ (e.g. defined by GO and (filtered) feasible genes on the chip).

sd_0 39
  parameter in the Significance Analysis of Microarrays statistic (modified t-test).

sd_g 39
  sample estimate for standard deviation of the expression difference for gene g.

$\mathrm{sd}_g^{\mathrm{shrink}}$ 40
  shrinkage estimate for the standard deviation of the expression difference for gene g.

$s_i^\star$ 82
  sampled gene set in simulation to obtain a preset profile.

S^{maSigFun} 73
  number of maSigFun prototype gene sets in the profile algorithm comparing simulation.

s^{maSigFun} 75
  a significant gene set identified by the maSigFun algorithm in the simulation study to compare the five competing profile algorithms.

$\tilde{s}^{\mathrm{maSigFun}}$ 76
  active gene set with spiked-in genes from a gene set with a significant maSigFun profile in the simulation study to compare the five competing profile algorithms.

$S_o^{(t)}$ 80
  gene set, containing those genes, which are designated to be not differentially expressed in a simulation turn (evaluation of smoothing algorithms).

$s_\star^{\mathrm{1S-GSA}}$ 75
  the prototype gene set for a significant profile identified by the 1S-GSA algorithm in the simulation study to compare the five competing profile algorithms.

$s_\star^{\mathrm{2S-GSA}}$ 75
  the prototype gene set for a significant profile identified by the 2S-GSA algorithm in the simulation study to compare the five competing profile algorithms.

$s_\star^{\mathrm{2T-GSA}}$ 75
  the prototype gene set for a significant profile identified by the 2T-GSA algorithm in the simulation study to compare the five competing profile algorithms .

S_{SIM} 74
  number of artificial gene sets in simulation study to compare the competing profile algorithms.

$s_\star^{\mathrm{maSigFun}}$ 75
  the prototype gene set for a significant profile identified by the maSigFun algorithm in the simulation study to compare the five competing profile algorithms.

$s_\star^{\mathrm{STEM}}$ 75
  the prototype gene set for a significant profile identified by the STEM algorithm in the simulation study to compare the five competing profile algorithms.

S^{STEM} 73
  number of STEM prototype gene sets in the profile algorithm comparing simulation.

$\tilde{s}^{\text{STEM}}$ 76
  active gene set with spiked-in genes from a gene set with a significant STEM profile in the simulation study to compare the five competing profile algorithms.

sw 61
  step width for the STEM algorithm.

T 23
  total number of time points.

t 23
  index of time points $(1, \ldots, T)$.

TN 43
  number of accepted true null hypotheses in a multiple testing setting.

TP 43
  number of rejected true alternative hypotheses in a multiple testing setting.

v_g 39
  sample estimate for the variance of the expression of gene g., ... 40
  estimate for the variance of the multiple expression measurements for gene g.

v^{median} 40
  median over the variance estimates v_g across all genes.

w 52
  a weight parameter in the INSM smoothing (either λ_{fill} or λ_{wipe}).

$\bar{x}_g$ 39
  arithmetic mean of the expression of gene g.

x_{tgm} 23
  preprocessed and hence logarithmized gene expression value (time point t, gene g, replicate m).

I want morebooks!

Buy your books fast and straightforward online - at one of the world's fastest growing online book stores! Environmentally sound due to Print-on-Demand technologies.

Buy your books online at
www.get-morebooks.com

Kaufen Sie Ihre Bücher schnell und unkompliziert online – auf einer der am schnellsten wachsenden Buchhandelsplattformen weltweit! Dank Print-On-Demand umwelt- und ressourcenschonend produziert.

Bücher schneller online kaufen
www.morebooks.de

OmniScriptum Marketing DEU GmbH
Heinrich-Böcking-Str. 6-8
D - 66121 Saarbrücken

Telefax: +49 681 93 81 567-9

info@omniscriptum.de
www.omniscriptum.de

Printed by Books on Demand GmbH, Norderstedt / Germany